EXPLANATION

1. Superior Upland — Hilly area of erosional topography on ancient crystalline rocks.

2. Continental Shelf — Shallow sloping submarine plain of sedimentation.

3. Coastal Plain — Low, hilly to nearly flat terraced plains on soft sediments.

4. Piedmont Province — Gentle to rough, hilly terrain on belted crystalline rocks becoming more hilly toward mountains.

5. Blue Ridge Province — Mountains of crystalline rock 3,000 to 6,000 feet high, mostly rounded summits.

6. Valley and Ridge Province — Long mountain ridges and valleys eroded on strong and weak folded rock strata.

7. St. Lawrence Valley — Rolling lowland with local rock hills.

8. Appalachian Plateaus — Generally steep-sided plateaus on sandstone bedrock, 3,000 to 5,000 feet high on the east side, declining gradually to the west.

9. New England Province — Rolling hilly erosional topography on crystalline rocks in southeastern part to high mountainous country in central and northern parts.

10. Adirondack Province — Subdued mountains on ancient crystalline rocks rising to over 5,000 feet.

11. Interior low plateaus — Low plateaus on stratified rocks.

12. Central Lowland — Mostly low rolling landscape and nearly level plains. Most of area covered by a veneer of glacial deposits, including ancient lake beds and hilly lake-dotted moraines.

13. Great Plains — Broad river plains and low plateaus on weak stratified sedimentary rocks. Rises toward Rocky Mountains at some places to altitudes over 6,000 feet.

14. Ozark Plateaus — High, hilly landscape on stratified rocks.

15. Ouchita Province — Ridges and valleys eroded on upturned folded strata.

16. Southern Rocky Mountains — Complex mountains rising to over 14,000 feet.

17. Wyoming Basin — Elevated plains and plateaus on sedimentary strata.

18. Middle Rocky Mountains — Complex mountains with many intermontane basins and plains.

19. Northern Rocky Mountains — Rugged mountains with narrow intermontane basins.

20. Columbia Plateau — High rolling plateaus underlain by extensive lava flows; trenched by canyons.

21. Colorado Plateau — High plateaus on stratified rocks cut by deep canyons.

22. Basin and Range Province — Mostly isolated ranges separated by wide desert plains. Many lakes, ancient lake beds, and alluvial fans.

23. Cascade-Sierra Nevada Mountains — Sierras in southern part are high mountains eroded from crystalline rocks. Cascades in northern part are high volcanic mountains.

24. Pacific Border Province — Mostly very young steep mountains; includes extensive river plains in California portion.

ENVIRONMENTAL GEOLOGY

EDWARD A. KELLER
University of California/Santa Barbara

With assistance from E. M. Burt,
Williams & Works, Grand Rapids, Michigan

Charles E. Merrill Publishing Company
A Bell & Howell Company
Columbus, Ohio

For
JACKIE

Published by
Charles E. Merrill Publishing Company
A Bell & Howell Company
Columbus, Ohio 43216

This book was set in Times Roman and Helvetica.
The Production Editor was Margaret Shaffer.
The cover was designed by Will Chenoweth.

International Standard Book Number: 0–675–08617–5

Library of Congress Catalog Card Number: 75–44887

4 5 6 7 8 9 10—81 80 79 78 77

Acknowledgments for cover photographs
U.S. Army Corps of Engineers
Ward's Natural Science Establishment, Inc.
ERTS photo
Hawaii Visitors Bureau
National Park Service

Printed in the United States of America

PREFACE

Environmental Geology is an in-depth treatment of relations between man and his geologic environment. Introductory chapters develop a philosophical framework that unites the cultural and physical environments while introducing the terminology necessary for the remainder of the book. Traditional subjects such as geologic hazards are introduced and discussed by interweaving geologic and human aspects of each hazard: flooding, landslides and subsidence, earthquakes, volcanoes, and coastal erosion. Other material includes hydrology and human use, waste disposal, geologic aspects of environmental health, resources and energy, land-use planning, site selection, landscape aesthetics, environmental impact, and environmental law. The main thrust of these topics is to emphasize the geologic aspect or geologic considerations that might arise under certain conditions in working in these areas.

The author is indebted to many individuals, companies, and agencies that supported the book by critical reviews or by providing photographs and illustrations.

Reviews of chapters by Samuel E. Swanson, Harold L. Krivoy, and James Dennis Rash are recognized, as are the reviews of the entire text by Douglas G. Brookins, John B. Droste, and Stanley P. Fisher.

The author is indebted to Maggie Shaffer for editing the text and making numerous suggestions for improvement. Special recognition is also given to the Bette Davis Company for typing the manuscript; Olive Pistor for assistance with government documents; drafting of illustrations by Mike Pearson; endless typing of letters by Deborah Blackmon and Mickey Gregory; and photographic assistance by Larry Ferguson. The index was prepared by Jacqueline J. Keller.

CONTENTS

Part One

PHILOSOPHY AND FUNDAMENTAL PRINCIPLES

Everything has a beginning and an end. Our earth began approximately 5 billion years ago when a cloud of interstellar gas known as a *solar nebula* collapsed, forming protostars and planetary systems; life on earth began about 2 billion years later, or 3 billion years ago. Since then, a multitude of different types of organisms have emerged, prospered, and died out, leaving only their fossils to mark their place in earth's history. Several million years ago, on one of the more recent pages in earth history, our ancestors set the stage for the eventual dominance of man. As certainly as our sun will eventually die, man, too, will disappear. The impact of man on earth history may not be significant, but to us living now, our children and theirs, our environment is significant indeed.

       Environmental geology is applied geology. Specifically, it is the application of geologic information to solving conflicts, minimizing possible adverse environmental degradation, or maximizing possible advantageous conditions resulting from man's use of the natural and modified environment. This includes evaluation of *natural hazards* such as floods, landslides, earthquakes, and volcanic activity to minimize loss of human life and property damage; evaluation of the *landscape* for site selection, land-use planning, and *environmental impact* analysis; and evaluation of *earth materials* (such as elements, minerals, rocks, soils, and water) to determine their potential use as resources or waste disposal sites and the effects on human health, and to assess the need for conservation practices. In a broader sense, environmental geology is that branch of earth science that emphasizes the entire spectrum of man's interactions with the physical environment.

Environment may be considered as the total set of circumstances that surround an individual or a community. It may be defined to include two parts: first, physical conditions such as air, water, gases, landforms, etc., which affect the growth and development of an individual or a community; and second, social and cultural aspects such as ethics, economics, aesthetics, etc., which affect the behavior of an individual or a community. Therefore, a complete introduction to environmental geology involves the consideration of philosophical and cultural aspects which influence how we perceive and react to our landscape, as well as the physical earth processes, resources, and landforms which may be more readily recognized by the observant earth scientist.

Chapters 1 through 3 provide the philosophical framework for the remainder of the book. Chapters 1 and 2 are designed to integrate the influence of cultural and physical activities of our total environment, and Chapter 3 introduces the physical environment through the geological cycle. The term *cycle* emphasizes that most earth materials, such as air, water, soil, minerals, and rock, although changed physically and chemically and transported from place to place, are constantly being reworked, conserved, and renewed by natural earth processes. Chapter 3 also introduces basic earth science terminology necessary to understand the remainder of the book.

CULTURAL BASIS FOR THE ENVIRONMENTAL CRISIS

The cultural aspect of the environmental crisis involves the entire way of life that we have transmitted from one generation to another. Therefore, if we are going to uncover the roots of our present condition, we must look to the past and consider various functional categories and social institutions that have developed. The functional categories of society that are especially significant in environmental studies are ethical, economic, political, aesthetic, and, perhaps, religious. The interaction between individuals and the institutions responsible for maintaining these functions are intimately associated with the way we perceive and respond to our physical environment.

Environmental Ethics

What started as the "quiet crisis" of the 1960s has evolved into what Stewart Udall, statesman and conservationist, refers to as the "crisis of survival" (1). More important than the certainty of a crisis is whether society believes that there is a crisis. In other words, is there a new awareness that is destined to change our life-style, morals, ethics, and institutions, or is the environmental revolution just another prestigious fad that interests the intellectual community?

Aldo Leopold, in telling the story of Odysseus, who, upon returning from the war in Troy, hanged a dozen slave women for suspected misbehavior during his absence, emphasizes the lack of ethics regarding property. The hangings involved no question of ownership; the women were property, and the disposal of property was a matter of expediency, much as it is today. Although concepts of right and wrong were present in Greece three thousand years ago, these ethical values did not extend to slaves (2). Since that time, ethical values have been extended to many other areas of human behavior; but, apparently, only within this century has the relation between man and his environment begun to emerge as a relation with moral considerations.

Ecological ethics involve limitations on social, as well as individual, freedom of action in the struggle for existence in our stressed environment (2). A land ethic assumes that man is ethically responsible not only to other individuals and society but also to the total environment, that larger community consisting of plants, animals, soil, atmosphere, etc. The environmental ethic proposed by Leopold affirms the right of all resources, including plants, animals, and earth materials, to continue existence and, at least in certain locations, their continued existence in a natural state. This ethic effectively changes man's role from conqueror of the land to that of citizen and protector of the environment. This obviously requires us to revere and love our land and not, for instance, to allow economics to determine all land use.

There exists a possible dichotomy or source of confusion between an ideal and a realistic land ethic. To give rights to the plants, animals, and landscape might be interpreted as granting to individual plants and animals the fundamental right to live. However, if man is to be part of the environment, he, along with the other members, must extract the energy necessary to survive. Therefore, although the land ethic assigns rights for animals such as deer, cattle, or chickens to survive as a *species,* it does not necessarily assign rights to an *individual* deer, cow, or chicken. The same argument may be given to justify the use of stream gravel for construction material, or to mine and use the other resources necessary for our well-being. However, unique landscapes with high aesthetic value, like endangered species, are in need of complete protection within our ethical framework.

Environmental ethics and moral responsibility are re-restated by Stewart Udall. Each generation has its own rendezvous with the land, for despite our fee titles and claims of ownership, we are all brief tenants on this planet. By choice or by default, we will carve out a land for our heirs. We can misuse the land and diminish the usefulness of resources, or we can create a world in which physical affluence and spiritual affluence go hand in hand (1).

The resounding message is that man is an integral part of the environment. He is no less than any other being, and he has a moral obligation to those beings who will follow. This obligation is to insure that they will also have the opportunity to experience the pleasure of belonging to and cooperating with the entire land community.

Economic and Political Systems

Arriving in late fall of 1620, after two months on the stormy North Atlantic, 73 men and 29 women from the *Mayflower* confronted what they considered a wild and savage land. The colonists were not equipped with the skills and knowledge necessary to adapt quickly to their new environment. Regardless of these shortcomings and despite their fear of the wilderness, they brought three things which assured their success in the New World. First, they brought a new technology. Reportedly, when the Pilgrims landed, they did not even have a saw, but they did have Iron Age skills necessary to insure relentless subjugating of the land and its earliest inhabitants, the Indians. In the long run, the ax, gun, and wheel asserted their supremacy. Second, the colonists brought with them the blueprints to remake the New World. They knew how to organize work, utilize work animals, and sell their surplus to overseas markets. Third, they brought with them a concept of land ownership completely different from that of the Indians whose bonds to the land were religious and held by kinship and nature rather than exclusive possession. The Indian had a deep affection for the land and had a notion of ownership different from that of the colonist, whose idea of ownership involved an absolute title to land regardless of who worked the land or how far away the owner was. After the Indians were displaced, land use or abuse depended entirely on the attitude of the owner.*

America now, as in its early years, suffers greatly from the Myth of Superabundance. This myth assumes that the land and resources in America are inexhaustible and that, therefore, management of resources is unnecessary. Management of the people and their society, however, had and continues to have the deep roots transposed from the Old World.

Stewart Udall writes that the land myth was instrumental in environmental degradation from the "birth of land policy" in the eighteenth century throughout the "raid on resources" which lasted into the twentieth century. Even young Thomas Jefferson, who in later life was to become aware of the value of conservation, stated that there was such a great deal of farmland that it could be wasted as he pleased. However, the real raid on resources probably began with the mountain men and their trapping of beaver in the 1820s. This was only the beginning and was followed by machines that for the first time were capable of large-scale removal of resources and landscape alteration. These inventions included the sawmills which precipitated the destruction of the American forests, and the "Little Giant" hoze nozzle that could tear up an entire hillside in the search for California gold.

The "Great Giveaway" of land which resulted in the destruction of forests and consequent soil erosion eventually ended, and in 1884 hydraulic mining was outlawed. However, the effects of these repugnant land-use practices can still be viewed today. Similar examples of the raid on resources can be given: the plights of the fur seal, the buffalo, and the passenger pigeon; and the Dust Bowl of the 1930s (1).

---

*Stewart Udall, *The Quiet Crisis,* pp. 25–27.

The seeds of conservation were planted in the latter part of the nineteenth century by men such as Carl Schurz, Secretary of the Interior, and John Wesley Powell, geologist and explorer. Their messages concerning conservation of resources and land-use planning, although largely ignored when first introduced, today stand as landmarks in perceptive and innovative conservation (1).

The historical roots of our landscape heritage, while not a pretty picture, are full of lessons to be learned. For example, we painstakingly learned that our resources are not infinite and that land and water management is necessary for meaningful existence. This conclusion has become even more significant over the years as American society continues to urbanize and consume resources at an ever-increasing rate.

The convergence of available resources with the needs of society, along with an ever-growing production of waste, has produced what is popularly referred to as the *environmental crisis*. This impending crisis in America, according to Lewis W. Moncrief, is a result of individual and institutional inability in our democratic system to organize technology, conservation, urbanization, and the capitalistic mission to the betterment of our landscape (3). He further contends that the present condition is characterized by three features that tend to restrict a quick solution to environmental problems: first, the absence of individual and personal moral direction concerning the way we treat our natural resources; second, the inability of our social institutions to make adjustments to reduce environmental stress; and third, an abiding faith in technology.

Overpopulation, urbanization, and industrialization, combined with little ethical regard for our land and inadequate institutions (or perhaps too many institutions stumbling over one another) to cope with environmental stress, may well be the immediate source of the crisis. The overpopulation, urbanization, and industrial factors, although part of the political and economical systems, currently tend to transcend those systems.

Political and economic theorists are often surprised to learn that disruption of the environment is as serious a problem in the USSR as in the United States. Goldman reports that the Soviets have greatly misused their natural resources. For example, most major Soviet cities have air pollution problems, and water pollution there has resulted in massive fish kills, in turn resulting in an increase in mosquitoes and malaria peril. The oil-covered Iset River was accidentally ignited in 1966, as was the Cuyahoga River in Ohio. Over 65 percent of the factories in Russia discharge their waste without treatment, and 60 percent of the cities and suburbs have no water treatment facilities. Land-use problems in Russia include a system of dams, reservoirs, and canals that have diverted such a large quantity of water that there is serious concern for the future of the Caspian Sea and its famous caviar fisheries. The reservoirs have also increased evaporation, which is disrupting natural moisture patterns and changing rainfall cycles. Furthermore, seepage from unlined canals has caused the water table to rise in normally dry areas, facilitating the deposition of harmful salts in the soil. In addition, the mining of beach deposits for construction material in conjunction with decreased supply of sediment to the beaches (the reservoirs are holding back the natural flow of sediment) has resulted in serious coastal erosion, for without the sand and gravel to impede the impact of waves, the coastline is subject to rapid erosion.

The factors responsible for environmental problems in the USSR, as in America, are population explosion and rapid industrialization. Furthermore, the Russian government, as sole owner of the productive resources, has not been any more successful than other countries in regulating or controlling environmental degradation. In fact, the national commitment to centralized control of industry has given rise to unique environmental problems resulting from uniform regulations for industry regardless of local conditions.

Goldman notes several other disadvantages of an urbanizing, socialized system. First, there are no private interests that can challenge government proposals. Thus, a well-intentioned program leading to possible unanticipated degradation to the environment might be introduced. Such a program would more likely be terminated in a society that allows public criticism. Second, there is environmental disruption due to a lack of private ownership. For example, the removal of gravel and subsequent erosion of the Black Sea coast is a result of the absence of private control. Because no one owns the gravel, it is free to anyone who can carry it away. Consequently, contractors in need of gravel need not worry about infringing on the rights of coastal property owners worried about aesthetic degradation erosion and loss of tourist trade. Third, because the Soviet government is the sole owner of the country's industry, it is generally unable to take the role of an impartial referee between consumer and industry. This condition is not likely to change as long as the USSR, which has only recently become industrialized, stresses rapid growth and production.

On the other hand, Goldman notes some aspects of socialism in the USSR that have by design, or more likely by fortuitous circumstances, resulted in lesser environmental disruption. For example, Russia has de-emphasized the production of consumer goods, resulting in less prolific production of easily disposed of items and less waste material. Furthermore, relatively low labor costs, which are not necessarily a planned characteristic of their system, facilitate trash collection and sewage disposal despite the general lack of a sewer system. This results because the collection of sewage "nite soil" is apparently offset by its value as fertilizer. The central power of the state may also be an advantage in situations where expediency is sought. For example, establishment of national parks may be a relatively quick, inexpensive process there compared to countries with private ownership of property.*

The conclusion from consideration of political systems is that rather than private enterprise, industrialization, urbanization, economic consideration, and lack of a land ethic are primarily responsible for environmental degradation. Therefore, the salvation of the landscape community involves social, economic, and ethical behavior on the part of individuals rather than the political systems as they exist today.

Optimistically, it appears that the emerging environmental ethics inherent in the spirit of individual actions and new legislation will facilitate the types of changes needed. These changes are possible because our democratic system with private ownership of resources and free enterprise has the flexibility necessary to allow meaningful change. However, it is emphasized that the system cannot be

---

*M.I. Goldman, *Environmental Disruption in the Soviet Union,* pp. 63-65.

expected to react until individual citizens are willing to practice environmental ethics and the political, economic, and legal institutions and processes are modernized and streamlined in keeping with the original constitutional intent. (There appears to be a tendency to solve problems by "legislating" them out of existence, while "insuring" no real progress by keeping antiquated or conflicting laws or institutions *ad infinitum,* sometimes by simply changing names.)

Aesthetic Preference and Judgment

Environmental intangibles such as the pleasures of outdoor experiences in isolation in nature are extremely difficult to evaluate. The hunter in the blind on a crisp autumn morning; the fisherman in icy mountain stream as the day gives way to darkness; the nature photographer about to culminate the search for an elusive subject on a lonely mountain top; the hikers showing their three-year-old son a snail shell; the picnickers just relaxing; or the motorist out for a Sunday drive in the country—all perceive to a lesser or greater extent various aspects of the landscape and react to it with various types of behavior. Their experiences and memories cannot readily be equated to economic value, but to the individual they may be priceless.

An understanding of beauty was originally studied in the branch of philosophy known as *aesthetics.* Today, little attention is given to philosophers, and beauty is defined by artists and art critics (5). There appears to be an important distinction between the philosopher who studies aesthetics to establish evaluative judgments, similar to verdicts and findings, and the artist and art critic who may be more concerned with appreciative aesthetic judgments which express preferences such as affection and antipathy (6). Therefore, given that there is a distinction between evaluation and preference, the former constitutes a more objective approach.

One of the perplexing problems associated with aesthetic evaluation is the impact of personal preference. For example, one person may appreciate a meandering river in an isolated swamp. Another may prefer a bubbling mountain stream, while a third would rather visit a public park in an urban area. Regardless of such preferences, if we are going to consider aesthetic factors in local, regional, and national land-use planning, we must develop a method of aesthetic evaluation for landscapes that is easy to understand and is quantitative, credible, and predictive.

Three basic criteria necessary to judge aesthetic quality have been recognized: unity, vividness, and variety (7). *Unity* refers to the quality of wholeness of the perceived landscape, not as an assemblage but as a single harmonious unit. *Vividness* refers to that quality of the landscape which reflects a visually striking scene. This is nearly synonymous with *intensity, novelty,* or *clarity. Variety* refers to how different one landscape is from another. *Diversity* and *uniqueness* also have a similar meaning. However, it is emphasized that greater diversity is not necessarily an indication of higher aesthetic value.

Impact of Religion

The role of religion in causing, perpetuating, or condoning environmental disruption and degradation is a vigorously debated issue. One school of thought holds that the

Judeo-Christian heritage of Western man is responsible for the way we treat the environment. A second school refutes this and argues that man's treatment of the land is a characteristic that transcends religious and cultural teaching.

Arguing that the Judeo-Christian heritage is responsible for the Western man's attitudes and behavior with respect to his environment, Lynn White, Jr., cites the following evidence. First, Christianity is the most anthropocentric religion the world has seen. It establishes a dualism of man and nature, insisting that it is God's will that man exploit nature. Second, by destroying pagan animism, Christianity, which previously tended to unite man with nature, made it possible for man to degrade the environment in a way completely indifferent to the rights or feelings of natural objects. Third, Western science, technology, and industrialization are a natural result of Judeo-Christian dogma of creation, which teaches that man was created after plants, animals, and fishes as their rightful monarch. Fourth, great Western scientists from the thirteenth century to the eighteenth explained their motivation in religious terms, indicating that science was seeking to understand God's mind by discovering how His creation operates rather than the previous emphasis on decoding the physical symbols of God's communication with man. In other words, before the thirteenth century, man studied nature because God created nature and consequently nature revealed the divine mentality. In the early church, therefore, nature was conceived as a symbolic system through which God spoke to man. In the West, this changed in the thirteenth century when scientists in the name of religious progress began investigating physical processes of light and matter (8).

The second school argue that the environmental crisis is not a religious problem. They criticize White's thesis on several grounds. First, prehistoric man with his use of fire and water also caused considerable environmental disruption. White also recognizes this. Second, before the birth of Christ, the early Greeks and Romans both imposed their will on the environment. Third, the triumph of Christianity over paganism brought no revolutionary change to the relation between society and nature. Fourth, although the ideals of some cultures may suggest that land is sacred, there is a considerable hiatus between ethical ideals and actual land-use practices (3, 9).

One can conclude that religious attitudes and beliefs are not a primary cause of the environmental crisis. This does not suggest that religious activities are not responsible for considerable environmental disruption. On the contrary, numerous examples such as timber shortages in China caused by the cremation of the dead introduced by Buddhism and deforestation in Japan associated with the construction of huge wooden Buddhist halls and temples suggest that religious activity can promote environmental problems.

Implication that one religion, one culture, or one political system is responsible for the way man treats the land cannot be rigorously defended. Therefore, White's model (Figure 1.1A) is not entirely correct; and, likewise, Moncrief's modification of White's model (Figure 1.1C) to include capitalism and democracy, with or without the Judeo-Christian start, is not a complete and accurate progression. Since environmental degradation apparently transcends both religious belief and political system, we must look further for the primary cause of our present condition. A simpler explanation (Figure 1.1B) assumes that our environmental problems are due to a pattern of human development which began when earliest man attempted to use tools to better his chances for survival. A product of harsh times, early man, like other animals, extended his niche as far as restraints allowed.

Therefore, each innovation not only asserted the individual but also assured that every man who followed had an easier time. As a result, increased populations increased demand on resources as well as demanding more innovations. This spiral has continued to the present, when there are signs that we may be on a collision course with our environment. In a small way, the condition may be analogous with what sometimes happens to deer when, through man's artificial management of animals, their numbers exceed the carrying capacity of the land. Deprived of their natural enemies, the deer herds increase until these "artificial deer" completely eat all available food, insuring that there is a serious shortage of winter feed and little reproduction of food plants for the following spring. Everything from wildflowers to trees is gradually impoverished, and the deer either become dwarfed from malnutrition or starve to death (2). Is man with his increasingly artificial environment doing to himself what he has imposed on the deer?

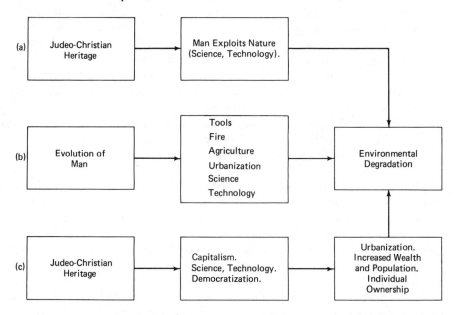

FIGURE 1.1 *Models showing possible paths leading to environmental degradation.*

SUMMARY AND CONCLUSIONS

Functional categories of society that are significant in environmental studies and the cultural basis for environmental degradation are ethical, economic, political, aesthetic, and, perhaps, religious.

       Our ethical framework appears to be slowly expanding and will eventually include the total environment in a land ethic. This ethic affirms the right of all resources, including plants, animals, and earth materials, to continue existence, and, at least in certain locations, their continued existence in a natural state (2).

The immediate cause of environmental degradation is over-population, urbanization, and industrialization combined with, as yet, little ethical regard for our land and inadequate institutions to cope with environmental stress. These problems are not unique to a particular political system; and, therefore, we conclude that the salvation of the landscape community necessitates changing social, economic, and ethical behavior that transcends political systems.

Aesthetic factors are now being considered in local, regional, and natural land-use planning, and scenery is considered as a natural resource. A problem remains in finding a method of aesthetic evaluation that is easy to understand and is quantitative, credible, and predictive. Until we do have a satisfactory methodology, it will remain difficult to balance aesthetic and economic costs and benefits.

The role of religion in causing, perpetuating, or condoning environmental degradation remains a much debated issue. Some authors argue that the Judeo-Christian heritage is responsible for Western man's attitudes and behavior toward the environment. The argument is that the Judeo-Christian teachings and practices destroyed pagan animism, which previously tended to unite nature and man, and thereby made it possible for man to degrade the environment with complete indifference. This argument cannot be rigorously defended. Prehistoric man and modern peoples believing in both Eastern and Western religions have exploited and disrupted their land. One can conclude that religious institutions have indeed been responsible for some environmental problems, but that the general tendency for degradation of the environment is a more universal problem that transcends religious teachings.

References

1. Udall, S. L. 1963. *The quiet crisis.* New York: Avon Books.
2. Leopold, A. 1949. *A Sand County almanac.* New York: Oxford University Press.
3. Moncrief, L. W. 1970. The cultural basis for our environmental crisis. *Science* 170: 508–12.
4. Goldman, M. I. 1971. Environmental disruption in the Soviet Union. In *Man's impact on environment,* ed. T. R. Detwyler, pp. 61–75. New York: McGraw-Hill.
5. Florman, S. C. 1968. *Engineering and the liberal arts.* New York: McGraw-Hill.
6. Zube, E. H. 1973. Scenery as a natural resource. *Landscape Architecture* 63: 126–32.
7. Litton, R. B. 1973. Aesthetic dimensions of the landscape. In *Natural Environments,* ed. J. V. Kantilla, pp. 262–91. Baltimore: John Hopkins University Press.
8. White, L., Jr. 1967. The historical roots of our ecological crisis. *Science* 155: 1203–7.
9. Yi-Fu, T. 1970. Our treatment of the environment in ideal and actuality. *American Scientist* 58: 244–249.

2

FUNDAMENTAL CONCEPTS

In this chapter we introduce concepts basic to the understanding and study of environmental geology. Although these concepts probably do not constitute a complete list, they provide the philosophical framework of this book. They are not to be memorized. An understanding of the general thesis of each concept will be a significant help in comprehending and evaluating philosophical and technical material throughout the remainder of the text.

Concept One

The earth is essentially a closed system.

A system may be considered as any part of the universe that is isolated in thought or in fact for the purpose of studying or observing changes that take place under various imposed conditions (1). Thus, examples of systems might include a planet, a volcano, or an ocean basin. Most systems contain various component parts which mutually adjust, and each part exerts a partial control on the others. For example, the earth may be considered a system with four parts: the *atmosphere;* the *hydro-sphere;* the *biosphere;* and the *lithosphere.* The mutual interaction of these parts is responsible for the surface features of the earth today. Furthermore, any change in the magnitude or frequency of processes in one part will affect the other parts. For example, a change in the magnitude of the processes which produce mountains may affect the atmosphere by releasing volcanic gas and causing regional changes in precipitation patterns as a new rain shadow is produced. This in turn will locally affect the hydrosphere as greater or less runoff reaches the ocean basins. Biospheric changes due to the changing environment can also be expected, and eventually the steeper slopes will also affect the lithosphere by facilitating increased erosion, which in turn will change the rate and type of sediments produced and thus the type of rocks produced from the sediments. These interactions among the variables in systems are not random and may be understood by examining each variable to deter-mine how it interacts with other variables and how it varies spatially over a site, area, or region. For example, in the hydrosphere, the spatial distribution of the

TABLE 2.1. *Residence times of some natural cycles.*

Earth Materials	Some Typical Residence Times
Atmosphere circulation	
Water vapor	10 days (lower atmosphere)
Carbon dioxide	5 to 10 days (with sea)
Aerosol particles	
Stratosphere (upper atmosphere)	Several months to several years
Troposphere (lower atmosphere)	One week to several weeks
Hydrosphere circulation	
Atlantic surface water	10 years
Atlantic deep water	600 years
Pacific surface water	25 years
Pacific deep water	1300 years
Terrestrial groundwater	150 years [above 2500 feet (760 m) depth]
Biosphere circulation [a]	
Water	2,000,000 years
Oxygen	2000 years
Carbon dioxide	300 years
Seawater constituents [a]	
Water	44,000 years
All salts	22,000,000 years
Calcium ion	1,200,000 years
Sulfate ion	11,000,000 years
Sodium ion	260,000,000 years
Chloride ion	Infinite

SOURCE: *The Earth and Human Affairs* by the National Academy of Sciences. Copyright © 1972 by the National Academy of Sciences (Canfield Press). By permission of Harper & Row, Publishers.
[a] Average time it takes these materials to recycle with the atmosphere and hydrosphere.

oceans with respect to the sun affects the evaporation process of ocean waters, which in turn affects atmospheric conditions by increasing or decreasing the amount of water in the atmosphere.

We know that the earth is not static. Rather, it is a dynamic, evolving system in which material and energy are constantly changing. Such dynamics might be considered evidence that the earth is an open system with no boundaries of energy or material. This interpretation is applicable as long as the sun continues to impart energy to the earth. However, considering natural earth cycles such as the water and rock cycles in which there is a continual recycling of earth materials, we can best think of the earth as a closed system or, in reality, a coalition of a large number of closed systems (2). For example, the rain that falls today will eventually return to the atmosphere, and the sediment deposited yesterday will eventually be transformed into solid rock. Therefore, although the earth is currently, and seemingly forever will be, an open system in terms of energy and material, it is essentially a closed system in terms of natural earth cycles.

As more and more demands are made on the earth and its limited resources, it has become increasingly important to understand the magnitude and frequency of the processes which maintain earth cycles. For example, if we hope to manage the water resources of a region, we must know the nature and extent to

TABLE 2.2. *Rates of some natural cycles.*

Earth Processes	Some Typical Rates
Erosion	
Average U.S. erosion rate[a]	2.4 inches (6.1 cm) per 1000 years
Colorado River drainage area	6.5 inches (16.5 cm) per 1000 years
Mississippi River drainage area	2.0 inches (5.1) cm) per 1000 years
N. Atlantic drainage area	1.9 inches (4.8 cm) per 1000 years
Pacific slope (Calif.)	3.6 inches (9.1 cm) per 1000 years
Sedimentation[b]	
Colorado River	281 million metric tons per year
Mississippi River	431 million metric tons per year
N. Atlantic coast of U.S.	48 million metric tons per year
Pacific slope (Calif.)	76 million metric tons per year
Tectonism	
Sea-floor spreading	
N. Atlantic	1 inch (2.5 cm) per year
E. Pacific	3 to 4 inches (7 to 10 cm) per year
Faulting	
San Andreas (Calif.)	0.5 inch (1.3 cm) per year
Mountain uplift	
Cajon Pass, San Bernardino Mts. (Calif.)	0.4 inch (1 cm) per year

SOURCE: *The Earth and Human Affairs* by the National Academy of Sciences. Copyright © 1972 by the National Academy of Sciences (Canfield Press). By permission of Harper & Row, Publishers.
a Thickness of the layer of surface of the continental United Stated eroded per 1000 years.
b Includes solid particles and dissolved salts.

which natural processes will supply groundwater and surface water. Or, if we are concerned with disposal of dangerous chemicals in a disposal well, we must know how the disposal procedure will interact with natural cycles to insure that we or our heirs will not be exposed to hazardous chemicals. This becomes especially critical in dealing with radioactive wastes which must be contained from several centuries to as long as one quarter million years. Therefore, it is exceedingly important to recognize earth cycles and determine the length of time involved in various parts of specific cycles. Tables 2.1 and 2.2 list the rates of several processes and typical residence times for some natural cycles.

Concept Two

>   *The earth is the only suitable habitat we have, and its resources are limited.*

The place of man in the universe is well stated in the *Desiderata:* "You are a child of the universe, no less than the trees and the stars; you have a right to be here. And whether or not it is clear to you, no doubt the universe is unfolding as it should" (3).

Leo F. Laporte, senior author of *The Earth and Human Affairs,* considers that the context of Concept Two includes two fundamental truths: first, that this earth is indeed the only place to live that is now accessible to us; and second, that our resources are limited, and while some resources are renewable, many are not. Therefore, we will eventually need large-scale recycling of many materials, and a large part of our solid and liquid waste-disposal problems could be alleviated if these wastes were recycled. In other words, many things now considered pollutants may be considered resources out of place.

There are at least two dichotomous views on natural resources. One school holds that finding resources is not so much a problem as is finding ways to utilize them. In other words, the entire earth, including the ocean and atmosphere, has raw materials that can be made useful if we can develop the necessary ingenuity and skill (4). The basic assumption is that as long as there is freedom to think and innovate, man will be able to produce sufficient energy and locate sufficient resources to meet his needs. Evidence to support this line of reasoning includes that first, efficient and intelligent use of materials has historically been a successful venture; second, we know more about extracting minerals and fuel than we did in the past and so can find new resources faster and mine lower-grade mineral deposits; and third, new work with atomic power and recycling of resources will more than meet the needs of the future.

The second school holds that "cornucopian premises" as outlined above are fallacious on grounds that an exponential increase of people and mineral products on a finite resource base is impossible. Furthermore, Preston Cloud states that we are in a resource crisis because of first, improvements in medical technology contributing to the overpopulation of the earth; second, an unrealistic view of the necessity of an ever-increasing gross national product based on obsolescence and waste; third, the finite nature of the earth's accessible minerals; and fourth, increased risk of irreversible damage to the environment as a result of overpopulation, waste, and the necessity of larger and larger mining operations to obtain ever smaller proportions of useful minerals (5).

The history of man can be traced back only several million years. Geologically, this is a very short time. The dinosaurs, for example, ruled the land for more than 100 million years. The evidence from earth history suggests that more species have become extinct than have survived! What then will be the history of man and who will write it? Hopefully, we will be something more in the geologic record than a good index fossil indicating a short time in earth history when man was abundant.

Concept Three

Today's physical processes are modifying our landscape and have oper-
ated throughout much of geologic time. However, the magnitude and
frequency of these processes are subject to natural and man-induced
change.

The concept that an understanding of present processes which are forming and modifying our landscapes will facilitate the development of inferences concerning the geologic history of a landscape is known as the *doctrine of uniformitarianism.* Simply stated as the "present is the key to the past," uniformitarianism was first suggested by James Hutton in 1785, elegantly restated by John Playfair in 1802, and popularized by Charles Lyell in the early part of the nineteenth century. Today it is heralded as one of the fundamental concepts of the earth sciences.

Uniformitarianism does not demand or even suggest that the magnitude and frequency of natural processes remain constant with time. Furthermore, it is obvious that the principle cannot be extended back throughout all of geologic time because the processes operating in the oxygen-free environment during the first 2 billion years of earth history were quite different from today's processes. However, as long as the past continents, oceans, and atmosphere were similar to those of today, we can infer that the present processes also operated in the past. For example, if we have studied present alpine glaciers and the characteristic erosional and depositional landforms associated with alpine glaciation, then we can infer that valleys with similar landforms were at one time glaciated even if no glacial ice is present today. In like manner, if one finds ancient gravel deposits with all the characteristics of stream gravel on the top of a mountain, then uniformitarianism can be used to suggest that a stream must have flowed there at one time. In other words, what was originally a stream valley has been changed by differential erosion to a mountain top. This is known as *inversion of topography.* Phenomena such as this would be very difficult to determine correctly if it were not for the principle of uniformitarianism.

It is important to understand the effects of man on increasing or decreasing the magnitude and frequency of natural earth processes. For example, rivers will flood regardless of man's activities, but the magnitude and frequency of flooding may be greatly increased or decreased due to man's activity. Therefore, in order to predict the long-range effects of a certain process such as flooding, we must be able to determine how man and his future activities will change the rate of the process. In this case, the present may have to be the key to the future. We can assume that the same processes will operate but that the rates will vary as the environment adjusts to man's activity. Furthermore, it must be concluded that ephemeral landforms such as beaches and lakes will appear and disappear in response to natural processes, and man's influence may be small in comparison.

Although the effects of man may be small on a global scale, they are very pronounced in a local area. For example, one year of erosion at a construction site may exceed many decades of erosion from an equivalent tract of woodlands or even agricultural land (6). This erosion results from exposure of the soil following the removal of vegetation. Therefore, if we are going to maximize the value of geologic knowledge in land-use planning, we must be able to use our understanding of natural earth processes in both a historical and a predictive mode. For example, when environmental geologists examine a recent mudflow (flowage of saturated heterogeneous debris) deposits in an area designated to become a housing development, they must use uniformitarianism to infer where there will be future mudflows, as well as to predict what effects urbanization will have on the magnitude and frequency of future flows.

Concept Four

There have always been earth processes that are hazardous to man. These natural hazards must be recognized and avoided where possible, and their threat to human life and property must be minimized.

Our discussion of uniformitarianism established that present processes have been operating a good deal longer than man has been on the earth. Therefore, man has always been obligated to contend with processes that tend to make his life difficult. Surprisingly, however, modern man appears to be a product of the Ice Age, one of the harshest of all environments.

Early in man's history, his struggle with natural earth processes was probably a day-to-day experience. However, his numbers were neither great nor concentrated, and, therefore, his losses from hazardous earth processes were not very significant. As man developed and learned to produce and maintain a constant food supply, his numbers increased and probably also the effects of hazardous earth processes. This resulted because population centers probably became local centers of pollution and disease. Furthermore, the concentration of population and resources increased the impact of periodic earthquakes, floods, and other natural disasters. This trend has continued until today many people either are living in areas that are likely to be damaged by hazardous earth processes or are susceptible to adverse impact of such processes in adjacent areas.

Natural earth processes are called *exogenetic* if they operate at or near the surface of the earth and *endogenetic* if they operate within or below the earth's crust. Exogenetic processes include weathering; mass wasting; and either erosion or deposition by such agents as running water, wind, or ice. Volcanic activity and diastrophism (processes which produce mountains, continents, ocean basins, etc.)

are common endogenetic processes. The work of organisms, including man, are primarily exogenetic processes; however, man is now able to cause some endogenetic processes such as earthquakes.

Many processes continue to cause loss of life and property damage, including flooding of coastal or floodplain areas; earthquakes; volcanic activity; mass-wasting phenomena such as landslides and mudflows; and weathering. The magnitude and frequency of these processes depend on such factors as the climate, geology, and vegetation of the region. For example, the effects of running water as an crosional or depositional process depend on the intensity of the rainfall; the frequency of the storms; how much and how fast the rainwater is able to infiltrate into the rock or soil; the rate of evaporation and transpiration of water back into the atmosphere; the nature and extent of the vegetation; and topography. Considering these factors, Peltier concluded that there presently are three areas where the rate of erosion by running water is minimized: first, mid-latitude desert regions of low rainfall; second, arctic and subarctic regions of low rainfall; and third, tropical regions characterized by abundant plant cover (7). The areas of maximum erosion by running water are the mid-latitude regions with more abundant rainfall and less vegetation cover than found in tropical regions (Figures 2.1 and 2.2). Similar reasoning can be used to determine areas of maximum weathering (Figure 2.3) and maximum mass wasting (Figure 2.4). The endogenetic processes associated with volcanoes, earthquakes, and tsunamis (very large sea waves, usually incorrectly referred to as *tidal waves*) are located primarily in response to geologic conditions. For example, the "ring of fire" consisting of the circum-Pacific area is well known for its high incidence of earthquakes and volcanic activity resulting from dynamic, global geologic processes originating deep within the earth.

From our discussion of natural earth processes we can conclude that many processes can be recognized and predicted by considering climatic, biologic, and geologic conditions. After earth scientists have identified potentially hazardous processes, they should make the information available to planners and decision

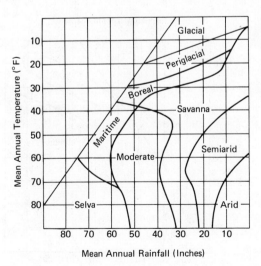

FIGURE 2.1 *Morphogenetic regions. (Source: L. C. Peltier. Reproduced by permission from the* ANNALS *of the Association of American Geographers, Volume 40, 1950.)*

FIGURE 2.2 *Intensity of erosion by running water. (Source: L. C. Peltier. Reproduced by permission from the* ANNALS *of the Association of American Geographers, Volume 40, 1950.)*

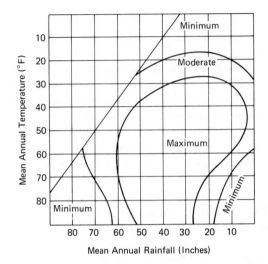

FIGURE 2.3 *Intensity of weathering. (Source: L. C. Peltier. Reproduced by permission from the* ANNALS *of the Association of American Geographers, Volume 40, 1950.)*

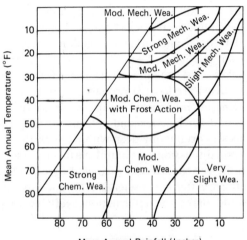

FIGURE 2.4 *Intensity of mass wasting. (Source: L. C. Peltier. Reproduced by permission from the* ANNALS *of the Association of American Geographers, Volume 40, 1950.)*

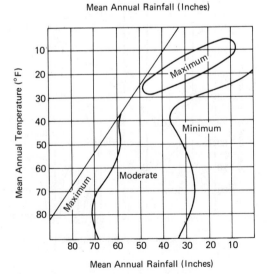

makers. Then various alternatives can be formulated to avoid or minimize any threat to human life or property.

Concept Five

Land- and water-use planning must strive to obtain a balance between economic considerations and the less tangible variables such as aesthetics.
Scenery is now considered a natural resource, and the aesthetic evaluation of a site or landscape before modification is becoming an important part of the environmental impact statement. It is refreshing to note that consideration is now being given to the less tangible variables such as aesthetics, as well as the more traditional benefits-cost analysis. Until this revolution, the justification for a project was weighed by comparing the financial benefits over a period of time with the cost. The assumption now formulated is that there are varying *scenic* values, just as there are varying economic values, associated with a landscape and proposed modification (8).

Most of the research on landscape aesthetics has been concerned with unique landscape nearly untouched by the activity of man. However, since most of us now are living in an urban environment, we should evaluate the aesthetic resources of these areas. Such evaluation would facilitate land use by identifying the various alternatives. During the evaluation state, this might temporarily interfere with the supply and demand aspects of urbanization, but the result might be worth the extra trouble.

The balancing of economic criteria with aesthetic criteria is ambitious and optimistic as well as very difficult. The problem is, On what basis can the two be compared? The one logical solution is a hierarchical ranking of the economic alternatives compared to a similar ranking of the aesthetic evaluation. Trade-offs and compromise can then be considered. Pragmatically, the problem of comparing economic consideration with aesthetic evaluation is not particularly difficult, provided we can first develop a generally agreed upon rating scale for aesthetic evaluation; second develop a reliable quantitative method to analyze the data and hierarchically rank the alternatives; and third develop techniques to map our scenic resources.

Concept Six

The effects of land use tend to be cumulative, and, therefore, we have an obligation to those who follow.
Several million years ago, when early hominids roamed the grasslands, marshy deltas, and adjacent forests of ancient Lake Rudoff along the Great Rift Valley system of East Africa, man was completely dependent upon his immediate environment. His effect on that environment was probably insignificant as he hunted game and was in turn hunted by predators. This relationship between man and his environment probably existed until about 800,000 years ago when he developed his skill in the use of fire.

The use of fire brought new effects different from earlier impacts on the environment. First, fire was capable of affecting large areas of forest or grasslands. Second, it was a repetitive process capable of damaging the same area at rather frequent intervals. Third, it was a rather selective process in that certain species were locally exterminated while other species that exhibited a resistance to or rapid recovery from fire were favored (9). It is doubtful that early man's use of fire for protection and hunting had any long-term effect on the environment. However, as man became more and more dependent on an increasing variety of resources for his clothes, lodging, and hunting, he also increased his capacity to observe and test his environment. This early experimentation probably led to the use of plants and primitive agriculture about 7000 B.C.

The emergence of agriculture was the first instance of an artificial land use capable of modifying the natural environment. It also set the stage for the development of a more or less continuously occupied site or cluster of sites which introduced further modification of the environment such as shelter for living space, primitive latrines, and protective barriers against predators and other people. Furthermore, these early sites probably became the first areas to experience pollution problems resulting from disposal of waste and soil erosion problems resulting from removal of indigenous vegetation (9). The innovation of agriculture also supplied the necessary nutrients for an increasing population that necessitated the clearing of additional land. This activity certainly influenced the ecological balance of the area as some species were domesticated or cultivated and others were removed as pests. Therefore, it is not surprising to note that the increase in human population is paralleled by an increase in the number of extinctions in birds and mammals (Figure 2.5).

The significant point of the entire developmental process of man through time is that as cities and farms increase, there is an increased demand for diversification of land use, the effects of which tend to be cumulative with time (10). If this is the case, then from an ethical and moral standpoint, we need to examine the effects of land use in a historical framework, if only to insure that our children and their children can survive in the environment they inherit. This is especially critical when it has been determined that at least since the beginning of civilization, 6000 years ago, and perhaps as far back as 15,000 years ago, virtually the entire surface of the earth has been altered by human activity. In other words, there is little, if any, land that can be considered original or untouched (2). Furthermore, man's ability to cause further changes is increasing at a rapid rate. In defense of man's activity, it can be noted that in comparison to the energy of mountain building, volcanic activity, and erosion power of streams, man's impact is small. However, it is not insignificant, especially in the large metropolitan areas where there are many negative aspects of urbanization.

The lessons in land use from the Old World are quite explicit. Where sound conservation practices were used, there were successful adjustments of population. However, where wasteful exploitation of resources was practiced, the results varied from gullied fields and alluvial plains in rocky hills and steep slopes, to silted-up irrigation reservoirs and canals, to ruins of prosperous cities. Three examples from Lowdermilk's study will emphasize this point (10).

Ancient Phoenicia and Slope Farming   About 5300 years ago, the Phoenicians migrated from the desert to settle along the eastern coast of the Mediterranean Sea

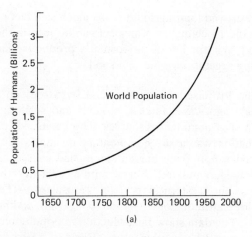

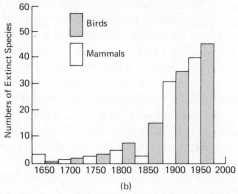

FIGURE 2.5 *(a) Increase in the human population paralleled by (b) an increase in the extinction of birds and animals. [Reproduced, by permission, from V. Ziswiler,* Extinct and Vanishing Species *(New York: Springer-Verlag, 1967).]*

and established the coastal towns of Tyre and Sidon, Beyrouth and Byblos. The land is mountainous with a relief of 10,000 feet and, at that time, was heavily forested with the famous cedars of Lebanon. These trees became the timber supply for the alluvial plains of the Nile and Mesopotamia. As the limited, flat land along the coast was populated to the carrying capacity, the people moved to the slopes, and as the slopes were cleared and cultivated, they were subject to soil erosion. Today there are a large number of terrace walls in various states of repair, indicating that the ancient Phoenician farmers attempted to control erosion with rock walls across the slope as many as 40 to 50 centuries ago.

The cedars of Lebanon retreated under the ax, until today very little of the original forest of approximately 1000 square miles is left. Evidence suggests that given present climatic conditions, the forests would grow where soil has escaped the process of erosion. Today the bare limestone slopes strewn with remnants of former terrace walls are testimony to the results of erosion and the decline and loss of the resources of a country.

Dead Cities of Syria   There exist in northern Syria a number of formerly prosperous cities that are now for all practical purposes dead. Before the invasion of the Persians and Arabs, cities such as El Bare prospered from the conversion of forest to farmland, which resulted in exports of olive oil and wine to Rome. After the invasion, which destroyed their agriculture, as much as 6 feet of soil eroded from the slopes, and today the 13 centuries of neglect can be seen in the nearly complete destruction of the land. What is left of the formerly productive land is a man-made desert generally lacking vegetation, water, and soil.

Palestine Story The Promised Land described by Moses on Mount Neba approximately 3000 years ago was a land of streams and springs, a land of wheat, barley, and vines, a land of abundant resources. The Promised Land at the time of Lowdermilk's investigation was a sad commentary on man's use of the land. Soil erosion and accelerated runoff from barren slopes had caused a good deal of the hills to become greatly depopulated. For example, consider the Wadi Musrara watershed which drains the western slope from Jerusalem to Tel-Aviv. The area can be divided into three altitude zones: plains (0–100 m); foothills (100–300 m); and hills (over 300 m). The data show the reduction in population of ancient village sites with topography since the seventh century (Figure 2.6). The breakdown of old terrace walls on steeper slopes and subsequent erosion of the soil are probably sufficient to explain the greater frequency of abandoned sites in the hills.

Although the land in Palestine cannot be restored to its original productivity, the efforts to redeem the land in recent years is a good example of what can be done at great costs. Lowdermilk reported that as of 1960, Israel had more than doubled its cultivated land to a million acres, drained 44,000 acres of marshland, and established range cover on vast amounts of uncultivated land for its livestock industry. The results of their efforts are significant in that the country had at that time nearly attained agricultural self-sufficiency with an export-import balance in food (11).

History of areas long occupied by man suggest that soil erosion is a serious problem that has destroyed land and retarded the progress of civilization. Therefore, conservation of our soils must remain a national interest for, as stated by Lowdermilk, "One generation of people replaces another, but productive soils destroyed by erosion are seldom restorable and never replaceable."

FIGURE 2.6 *Relationship between abandoned village sites and topography in the Wadi Musrara watershed. [Data from W. C. Lowdermilk,* Lessons from the Old World to the Americans in Land Use *(Smithsonian Report, 1943).]*

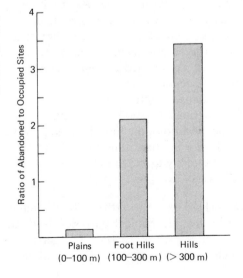

Concept Seven

> *The fundamental component of every person's environment is the geologic factor, and understanding of this environment requires a broad-based comprehension and appreciation of the earth sciences and other related disciplines.*

Concept Seven arises from the fact that all geology is environmental, and since we all live on the surface of the earth, we are both directly and indirectly affected by geologic processes (12). Therefore, an understanding of our complex environment requires considerable knowledge of such disciplines as *geomorphology,* the study of landforms and surface processes; *petrology,* the study of rocks and minerals; *sedimentology,* the study of environments of depositions of sediments; *tectonics,* the study of processes which produce continents, ocean basins, mountains, and other large structural features; *hydrogeology,* the study of surface and subsurface water; *pedology,* the study of soils; *economic geology,* the application of geology to locating and evaluating mineral materials; and *engineering geology,* the application of geologic information to engineering problems. Beyond this, the serious earth scientist should also be aware of the contributions to environmental research from areas such as physical, cultural, economic, and urban geography; biology; conservation; atmospheric science; chemistry; environmental law; architecture; and engineering. Environmental geology is the domain of the generalist with strong interdisciplinary interest. This in no way refutes the significant contributions of specialists in various aspects of environmental studies, or the importance of the generalist's consideration of specific problems or specialty areas for research. It merely suggests that although our research interest may be specialized, we should be generalists in terms of awareness of other disciplines and their contribution to environmental geology. It is further emphasized that many projects may be studied best by an interdisciplinary team of scientists (but teams must be well disciplined).

The importance of the interdisciplinary nature of environmental geology becomes apparent when the nature of environmental problems is explored. Most projects are complex and involve many different facets that may be generalized into three categories: physical; biological; and of human use and interest. These are essentially the same categories used by Leopold to evaluate river valleys, and their extension to other areas of environmental research seems appropriate (13). The category of physical factors includes such considerations as physical geography; geologic processes; hydrologic processes; rock and soil types; and climatology. Biologic factors include consideration of the nature of plant and animal activity; changes in biologic conditions or processes; and spatial analysis of biologic information. Human use and interest factors include such attributes as land use; economics; aesthetics; interaction between man's activity and the physical and biological realms; and environmental law.

Obviously, no one project or research interest will involve all possible factors in each of the three categories, and there is considerable interaction between the categories. Projects such as waste-disposal operations, highway construction, mass transit systems, urban land-use planning, and mining of resources may be concerned with all of the categories. For example, the planning, construction, and operation of a sanitary landfill site is concerned with physical factors such as physical location, topography, soil type, and hydrologic conditions; biologic processes, which determine the rate of decay of the organic refuse, as well as any contamination of the biologic realm in the vicinity of the site; and human interests such as compliance to laws, regulations, and good engineering practice.

SUMMARY AND CONCLUSIONS

Seven fundamental concepts basic to the understanding of environmental geology were introduced:

1. The earth is essentially a closed system.

2. The earth is the only suitable habitat we have, and its resources are limited.

3. Today's physical processes are modifying our landscape and have operated throughout much of geologic time. However, the magnitude and frequency of these processes are subject to natural and man-induced change.

4. Earth processes that are hazardous to man have always existed. These natural hazards must be recognized and avoided where possible, and their threat to human life and property must be minimized.

5. Land- and water-use planning must strive to obtain a balance between economic considerations and the less tangible variables such as aesthetics.

6. The effects of land use tend to be cumulative, and, therefore, we have an obligation to those who follow.

7. The fundamental component of every person's environment is the geologic factor, and understanding of this environment requires a broad-based comprehension and appreciation of the earth sciences and other related disciplines.

Although they probably do not constitute a complete list, these concepts set a philosophical framework from which to investigate and discuss environmental geology.

References

1. Ehlers, E. G. 1968. *Phase equilibria: laboratory studies in mineralogy*. San Francisco: W. H. Freeman.
2. National Research Council. 1971. *The earth and human affairs*. San Francisco: Canfield Press.
3. *Desiderata*. 1692.
4. Holman, E. 1952. Our inexhaustible resources. *Bulletin of the American Association of Petroleum Geologists* 6: 1323–29.
5. Cloud, P. E., Jr. 1968. Realities of mineral distribution. In *Man and his physical environment*, eds. G. D. McKenzie, and R. O. Utgard, pp. 194–207. Minneapolis, Minnesota: Burgess.
6. Wolman, M. G., and Schick, A. P. 1967. Effects of construction on fluvial sediment, urban and suburban areas of Maryland. *Water Resources Research* 3: 451–64.
7. Peltier, L. C. 1950. The geographic cycle in periglacial regions as it is related to climatic geomorphology. *Annals of the Association of American Geographers* 40: 214–36.

8. Zube, E. H. 1973. Scenery as a natural resource. *Landscape Architecture* 63: 126–32.

9. Nicholson, M. 1970. Man's use of the earth: historical background. In *Man's impact on environment,* ed. T. R. Detwyler, pp. 10–21. New York: McGraw-Hill.

10. Lowdermilk, W. C. 1943. *Lessons from the Old World to the Americans in land use.* Smithsonian Report for 1943, pp. 413–28.

———. 1960. The reclamation of a man-made desert. *Scientific American* 202: 54–63.

12. Oakeshott, G. B. 1970. Controlling the geologic environment for human welfare. *Journal of Geological Education* 18: 193.

13. Leopold, L. B. 1969. *Quantitative comparison of some aesthetic factors among rivers.* U.S. Geological Survey Circular 620.

References

3

EARTH MATERIALS
AND PROCESSES

Geologic Cycle

Throughout the 5 billion years of earth history, the materials on or near the earth's surface have been created, maintained, and destroyed by numerous physical, chemical, and biochemical processes. Except during the early history of our planet, the processes which produce the earth materials necessary for our survival have periodically reproduced new materials. Collectively, the processes are referred to as the *geologic cycle* (Figure 3.1), which is really a group of subcycles (1). Two of the more important subcycles are the *tectonic* and *hydrologic* cycles.

The hydrologic cycle is the movement of water from the oceans, to the atmosphere, and back to the ocean, by way of evaporation, runoff in streams and rivers, and groundwater flow. Only a very small amount of the total water in the ocean is active in the hydrologic cycle at any one time, and yet this small amount of water is tremendously important in facilitating the movement and sorting of chemical elements in solution (geochemical cycle), sculpturing the landscape, weathering

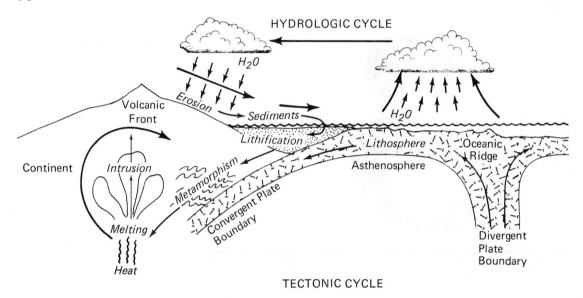

FIGURE 3.1 *The geologic cycle composed of subcycles, including the hydrologic, the rock, the tectonic, and the geochemical cycles.*

rocks, transporting and depositing sediments, and providing our water resources. Therefore, we should learn more about the hydrologic cycle to better understand and utilize this important resource.

The tectonic processes are driven by forces deep within the earth. They deform the earth's crust, producing external forms such as ocean basins, continents, and mountains. These processes are collectively known as the *tectonic cycle*. We now know that the outer layer of the earth, containing the continents and oceans, is 40 to 60 miles thick. It is called the *lithosphere,* and it is not a continuous, uniform layer. Rather, the lithosphere is broken into several large parts called *plates* which are moving relative to one another (Figure 3.2) (2). As the lithospheric plates move over the asthenosphere, which is thought to be a more or less continuous layer of little strength below the lithosphere, the continents also move (Figure 3.3) (3). This moving of continents is called *continental drift*. It is believed that the most recent episode of drift started about 200 million years ago, when a supercontinent called *Pangaea* broke up.

The boundaries between plates are geologically active areas, and most earthquakes and volcanic activities occur there. Three types of boundaries exist: divergent, convergent, and transform fault (4). *Divergent boundaries* occur at oceanic ridges where plates are moving away from each other, producing new lithosphere. *Convergent boundaries (subduction zones)* occur when one plate dives beneath the leading edge of another plate. However, if both leading edges are composed of relatively light continental material, it is more difficult for subduction to start, and a special type of convergent plate boundary called a *collision boundary* may develop. This produces linear mountain systems such as the Alps and the

Himalayas. *Transform fault* boundaries occur where one plate slides past another, as, for example, the San Andreas fault in California (Figure 3.2).

Two other subcycles of the geologic cycle are the *geochemical* and *rock* cycles. Geochemistry is the study of the distribution and migration of elements in earth processes, and the geochemical cycle is the migratory path of elements during geologic changes. This cycle involves the chemistry of the lithosphere, asthenosphere, hydrosphere, and biosphere. The rock cycle (Figure 3.4) is a sequence of processes which produce the three rock families: igneous, sedimentary, and metamorphic. The geochemical and rock cycles are closely related to one another and intimately related to the hydrologic cycle which provides the water necessary for many chemical and physical processes. The tectonic cycle is also intimately related to the other cycles as it provides water from volcanic processes as well as heat and energy to form and change many earth materials.

Our discussion of the geologic cycle has established that earth materials such as minerals, rocks, soil, and water are constantly being created, changed, and destroyed by internal and external earth processes in numerous subcycles. The remainder of this chapter will be discussions of the different earth materials and various aspects of their occurrence and geology that have environmental significance.

Minerals

Minerals are naturally occurring, solid, crystalline substances with physical and chemical properties that vary within known limits (see Plate 3.A). Although there are over 2000 minerals, only a few are necessary to identify most rocks. Nearly 75 percent by weight of the earth's crust is oxygen and silicon. These two, in combination with aluminum, iron, calcium, sodium, potassium, and magnesium, account for the minerals that make up about 95 percent of the earth's crust. These minerals, called the *silicates,* are some of the most important rock-forming minerals. The three most important rock-forming silicate minerals or mineral groups are quartz, feldspar, and ferromagnesium.

Quartz, one of the single most abundant minerals in the crust of the earth, is a hard, resistant mineral composed of silicon and oxygen. It is often white but, due to impurities, may also be rose, purple, black, or another color. It is usually recognized by its hardness, greater than that of glass, and the characteristic way it fractures—conchoidally (like a clam shell). Because it is very resistant to natural processes that lead to the breakdown of most minerals, quartz is the common mineral in river and most beach sands.

The feldspars, the most abundant and perhaps the most important group of rock-forming minerals in the crust of the earth, are aluminosilicates of potash, soda, and lime. They are generally white, gray, or pink and are fairly hard. Feldspars are very abundant and are important commercial minerals in the ceramics and glass industries. Feldspars weather chemically to form clays, which are hydrated aluminosilicates. This has important environmental implications. All rocks are fractured, and water, which facilitates chemical weathering, enters the fractures.

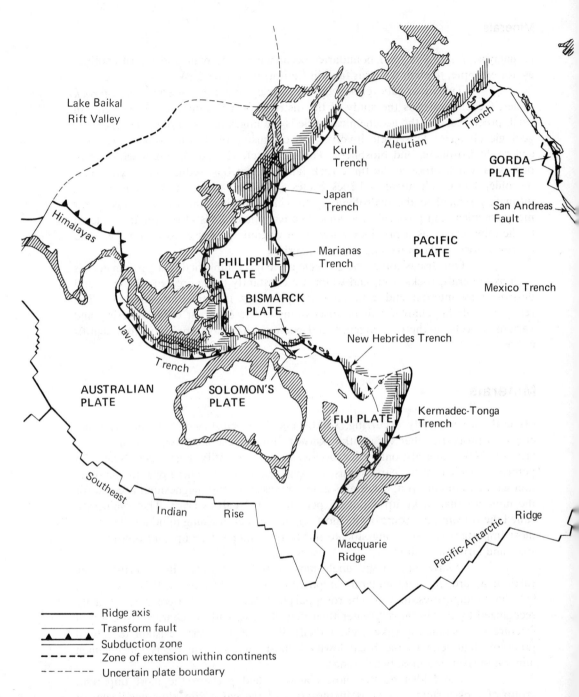

Lake Baikal
Rift Valley

Kuril
Trench

Japan
Trench

Aleutian   Trench

GORDA
PLATE

San Andreas
Fault

PACIFIC
PLATE

Marianas
Trench

PHILIPPINE
PLATE

Himalayas

BISMARCK
PLATE

Mexico Trench

New Hebrides Trench

Java

Trench

AUSTRALIAN
PLATE

SOLOMON'S
PLATE

FIJI PLATE

Kermadec-Tonga
Trench

Southeast

Indian   Rise

Macquarie
Ridge

Pacific-Antarctic   Ridge

——————— Ridge axis
▲▲▲▲ Transform fault
▲▲▲▲ Subduction zone
– – – – – Zone of extension within continents
‒ ‒ ‒ ‒ ‒ Uncertain plate boundary

FIGURE 3.2 *Lithospheric plates that form the earth's outer
layer. Three types of plate junctions are shown: the* ridge axis,
forming divergent boundaries; the subduction zones, *forming
convergent plate boundaries; and the* transform fault plate
boundaries, *where one plate is sliding by another. (From
"Plate Tectonics" by John F. Dewey. Copyright © May 1972 by
Scientific American, Inc. All rights reserved.)*

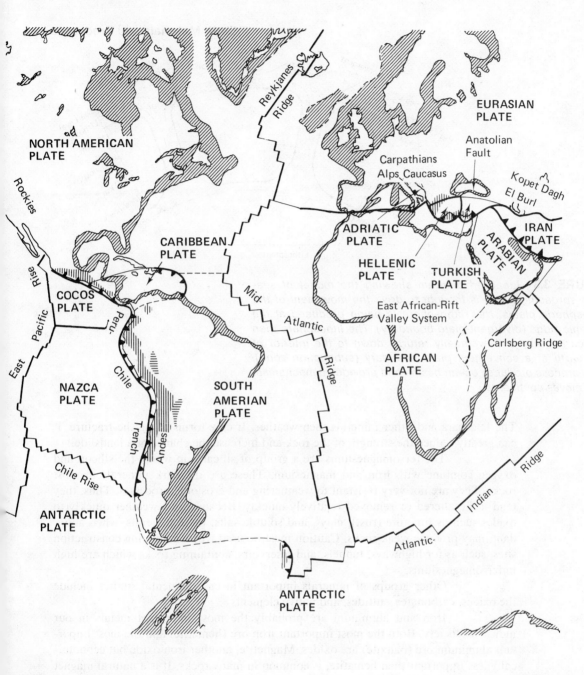

NORTH AMERICAN PLATE

Rockies

Reykjanes Ridge

EURASIAN PLATE

Anatolian Fault

Carpathians
Alps Caucasus

Kopet Dagh
El Burl

CARIBBEAN PLATE

ADRIATIC PLATE

IRAN PLATE

ARABIAN PLATE

COCOS PLATE

East Pacific Rise

Peru-

Chile

HELLENIC PLATE

TURKISH PLATE

Mid-

Atlantic

East African-Rift
Valley System

AFRICAN PLATE

Carlsberg Ridge

NAZCA PLATE

Trench

Andes

SOUTH AMERIAN PLATE

Ridge

Chile Rise

Chile Rise

Indian Ridge

ANTARCTIC PLATE

Atlantic-

ANTARCTIC PLATE

	Areas of Intermediate-Focus Earthquakes
	Areas of Deep-Focus Earthquakes
	Continental Crust

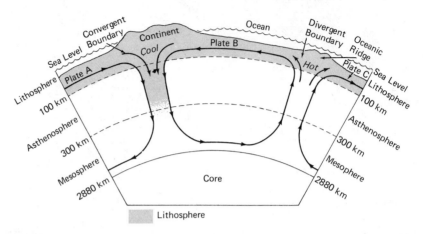

Lithosphere

FIGURE 3.3 *Idealized diagram showing the model of sea-floor spreading which is thought to drive the movement of the lithospheric plates. New lithosphere is being produced at the oceanic ridge (divergent plate boundary). The lithosphere then moves laterally and eventually returns down to the interior of the earth at a convergent plate boundary (subduction zone). This process produces ocean basins and provides a mechanism that moves continents.*

The feldspars and other minerals then weather. If clay forms along the fracture, it can greatly reduce the strength of the rock and increase the chance of a landslide.

The ferromagnesiums are a group of silicates in which the silicon and oxygen combine with iron and magnesium. These are the dark minerals in most rocks. They are not very resistant to weathering and erosional processes. Thus, they tend to be altered or removed relatively quickly. Because they weather quickly to oxides such as limonite (rust), clays, and soluble salts, these minerals, when abundant, may produce weak rocks. Caution must be used when evaluating construction sites, such as for highways, tunnels, and reservoirs, containing rocks which are high in ferromagnesiums.

Other groups of minerals important in environmental studies include the oxides, carbonates, sulfides, and native elements.

Iron and aluminum are probably the most important metals in our industrial society. Both the most important iron ore (hematite) and the most important aluminum ore (bauxite) are oxides. Magnetite, another iron oxide but economically less important than hematite, is common in many rocks. It is a natural magnet (lodestone) that will attract and hold iron particles. Where particles of magnetite are abundant, they may produce a black sand in streams or beach deposits.

Environmentally, the most important carbonate mineral is calcite, which is calcium carbonate. This mineral is the major constituent of limestone and marble, two very important rock types. Weathering by solution of this mineral from such rocks often produces *caverns,* surface pits *(sinkholes),* and other unique features. Cavern systems carry groundwater, and water pollution problems in urban areas over limestone bedrock are well known. In addition, construction of highways,

Varieties of feldspar, the most common rock-forming mineral in the earth's crust.

Two of the many possible varieties of quartz. Quartz is an important rock-forming mineral and is very resistant to weathering.

Ferromagnesian, a rock-forming mineral (augite).

Ferromagnesian, a rock-forming mineral (biotite).

Calcite, the abundant mineral in the rock limestone. Calcite is environmentally important in the rock limestone. It is environmentally important because limestone terrain may be characterized by caverns, sinkholes, and underground drainage often associated with pollution and construction problems.

Two varieties of halite (salt), clear and impure. The mineral halite is environmentally important because it is the main constituent of the rock known as rocksalt in which we may permanently dispose of nuclear waste.

Pyrite (fool's gold) is iron sulfide. It is environmentally important because in coal-mining areas the mineral is common and reacts with water and oxygen to form sulfuric acid associated with acid mine drainage.

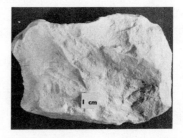

This is only one of many clay minerals. The clay minerals are environmentally important because when they are present in the soil, they give rise to such undesirable properties as low strength, high water content, high shrink-swell potential, and poor drainage.

PLATE 3.A *Some of the common rock-forming or environmentally significant minerals. The white square used for scale is one centimeter, giving an area of one square centimeter.*

reservoirs, and other engineering structures are a problem where caverns or sink-holes are likely to be encountered.

The sulfide minerals, such as pyrite, or iron sulfide (fool's gold), are sometimes associated with environmental degradation. This occurs most often when roads, tunnels, or mines cut through rocks such as coal containing sulfide minerals. The minerals oxidize to form such compounds as ferric hydroxide and sulfuric acid. This is a major problem in the coal regions of Appalachia.

Native elements such as gold, silver, copper, and diamonds have long been sought as valuable minerals. They usually occur in rather small accumulations but occasionally are found in sufficient quantities to justify mining. As we continue to mine these in ever lower-grade deposits, the environmental impact will continue to increase.

Rocks

Rocks are aggregates of one or more minerals (see Plate 3.B), and the rock cycle is truly the largest of the earth cycles. For this cycle to operate, the tectonic cycle is required for energy, the geochemical cycle for materials, and the hydrologic cycle for water. The water is used in the processes of weathering, erosion, transportation, deposition, and lithification of sediments.

The general model of the rock cycle is that the three rock families—igneous, metamorphic, and sedimentary—are involved in a worldwide recycling process. The elements of the cycle are shown in Figure 3.4. Internal heat from the tectonic cycle drives the rock cycle and produces igneous rocks from molten materials. These may crystallize beneath or on the earth's surface. Rocks at or near the surface break down chemically and physically by weathering processes to form sediments that are transported by wind, water, and ice. The sediments accumulate in depositional basins, such as the ocean, where they are eventually transformed into sedimentary rocks. After the sedimentary rocks are buried to sufficient depth, they

FIGURE 3.4 *Idealized diagram showing the rock cycle.*

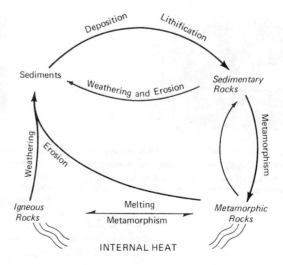

Granite (intrusive igneous rock). The dark mineral is fer-romagnesian; the white is feldspar; and the gray is quartz.

Basalt (extrusive igneous rock). Note the fine-grained size compared to that of granite.

Sandsone (detrital sedimentary rock). The individual sand grains can be seen on close inspection.

Shale (detrital sedimentary rock). Fine-grained texture and presence of clay minerals in this rock may cause problems when shale is used as a foundation material for engineering structures.

Conglomerate (detrital sedimentary rock), coarse-grained rock composed of gravel, pebbles, and finer sediment.

Limestone (biochemical or chemical sedimentary rock). Note the fragments of shells and other marine life (fossils).

Gneiss (foliated metamorphic rock). Notice the dark bands of ferromagnesian minerals and lighter bands of feldspar.

Slate (foliated metamorphic rock). Fine-grained texture and foliation give rise to flat planar surfaces characteristic of slate.

PLATE 3.B *Some common rock types. The scale is shown by the white square one centimeter on a side, giving an area of one square centimeter.*

may be altered by heat, pressure, or chemically active fluids to produce metamorphic rocks, which may then melt to start the cycle over.

Possible variations of the idealized sequence are indicated by the arrows in Figure 3.4. For example, either igneous or metamorphic rocks may be altered by metamorphism into a new metamorphic rock without ever being exposed to weathering or erosion processes.

The tectonic cycle provides several environments for the rock cycle (Figure 3.1). The general idea is that new lithosphere is produced at the oceanic ridges. This material moves laterally away from the ridges, which are divergent plate boundaries, until it dives beneath another plate to form a convergent plate boundary. These environments and associated rock-forming processes are shown in Figure 3.1. Transform fault and collision boundaries also produce rock-forming processes that fit in the generalized rock cycle. Recycling of rock and mineral material is the most important aspect of the rock cycle, while the processes which drive and maintain the cycle are important in determining the properties of resulting rocks. Therefore, interest is more than academic because it is upon these earth materials that we build our homes, industries, roads, and other structures. Furthermore, understanding of the various aspects of this cycle, such as the igneous-rock-forming processes or the soil-forming processes, will facilitate best use of these resources.

Other aspects of the rock cycle are responsible for concentrating as well as dispersing materials. These aspects are extremely important in mining minerals. If it were not for igneous, sedimentary, and metamorphic processes that concentrate minerals, it would be difficult indeed to extract resources. Thus, man takes resources that are concentrated by one aspect of the rock cycle, transforms these resources through industrial activities, and then returns them to the cycle in a diluted form where they are further dispersed by continuing earth processes (5). Once this process has taken place, the resource cannot be concentrated again within a useful frame of time. Take, for example, the lead in automobile fuel. Lead is mined in a concentrated form; transformed and diluted in fuel; and further dispersed by traffic patterns, air currents, and other processes. Eventually, it may become abundant enough to contaminate soil and water but never sufficiently concentrated to be recycled. Similar examples may be cited for many other resources used in paints, solvents, and other industrial products.

Important aspects of man's use of earth materials produced by the rock cycle include properties which affect engineering design. Therefore, the terminology of applied geology is somewhat different from that used in traditional geologic investigations. For example, the term *rock* can be reserved for earth materials that cannot be removed without blasting, and *soil* can be defined as those earth materials that can be excavated with normal earth-moving equipment. Therefore, a very friable (loosely compacted or cemented) sandstone may be considered a soil, but a well-compacted clay may be called a rock. Although confusing at first, this pragmatic approach has advantages over conventional terminology in that more useful information is conveyed to planners and designers. For example, clay is generally unconsolidated and easily removed. Thus, a contractor might assume, without further information, that clay is a soil and bid low for an excavation job on the belief that it can be removed without blasting. However, if the clay turns out to be well compacted, he may have to blast it, which is much more expensive. This kind of error is avoided if a proper preliminary investigation is made and the contractor forewarned that such a clay should be considered as rock.

Strength of Rocks   The strength of earth materials varies with composition, texture, and location. Thus, weak rocks, such as those containing many altered ferromagnesium minerals, may creep or nearly flow under certain conditions and be very difficult to tunnel through. On the other hand, granite, a very common igneous rock, is generally a strong rock needing little or no support. However, if granite were placed deep within the earth, it might also flow and be deformed. Hence, the strength of a rock may be quite different under different types and different amounts of stress.

Stress is defined as a force per unit area generally measured in pounds per square inch (lb/in²). Three types of stress, *compressive, tensile,* and *shear,* are shown in Figure 3.5. *Strain* is defined as deformation induced by a stress. A material under stress may deform in an *elastic* or a *plastic* manner. Elastic deformation is like a rubber band; that is, the deformed material returns to its original shape after the stress is removed. Plastic deformation is characterized by permanent strain; that is, the material does not return to its original shape after the stress is removed. Materials that rupture before any plastic deformation are *brittle,* and those that rupture after elastic and plastic deformation are *ductile* (Figure 3.6) (6).

The strength of an earth material is usually described as the *compressive, shear,* or *tensile stress* necessary to break a sample of it. However, it must be remembered that rocks are neither homogeneous nor isotropic, and both conditions are necessary before completely reliable strength values can be assigned to any material. Thus, it is necessary to assign a safety factor (SF) to insure that a rock will

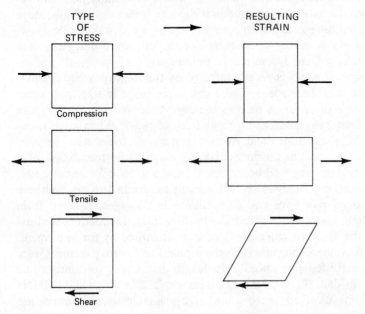

FIGURE 3.5 *Common types of stress and resulting strain.*

FIGURE 3.6 *Typical stress-strain diagram. [Reproduced, by permission, from M. P. Billings,* Structural Geology *(New York: Prentice Hall, 1954).]*

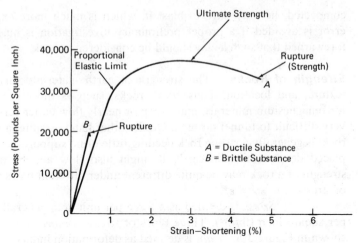

perform as expected; that is, the rock is loaded only to a fraction of its assumed strength. For example, if a rock has a compressive strength of 10,000 lb/in², then with a safety factor of 10, the rock should be loaded only to 1000 lb/in². It is emphasized that the testing necessary to establish strengths of rocks can be quite expensive. Therefore, this information is obtained only for large engineering structures. For small structures, the experienced earth scientist can render a judgment consistent with good engineering practice.

The lack of extensive testing to determine the strength of rocks at a site does not extend to a similar treatment of soils. The cost of a good soil survey to determine the strength characteristic of soils is well worth the initial cost compared to the possible damages and costs resulting from not considering these properties. Even small structures, such as apartment buildings with a swimming pool and retaining wall, are subject to considerable structural damage (cracking of walls, steps pulling away from the buildings, etc.) by differential settling on unstable soils. However, if proper use of a soil survey is made before buildings are designed, then damage caused by adverse soil conditions can be minimized.

The strength of rocks is greatly affected by the frequency and orientation of fractures, joints, and shear zones. These structures range in size from small hairline fractures to huge shear zones such as the San Andreas fault zone (Figure 3.7) which scars the California landscape for hundreds of miles. When large structures are planned, these shear zones and fracture systems in rocks must be considered. One active shear zone in the immediate vicinity of the foundation of a proposed large masonry dam may well be sufficient reason to look for another site. It has been concluded that the Baldwin Hills Reservoir failure in 1963 in Southern California, which claimed five lives and $11 million in damages, resulted from gradual movement along shear zones (faults). The tile drain, foundation, and asphaltic membrane of the dam and reservoir floor were fractured by the movement. After the reservoir had drained, long cracks in the asphaltic concrete pavement were observed. The cracks extended continuously the length of the reservoir and in line with the deep gash in the dam (Figure 3.8). The dam was 232 feet high and 650 feet long. It incorporated the most advanced knowledge available when construction began in 1947. The fracture, along with several others, was discovered during

The delta of the Mississippi River, showing the cloud of silt flowing into the Gulf of Mexico. The colors result from filters used in this satellite picture. (ERTS photo)

The Susquehanna River cuts across the folds of the Appalachian Mountains in Pennsylvania. The colors result from filters used in this satellite picture. (ERTS photo)

construction and was examined and judged not dangerous to the stability of the dam. However, because of a previous experience with shear zones, the 1928 failure of the St. Francis dam, the design of the Baldwin Hills dam included some provisions for the fractures in the foundation material. Furthermore, a series of periodic inspections were initiated to insure the safety of the operation. Even with these precautions, the failure came quite suddenly. Were it not for fortunate circumstances which gave several hours warning and enabled officials to act quickly, the loss of life might have been much greater (7).

The problems and dangers associated with fractures, shear zones, and faults are not confined to "active" systems. Once a fracture has developed, it is subject to weathering, which may produce clay minerals. These minerals may be unstable and may become a lubricated surface facilitating landslides, or the clays may wash out and create open conduits for water to move through the rocks. Therefore, although active faulting is obviously a hazard to the stability of engineered structures, even small, inactive fractures, faults, and other shear zones must be inspected and carefully evaluated to maximize the safety of man-made structures.

Types of Rocks Our discussion of the generalized rock cycle established that there are three rock families: igneous, sedimentary, and metamorphic. Table 3.1 lists the common rock types. Although rocks are primarily classified according to both mineralogy and texture, it is texture which is most significant in environmental geology. Rock texture—the size, shape, and arrangement of grains—along with fractures, shear zones, and faults, determines the strength and utility of rock. This

FIGURE 3.7 *Aerial view of the San Andreas fault zone in the vicinity of Point Reyes, California. (Photo courtesy of the California Division of Mines and Geology.)*

(a)

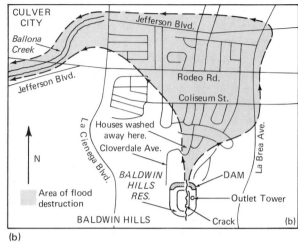

(b)

(c)

FIGURE 3.8 *Flood destruction due to the failure of the Baldwin Hills Reservoir [Photograph (a) courtesy of the California Department of Water Resources. Drawing (b) from Walter E. Jessup, "Baldwin Hills Dam Failure." Reprinted with permission from the February 1964 issue of* Civil Engineering— ASCE, *official monthly publication of the American Society of Civil Engineers, Vol. 34, 1964. Photograph (c) courtesy of the Los Angeles Department of Water and Power.]*

Rocks

43

should not be interpreted to mean that the mineralogy is not important. On the contrary, in specific cases it can be extremely significant. However, in considering groups of rocks such as the intrusive igneous rocks, mineralogy is of secondary importance to the texture and structure. For example, the granitic rocks can be given a number of different names based upon the mineralogy, but the engineering properties of all of them will be nearly the same. What is more important in practical applications is the texture and nature of fractures and related alteration zones.

TABLE 3.1 *Common rocks (engineering geology terminology)*

Type	Texture	Materials
Igneous		
Intrusive		
Granitic[a]	Coarse[b]	Feldspar, quartz
Ultrabasic	Coarse	Ferromagnesians, ± quartz
Extrusive		
Basaltic[c]	Fine[d]	Feldspar, ± ferromagnesians, ± quartz
Volcanic breccia	Mixed—coarse and fine	Feldspar, ± ferromagnesians, ± quartz
Welded tuff	Fine volcanic ash	Glass, feldspar, ± quartz
Metamorphic		*Parent material*
Foliated		
Slate	Fine	Shale or basalt
Schist	Coarse	
Gneiss	Coarse	Shale, basalt, or granite
Nonfoliated		
Quartzite	Coarse	Sandstone
Marble	Coarse	Limestone
Sedimentary		*Materials*
Detrital		
Shale	Fine	Clay
Sandstone	Coarse	Quartz, feldspar, rock fragments
Conglomerate	Mixed—very coarse and fine	Quartz, feldspar, rock fragments
Chemical		
Limestone	Coarse to fine	Calcite, shells, calcareous algae
Rocksalt	Coarse to fine	Halite

[a]Textural name for a group of coarse-grained, intrusive igneous rocks, including granite, diorite, and gabbro.
[b]Individual mineral grains can be seen with naked eye.
[c]Textural name for a group of fine-grained, extrusive igneous rocks, including rhyolite, andesite, and basalt.
[d]Individual mineral grains cannot be seen with naked eye.

Igneous rocks are rocks that have crystallized from a naturally occurring, mobile mass of quasi-liquid earth material known as *magma*. If magma crystallizes below the surface of the earth, the resulting igneous rock is called *intrusive*. Therefore, whenever intrusive igneous rocks such as granite are exposed at the surface of the earth, we may conclude that erosional processes have removed the original cover material. Magma is probably generated in the upper asthenosphere or the lithosphere. As it moves up, it displaces the rock it intrudes, often breaking off portions of the country rock (the material the magma moves into), and incorporates them into the mass of moving magma. These foreign blocks, known as *inclusions,* are good evidence of forcible intrusion. A few varieties of granite, the most common intrusive igneous rock, do not show evidence of being intruded and may form in place by a metasomatism which involves replacement rather than magmatic processes.

Whether metasomatic or magmatic, intrusive igneous rocks are generally strong and have a relatively high unconfined compressive strength varying from about 10,000 to 40,000 lb/in^2 (8). The strength of an individual granite depends upon the grain size and the degree of fracturing. In general, fine-grained granites are stronger than coarse-grained. Fresh, unweathered intrusive rocks, unless extensively fractured, are usually satisfactory for all types of engineering construction and operations (9). If they can be economically quarried, they are often good sources of construction material such as crushed stone for concrete aggregate, dimension stone for facing or veneer, broken stone for fill material, or riprap to protect slopes against erosion. However, even with the general favorable physical properties of intrusive rocks, it is not safe to assume that all of these rocks are satisfactory for their intended purpose. This is especially true for large, heavy structures whose stability depends upon a solid, safe foundation. In this case, an entire project may be jeopardized when the site is studied and evaluated as in good engineering practice and a shear zone or an alteration zone is identified. Therefore, although most intrusive rock *is* satisfactory for nearly all purposes, one should not *assume* that it is. Field evaluation, including mapping, drilling, and laboratory tests, is necessary before large structures are designed.

Extrusive igneous rocks form when magma reaches the surface and is blown out a volcano as pyroclastic debris or else flows out as lava. In either case, the resulting igneous rock is extrusive. The variety and composition of extrusive rocks are considerable, and, therefore, generalizations concerning their suitability for a specific purpose are difficult. Again, when one is working with these rocks, careful field examination, including detailed mapping and drilling, along with laboratory testing, is always necessary before large engineering structures are designed (9).

Extrusive rocks crystallized from lava flows may be stronger than granite (with an unconfined compressive strength exceeding 40,000 lb/in^2) (8). However, if the lava flows are mixed with volcanic breccia (angular fragments of broken lava and other material) or with thick vesicular or scoriaceous zones (produced as gas escapes from cooling lava), then the strength of the resulting rocks may be greatly reduced. Furthermore, after cooling and solidfying, lava flows often exhibit extensive columnar jointing (Figure 3.9) which may lower the strength of the rock. Solidified lava flows may also have large subterranean voids known as *lava tubes,* which may either collapse from the weight of the overlying material or carry

FIGURE 3.9 *Devils postpile. Characteristic columnar joints that have formed due to contraction during cooling of lava. (Photo by Cecil W. Stoughton, courtesy of the U.S. Department of the Interior, National Park Service.)*

large amounts of groundwater, either of which may cause problems during the planning, design, or construction phases of a project.

Pyroclastic debris ejected from a volcano produces a variety of extrusive rocks. The most common material is *tephra,* which is a comprehensive term for any clastic material ejected from a volcano. Volcanic ash consists of rock fragments and glass shards less than 4 mm in diameter. When it is compacted, cemented, or welded together, it is called *tuff.* Although the strength of a tuff depends upon how well cemented or welded it is, generally it is a soft, weak rock which may have an unconfined compressive strength of less than 5000 lb/in^2 (8). Some tuff may be altered to a clay known as *bentonite,* an extremely unstable material. When it is wetted, bentonite expands to many times its original volume. Pyroclastic activity also produces larger fragments which, when mixed with ash and cemented together, form *volcanic breccia* or *agglomerate.* The strength of this material may be comparable to that of intrusive rocks or may be quite weak, depending upon how the individual particles are held together. In general, however, volcanic breccia and agglomerates make poor concrete aggregates because of the large and variable amount of fine materials produced when the rock is crushed.

From the discussion of extrusive igneous rocks, and especially pyroclastic debris, it is easy to see that planning, design, and construction of engineering projects in or on these rocks can be complicated and risky. Such works should involve considerable application of detailed geologic information (9).

Sedimentary rocks form when sediments are transported, deposited, and then lithified by natural cement, compression, or other mechanism. Two types of sedimentary rocks exist: *detrital* sedimentary rocks, which form from broken parts of previously existing rocks; and *chemical* sedimentary rocks, which form from chemical or biochemical processes that remove material carried in chemical solution.

The detrital sedimentary rocks include shale, sandstone, and conglomerate. Of the three, shale includes about 50 percent of all sedimentary rocks and is by far the most abundant. Shale also causes considerable environmental problems, and the very existence of shale is a red flag to the applied earth scientist.

There are two types of shale: *compaction* shale and *cementation* shale. Compaction shale is held together primarily by molecular attraction of the fine clay particles. It is a very weak rock with the following environmental problems: First, depending upon the type of clay (mineralogy), it may have a high potential to absorb water and swell. Second, depending upon the bonding between the depositional layers (bedding planes), there may be a high potential to slide even on gentle slopes. Third, there is a high potential to slake; that is, contact with water will cause the surface to break away and curl up. This seriously retards the making of a firm bond between the rock and engineering structures such as foundations of buildings and dams. Fourth, due to the elastic nature of clays, these rocks tend to rebound if stress conditions change, making the rock a very poor foundation material. Fifth, these rocks have an unconfined compressive strength, as low as 25 lb/in^2.

Depending upon the degree and type of cementing material, cemented shales can be very stable, strong rock suitable for all engineering purposes. It is emphasized that the presence of shale is a danger sign, calling for a close look to determine the type and extent of the rock and its physical properties.

Sandstones and conglomerates are coarse-grained and make up about 25 percent of all sedimentary rock. Depending on the type of cementing material, these rocks may be very strong and stable for engineering purposes. Common cementing materials are silica, calcium carbonate, and clay. Of these, silica is the strongest; calcium carbonate tends to dissolve in weak acid; and clay may be unstable and tend to wash away. It is always advisable to carefully evaluate the strength and stability of cementing materials in the detrital sedimentary rocks.

Limestones make up about 25 percent of all sedimentary rocks and are by far the most abundant of the chemical sedimentary rocks. Limestone is composed almost entirely of the mineral calcite. Human use and activity generally do not mix well with limestone. Although it may have sufficient strength, it weathers easily to form subsurface cavern systems and solution pits. In such limestone areas, most of the streams may be diverted to subterranean routes. These routes may easily become polluted by contaminated runoff that enters the groundwater. The mineralogy of limestone is such that little or no natural purification of contaminated waters in contact with them occurs. In addition, construction may be hazardous in areas with abundant caverns and surface solution pits called *sinkholes*.

Another important chemical sedimentary rock is *rocksalt,* which is composed primarily of halite (sodium chloride). Rocksalt forms when shallow seas or lakes dry up. As water evaporates, a series of salts, one of which is halite, are precipitated. The salts may later be covered up with other types of sedimentary rocks as the area again becomes a center of deposition of sediments. It is from these sedimentary basins that salt is mined, often by solution mining, a process in which water is pumped down into the salt and flows or is pumped out again supersaturated with salt. The process leaves huge holes in the salt deposits. Because rocksalt is abundant, usually completely isolated from circulating groundwater, and located in areas with low earthquake risk, it has been suggested that old solution mining pits may be an acceptable place to dispose of radioactive wastes such as those from nuclear fuel reprocessing plants. Rocksalt is abundant in Kansas, New York, Michigan, and the Gulf Coast. Specific sites in these areas will probably be given serious consideration as disposal sites if pilot studies are successful (10).

Sedimentary structures, such as *folds* and *unconformities,* have important relationships to human interest and activity. Folds or bends (Figure 3.10) are produced when rocks are deformed under stress. Geologic events that fold rocks are

(a)

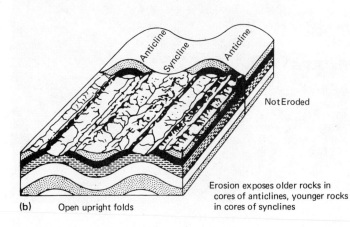

(b)   Open upright folds

Not Eroded

Erosion exposes older rocks in
cores of anticlines, younger rocks
in cores of synclines

FIGURE 3.10 *(a) Series of folds in the Rocky Mountains of southern British Columbia, and (b) idealized diagram showing the effects of erosion on folded rocks. (Photo courtesy of the Geological Survey of Canada.)*

usually associated with the tectonic cycle. Unconformities are buried erosion surfaces (Figure 3.11). Our present landscape is an erosion surface produced by the action of running water, wind, and other processes. If the sea were to rise and cover part of the landscape, sediments would be deposited over the surface and bury it, and an unconformity would be formed.

Folds and unconformities are important in environmental work for several reasons. First, they tend to influence the flow of groundwater in the rocks; that is, water usually migrates down the limbs of folds and along unconformities. Second, petroleum reservoirs are often found in folded sedimentary rocks and in association with unconformities. Third, folding produces typical fracture systems characterized by open tension fractures on the crest of anticlines and closed compression fractures in the troughs of synclines (Figure 3.12). This may be significant in the siting of reservoirs in areas of folded sedimentary rocks such as limestone where solutional weathering enlarges the fractures. Synclines are likely to hold the water better than anticlines and, therefore, may be favored sites (9). Fourth, erosion of folded rocks may expose a variety of rock types at different inclinations and hence may affect the design of an engineering structure for that area.

Metamorphic rocks are changed rocks. Heat, pressure, and chemically active fluids produced in the tectonic cycle may change the mineralogy and texture of rocks, in effect producing new rocks. Two types of metamorphic rocks exist: *foliated,* those in which the elongated or flat mineral grains have a preferential parallel alignment or banding of light and dark minerals; and *nonfoliated,* those without preferential alignment or segregation.

Foliated metamorphic rocks, such as slate, schist, and gneiss, have a variety of physical and chemical properties. Therefore, it is difficult to generalize as to the usefulness of such rocks for engineering projects. Slate is generally an excellent foundation material. Schist, when composed of soft minerals, is a poor foundation material for large structures. Gneiss is usually a hard, tough rock suitable for most engineering purposes.

FIGURE 3.11 *Unconformity (buried surface of erosion) near Monterey, California. The light rock is a steeply inclined shale that has been eroded and covered with more recent sediments that are nearly horizontal.*

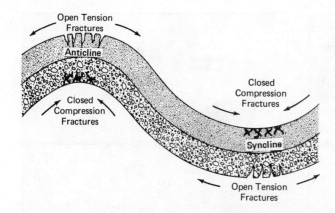

FIGURE 3.12 *Idealized diagram showing an anticline and a syncline and the accompanying tension and compression fractures formed during folding.*

The foliation planes of metamorphic rocks are potential planes of weakness. The strength of the rock, its potential to slide, and the movement of water through the rock all vary with the orientation of the foliation. For example, consider the construction of road cuts and dams in terrain where foliated metamorphic rocks are common. In the case of road cuts, foliation planes inclined toward the road result in unstable blocks which might fall or slide downslope toward the road. Also, groundwater will tend to flow down the foliation, causing a drainage problem at the road. The preferred orientation is where the foliation planes dip away from the road cut (Figure 3.13). For construction of dams, the preferred orientation is with nearly vertical foliation planes parallel to the axis of the structure (Figure 3.14) (8). This will minimize the chance of leaks and unstable blocks.

Important nonfoliated metamorphic rocks include *quartzite* and *marble*. Quartzite is a metamorphosed sandstone. It is a hard, strong rock suitable for many engineering purposes. The engineering properties of marble are similar to those of its parent rock (limestone), and cavern systems and surface pits should be expected. Vertical cavities appear to be more likely in marble than limestone, reflecting differences in joint patterns and groundwater movement.

On the night of March 12, 1928, over 500 lives were claimed and $10 million in property damage was done as ravaging flood waters raced down the San Francisquito Canyon near Sargus, California. The disaster did more to focus public attention on the need for geologic investigation as part of siting reservoirs than any previous event. The St. Francis Dam, 205 feet high with a main section 700 feet long and holding 38,000 acre-feet, had failed. The cause was clearly geologic. The adverse geologic conditions (Figure 3.15) included the following: First, the east canyon wall is metamorphic rock (schist) with foliation planes parallel to the wall. Before the failure, both recent and ancient landslides indicated the instability of the rock. Second, the rocks of the west side of the dam are sedimentary and form prominent ridges, suggesting that they were strong and resistant. Under semiarid conditions, this was true. However, when the rocks became wet, they disintegrated. This was not discovered and tested until after the dam had failed. Third, the contact

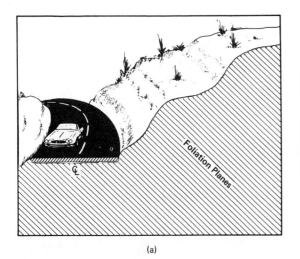
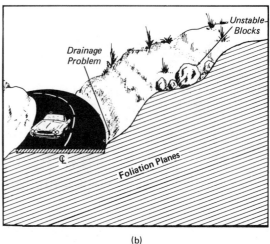

(a)   (b)

FIGURE 3.13   *Idealized diagram showing two possible orien-
tations of foliation in metamorphic rock and the effect on highway stability. Where the foliation is inclined away from the
road (a), there is less likelihood that unstable blocks will fall on
the roadway. Where the foliation is inclined toward the road
(b), unstable blocks of rock may cause a landslide.*

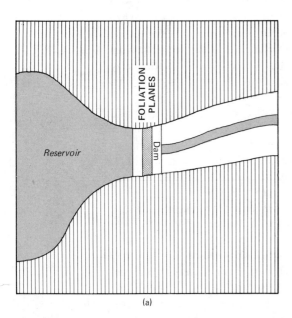
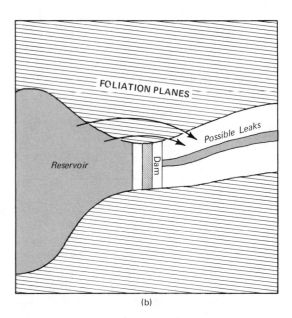

(a)   (b)

FIGURE 3.14   *Idealized diagram showing two possible orien-
tations of foliation in metamorphic rocks at a dam and reservoir
site. The most favorable orientation of the foliation is shown in
part (a), where the foliation is parallel to the axis of the dam.
The least favorable orientation is where the foliation planes
are perpendicular to the axis of the dam, as shown in (b)*

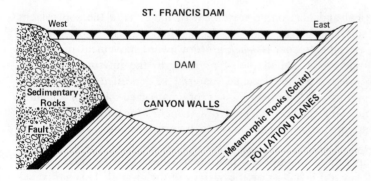

FIGURE 3.15 *Idealized diagram showing the geology of the St. Francis Dam site.*

between the two rock types is a fault with approximately a 5-foot thick zone of crushed and altered rock. The fault was on the 1922 fault map of California. It either was not recognized or was ignored. The causes of the dam failure were a combination of slipping of the metamorphic rock, disintegration and sliding of the sedimentary rock, and leakage of water along the fault zone which washed out the crushed rock (7, 8). These three causes together destroyed the bond between the concrete and the rock and precipitated failure. This tragic event clearly illustrated the need to carefully investigate the properties of earth materials before the construction of large engineering structures. Such investigations are now standard procedures.

Soils

Intimate interactions of processes between the rock and hydrologic cycles produce weathered rock materials which are basic ingredients of soils. Weathering is the physical and chemical breakdown of rocks and is the first step in the soil-forming process. The more insoluble weathered material may remain essentially in place and be further modified by organic processes to form a residual soil, as, for example, the red soils of the Piedmont in the southeastern United States. If the weathered material is transported by water, wind, or ice (glaciers), and then deposited and further modified by organic processes, then a transported soil forms. The fertile soils formed from glacial deposits in the Midwest are an example.

Soils may be defined in two ways. Soil scientists define soil as solid earth material that has been altered by physical, chemical, and organic processes, such that it can support rooted plant life. Engineers define soil as any solid earth material that can be removed without blasting. Both of these definitions are important in environmental studies.

Soil can be considered as an open system. As such, it is a function primarily of climate and topography and, secondarily, parent material (material the soil is formed from), maturity (time), and organic activity. Most of the differences we see in soils are the effects of climate and topography. The type of parent rock, the organic processes, and the amount of time the soil-forming processes have operated are of only secondary importance.

The vertical and horizontal movement of material in the soil system produces a distinct soil layering or soil profile divided into zones or horizons (Figure 3.16). An important process is *leaching* (downward movement of soluble material) from the upper zone, called the *A horizon,* to the intermediate zone, known as the *B horizon,* where the leached material is deposited. In different climates, this may produce a "hardpan" layer of compacted clay, or calcium cabonate cemented clays, or sands, etc., known as *caliche.* In addition, materials may also move upward in the soil in response to natural variations of the water table or when extensive irrigation raises the level of groundwater. When accompanied by extensive evaporation, this upward movement causes salts to move up and be deposited in the upper soil layers or on the surface of the ground. This happened in the Indos River Valley in Pakistan, where about 23 million acres were irrigated. The water table rose and extensive surface evaporation deposited salts, making the soil unsuitable for crops (11). The problem can be, and has been in some places, corrected by installing a drainage system for the land and then providing an abundance of water to leach the salts out. This is done in the Great Valley of California where salts are a problem.

Soil classification and terminology for environmental studies is a problem because in many cases we are interested in both the soil processes and the materials as they relate to human use and activity. Therefore, a genetic classification that includes engineering properties is most appropriate. However, none exists. Therefore, we will discuss both the concept of *zonal soils,* which groups soils in a

FIGURE 3.16 *Idealized diagram showing soil horizons and associated zones.*

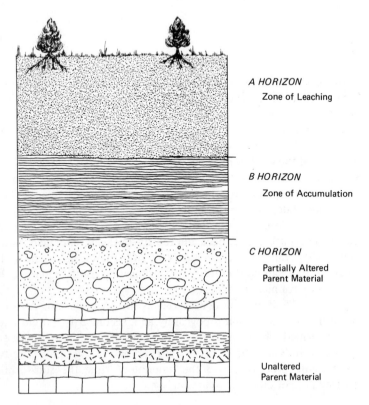

A HORIZON
Zone of Leaching

B HORIZON
Zone of Accumulation

C HORIZON
Partially Altered
Parent Material

Unaltered
Parent Material

genetic mode, and the *engineering classification,* which groups soils by material types and engineering properties.

The concept of zonal soils is based on the observation that many soil profiles are closely related to the climatic zone in which they are found. Hence, soils in which the profile is an adjustment to the climatic zone are called *zonal soils.* Recent surface materials such as sand dunes or flood deposits that do not have a distinctive soil layering are referred to as *azonal soils.* In addition, some soils have characteristics determined primarily by local conditions such as excessive water due to flooding, high groundwater table or poor drainage, or perhaps an abundance of salts or lime. These soils are referred to as *intrazonal soils* (12).

All soils above the water table have three distinct phases: the solid material; the liquid (mostly water); and the gas, which is mostly air. The usefulness of a soil to man is greatly affected by the variations in the proportions and structure of the three phases. The types of solid materials, the size, and the water content are probably the most significant variables determining engineering properties. For fine-grained soils, properties such as the liquid limit (LL) and the plastic limit (PL) are used in classifying soils for engineering purposes. The liquid limit is defined by the water content at which the soil behaves as a liquid; and the plastic limit (PL) is defined by the water content below which the soil no longer behaves as a plastic material.

The unified soil classification system is widely used in engineering practice. Because all natural soils are mixtures of coarse particles such as gravel and sand, fine particles such as silt and clay, and organic material, this classification is divided into these three groups. Each group is based on the predominant size or the abundance of organic material (Table 3.2). Coarse soils are those in which greater

TABLE 3.2 *Unified soil classification system*

Major Divisions				Group Symbols	Soil Group Name
COARSE-GRAINED SOILS — Over half of material larger than 0.074 mm	GRAVELS	CLEAN GRAVELS	Less than 5% fines	GW	Well-graded gravel
				GP	Poorly graded gravel
		DIRTY GRAVELS	More than 12% fines	GM	Silty gravel
				GC	Clayey gravel
	SANDS	CLEAN SANDS	Less than 5% fines	SW	Well-graded sand
				SP	Poorly graded sand
		DIRTY SANDS	More than 12% fines	SM	Silty sand
				SC	Clayey sand
FINE-GRAINED SOILS — Over half of material smaller than 0.074 mm	SILTS Non-plastic			ML	Silt
				MH	Micaceous silt
				OL	Organic silt
	CLAYS Plastic			CL	Silty clay
				CH	High plastic clay
				OH	Organic clay
Predominantly Organics				PT	Peat and muck

than 50 percent of the particles (by weight) are larger than 0.074 mm in diameter. Fine soils are those with less than 50 percent of the particles greater than 0.074 mm (8). Organic soils have a high organic content and are identified by their black or gray color and sometimes by a hydrogen sulfide (rotten egg) odor. A useful way of estimating the size of the soil particles in the field is as follows: It is sand or larger (coarse soil), if you can see the individual grains; silt (fine soil) if you can see the grains with a 10X hand lens; and clay (fine soil) if you cannot see the grains with such a hand lens.

Soils greatly affect the determination of the best use of the land, and a soil survey is an important part of planning for nearly all engineering projects. Such a survey should include a soil description; soil maps showing the horizontal and vertical extent of soils; and tests to determine grain size, moisture content, and strength. The purpose of the survey is to provide the information necessary for identification of potential problem areas before construction.

The more important engineering properties of soils for planning are strength, sensitivity, compressibility, erodability, permeability, corrosion potential, and ease of excavation.

Strength of soils is usually measured by the unconfined compressive strengths. Generalizations or numerical averages are often misleading because earth materials are often composed of several different mixtures, zones, or layers of materials with different physical and chemical properties. Nevertheless, clays and organic soils tend to have lower strengths than the coarser soils.

Sensitivity of soils reflects changes in the soil strength due to disturbances such as vibrations or excavations. Sand and gravel soils with no clay are the least sensitive. As fine material becomes abundant, soils become more and more sensitive. Some clay soils may lose 75 percent or more of their strength following disturbance (8).

Compressibility of soils is a measure of the tendency to consolidate, or decrease in volume. It is partly a function of the elastic nature of the soil particles and is directly related to settlement of structures (9). Excessive settlement will crack foundations and walls. Coarse materials such as gravels and sands have a low compressibility, and settlement will be considerably less than that of fine-grained or organic soils which have a high compressibility.

Erodability of soils refers to the ease with which soil materials can be removed by wind or water. Easily eroded materials (soils with a high erosion factor) include unprotected silts, sands, and other loosely consolidated materials. Cohesive soils, greater than 20 percent clay or naturally cemented soils, are not easily moved and therefore have a low erosion factor.

Permeability is a measure of the ease with which water can move through a material. Clean gravels or sands have the highest permeabilities and may transmit thousands to tens of thousands of gallons per day per square foot (GPD/ft^2). Mixtures of clean gravel and sand have permeabilities hundreds to thousands of gallons per day per square foot. As fine particles in such a mixture increase, permeability decreases. Clays generally have a permeability of less than one gallon per day per square foot.

Corrosion is a slow weathering or chemical decomposition that proceeds from the surface in. All objects buried in the ground, as, for example, pipes, cables, anchors, fenceposts, etc., are subject to corrosion. How corrosive a particular soil is depends upon first the chemistry of both the soil and the buried material, and second the amount of water available. It has been observed that the more easily a soil carries an electrical current (low resistivity), the higher its corrosion potential. Therefore, measurement of soil resistivity is used as a way to estimate the corrosion hazard (13).

Ease of excavation of soils pertains to the procedures, and hence equipment, required to remove it in construction. There are three general categories reflecting excavation techniques. *Common excavation* is accomplished with an earth mover, backhoe, or dragline. This equipment essentially removes the soil without having to scrape it first. *Rippable excavation* requires breaking the soil up with a special ripping tooth or teeth before it can be removed. *Blasting* or *rock cutting* is the third, and often most expensive, category.

The above discussion establishes that some materials are more desirable than others for specific uses. The planner concerned with land use will not make soil tests to evaluate engineering properties of soils. However, he will be better prepared to design with nature and take advantage of geologic conditions if he understands the basic terminology and principles of earth materials. For example, our discussion of engineering properties established that, generally, first, clay soils, because of their low strength, high sensitivity, high compressibility, and low permeability, should be avoided in projects involving heavy structures or structures with minimal allowable settling, or projects needing well-drained soils; and second, soils having high corrosive potentials or requiring other than common excavation should be avoided if possible. If such soils cannot be avoided, extra care, special materials and/or techniques along with higher-than-average first costs (planning, design, and construction) must be expected. The second costs (operation and maintenance) may be greater too.

Water

The hydrologic or water cycle (Figure 3.17) is driven by solar energy and supplies nearly all of our water resources. Of the water on earth, approximately 97 percent is in the oceans. A small percentage is locked up in glacier icecaps, and only a small fraction of 1 percent is in the entire atmosphere. Nevertheless, the freshwater phase of the hydrologic cycle is dependent upon this small portion of total water resources.

Of the many types of water that can be described, six are considered here. The first is *meteoric* water, the water in or derived from the atmosphere. The second is *connate* water, water which is no longer in circulation or contact with the present water cycle. Connate water is usually saline water trapped during the deposition of sediments and may be considered as "fossil water." The third is *juvenile* water, water which is derived from the interior of the earth but has not previously existed as atmospheric or surface water. The slow release of juvenile water, usually by volcanic activity today, is believed to be the source of most of the water in the hydrologic cycle. The fourth is *surface* water, the waters above the surface of the lithosphere, river, lake, ocean, etc., waters. The fifth is *subsurface* water, all of the waters within the lithosphere. These include soil, capillary, connate, etc., waters, and

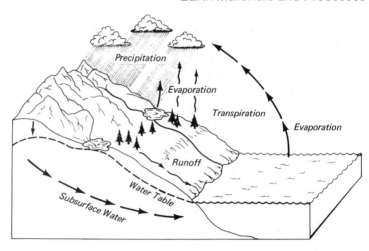

FIGURE 3.17   *Idealized diagram of the water cycle.*

groundwater. *Groundwater,* the sixth, is that part of the subsurface waters within the zone of saturation. These waters may be fresh or saline. Fresh water is an important part of our potable water supply.

On land, meteoric and surface water processes, and, to a lesser extent, groundwater processes, are responsible for erosion and deposition of earth materials that form most of our landscape. Even in arid areas, it is the work of running water that produces most of the landforms. In addition, the same water processes interact and provide the water resources necessary for our existence.

The hydrologic cycle shown in Figure 3.17 is for a region of humid climate. There, processes of precipitation balance with those of evaporation, transpiration, runoff, and groundwater flow, so the entire system forms a water budget. However, in arid areas the situation is simplified. There, evaporation may exceed precipitation, and the flow of water is essentially one way from the land up to the atmosphere (1).

Processes involved with the movement of surface waters and groundwaters are intimately related. In the contemporaneous United States, approximately 30 percent of the precipitation enters into the surface-subsurface flow system. Of this, approximately 1 percent reaches the ocean by way of groundwater flow (1). This, however, is a gross simplification because there are so many possible interreactions where surface water enters into the groundwater flow, and conversely. The rate of surface runoff or infiltration into the groundwater system depends upon the distribution and amount of precipitation; the type of soils and rocks; the slope of the land; the amount and type of vegetation; and the amount of rejected recharge, or the water that cannot enter the ground because the soil is already saturated.

Water that infiltrates the ground first enters the belt of soil moisture in the zone of aeration (Figure 3.18). It may then move through the intermediate belt, which is seldom saturated, and enter the groundwater system of the zone of saturation. The upper surface of this zone is the water table. The capillary fringe just above the water table is a narrow belt where water is drawn up by a repulsion due to surface tension of the water.

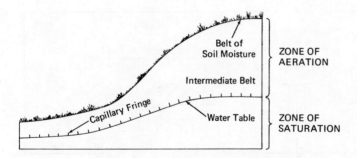

FIGURE 3.18 *Idealized diagram showing the zones in which groundwater occurs.*

     The rate of movement of groundwater depends upon the nature and extent of the primary (or intergranular) and secondary (or fractural) permeability of the soil or rocks and upon the hydraulic (pressure) gradient, which is related to the slope of the water table. The percentage of void or empty space in soil and rock is called the *porosity*. The measure of the ability of water to move through a particular material is called *permeability*. Some of the most porous materials, such as clay, have a very low permeability (Table 3.3) because the small, flat clay particles can have a great deal of pore space, but the individual small openings tenaciously hold the water.

     A zone of earth material capable of producing groundwater at a useful rate from the well is called an *aquifer*. A zone of earth material that will hold water but not transmit it fast enough to be pumped from a well is called an *aquiclude*. These zones may form an essentially impermeable layer through which little water moves. Clay soils, shale, and igneous or metamorphic rock with little interconnected porosity and/or fractures are likely to form aquicludes. Gravel, sand soils, sandstone, granite, and metamorphic rocks with high porosity or open fractures, on the other hand, will be quite good aquifers if groundwater is available.

     Aquifers are said to be *unconfined* if there is no impermeable layer restricting the upper surface of the zone of saturation. If a *confining* layer is present,

Material	Porosity (%)	Permeability (gallons/day/ft²)
Unconsolidated		
Clay	45	1
Sand	35	800
Gravel	25	5000
Gravel and sand	20	2000
Rock		
Sandstone	15	700
Dense limestone or shale	5	1
Granite	1	0.1

TABLE 3.3 *Porosity and permeability of selected earth materials.*

SOURCE: Modified after Linsley, Kohler, and Paulhus, *Hydrology for Engineers* (New York: McGraw-Hill, 1958).

FIGURE 3.19 *Diagram showing how an artesian well system may develop.*

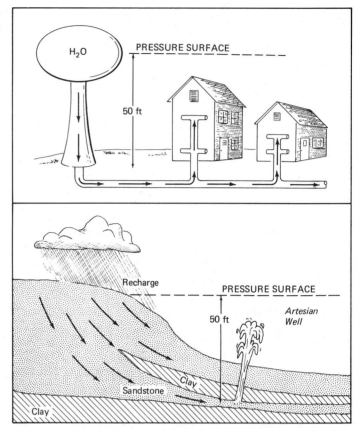

then the water below may be under pressure, forming artesian conditions. Water in artesian systems tends to rise to the height of the recharge zone. This condition is analogous to the function of the water tower producing water pressure for houses (Figure 3.19). Both confined and unconfined aquifers may be found in the same area (Figure 3.20).

When water is pumped from a well, a cone of depression in the water table or artesian pressure gradient forms (Figure 3.21). Overpumping of an aquifer causes the water level to lower continuously with searching air equilibrium. This necessitates lowering the pump settings or drilling deeper wells. These adjustments may work, depending on the hydrologic conditions, but they are not always successful. For instance, continued deepening to correct for overpumping of wells tapping igneous and metamorphic rocks is limited. Water from such wells is pumped from open fracture systems that tend to close or diminish in number and size with increasing depth.

Two types of streams exist—*effluent* and *influent* (Figure 3.20). Effluent streams tend to be perennial, and stream flow is maintained during the dry season by groundwater seepage into the channel. Influent streams are everywhere above the groundwater table and flow in direct response to precipitation. Water from the channel moves down through the water table, forming a recharge mound.

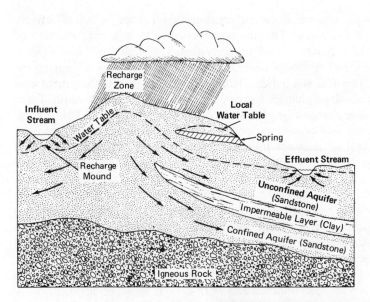

FIGURE 3.20 *Idealized diagram showing an unconfined aquifer, a local (perched) water table, and influent and effluent streams.*

These streams may be intermittent or ephemeral in that they flow only part of the year.

Streams and rivers are the basic transportation systems of the part of the rock cycle involved with erosion and deposition of sediments. They are also a primary erosion agent in the sculpture of landscape. Streams and rivers are open systems that generally maintain a dynamic equilibrium between the work done and the load imposed (14). To accomplish this, the stream must maintain a delicate balance between the flow of water and the movement of sediment. Since the stream cannot increase or decrease the amount of water or sediment it receives, adjustments

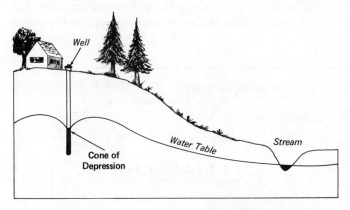

FIGURE 3.21 *Cone of depression in water table resulting from pumping water from a well.*

are made in terms of the channel slope and cross-sectional shape, which effectively change the velocity of the water. The change of velocity in turn may increase or decrease the amount of sediment carried in the system. The stream will tend to have a slope to provide just the velocity of water necessary to do the work of moving the sediment load (15). Since this is a delicate balance, any change in the sediment load or discharge will initiate slope changes to bring this system into balance again. Take, for example, a land-use change from forest to agricultural row crops. This change will cause increased soil erosion and an increase in the load supplied to the stream. The stream initially will be unable to transport the entire load, and sediment will be deposited, increasing the slope which will in turn increase the velocity of water and allow the stream to move more sediment. Slope will increase by deposition in the channel until the velocity is increased sufficiently to carry the new load. A new dynamic equilibrium may be reached, provided the rate of sediment increase levels out and the channel can adjust the slope and channel shape before another land-use change. If the reverse situation occurs, that is, if farmland is converted to forest, this sediment load will decrease and the stream will react by eroding the channel to lower the slope which in turn will lower the velocity of the water. This will continue until an equilibrium between the load imposed and work done is achieved again.

The above sequence of change is happening in parts of the Piedmont of the southeastern United States. There, land that was forest in the early history of the country was cleared for farming, producing accelerated soil erosion and subsequent deposition of sediment in the stream (Figure 3.22). The land is now reverting to pine forests, and this, in conjunction with soil conservation measures, has reduced the sediment load delivered to the streams. Thus, the once muddy streams choked with sediment are now clearing slightly and eroding their channels. Whether this trend continues depends upon future conservation measures and land use.

Consider now the effect of constructing a dam on a stream. Considerable changes will take place both upstream and downstream of the reservoir. Upstream, the effect will be to slow the stream down, causing deposition of sediment which will increase the gradient. This will continue until the channel is steep enough to again provide the water velocity necessary to transport the sediment supplied from further upstream. Downstream, the water coming out below the dam will have little sediment; but the stream, with a new high velocity, will erode the channel to pick up sediment, thus decreasing the slope until new equilibrium conditions are reached (Figure 3.23).

FIGURE 3.22 *Accelerated sedimentation and subsequent erosion due to land-use changes at the Mauldin Millsite on the Piedmont of middle Georgia. [Reproduced, by permission, from S. W. Trimble, "Culturally Accelerated Sedimentation on the Middle Georgia Piedmont," Master's thesis (Athens: University of Georgia, 1969).]*

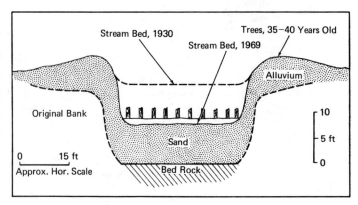

    Channel pattern is the configuration of the channel in plan view, as from an airplane. Two main patterns exist (Figure 3.24): *braided* channels characterized by numerous bars and islands which divide and reunite the channel; and *unbraided* channels. Because long, straight reaches are relatively rare, channels that do not braid are designated as *sinuous*. However, sinuous channels often contain relatively short, straight reaches. Individual bends in sinuous channels are called *meanders;* these bends migrate back and forth across the stream valley, depositing sediment on the inside of bends, forming point bars, and eroding the outside of bends. This process is prominent in the constructing and maintaining of the flood-

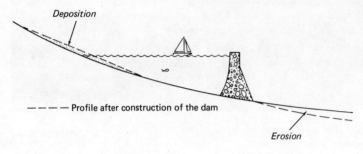

FIGURE 3.23 *Upstream deposition and downstream erosion due to construction of a dam and a reservoir.*

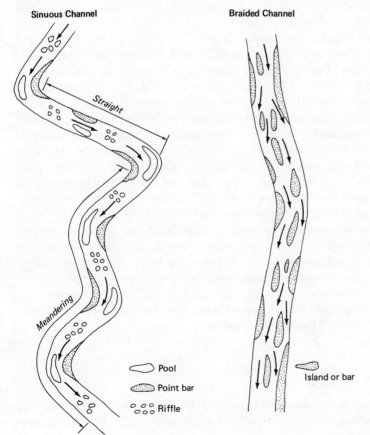

FIGURE 3.24 *Channel patterns.*

FIGURE 3.25 *Well-developed pool-riffle sequence in Sims Creek near Blowing Rock, North Carolina.*

plains. Although part of the floodplain is covered by floods, overbank deposition is of secondary importance in their development.

Sinuous channels often contain a series of regularly spaced *pools* and *riffles* (Figure 3.25). Pools at low flow are deep areas characterized by relatively slow movement of water. They are produced by scour at high flow. Riffles at low flow are shallow areas recognized by relatively fast water. Riffles are produced by depositional processes at high flow. The pool-riffle sequence is repeated approximately every 5 to 7 times the channel width. Streams with well-developed pools and riffles tend to have considerable gravel in the streambed and a relatively low slope (0.002 to 0.008). Streams with finer bed material or steep slopes tend to lack regularly spaced pools and riffles. Bedrock channels may also develop pools and riffles provided the channel slope is sufficiently low and the rock is relatively homogeneous with respect to erosion. These forms are important in environmental studies because they provide a variety of flow conditions characterized by deep, slow-moving water alternating with shallow, fast-moving water which facilitates desirable biologic productivity. They also provide visual and other sensuous varieties which increase the aesthetic amenity as well as the recreational potential by providing better fishing and boating conditions. In some instances, pools and riffles also help to stabilize the channel. For example, a straight reach with pool and riffles in which the deep part of the channel alternates from bank to bank may be morphologically more stable than a straight channel lacking these forms.

Wind and Ice

Wind- and ice-related processes are responsible for the erosion, transport, and deposition of tremendous quantities of surficial earth materials. Furthermore, these processes both modify and create a substantial number of landforms in environmentally sensitive areas (coastal, desert, arctic, and subarctic).

Wind   Windblown deposits are generally subdivided into two groups: *sand deposits,* mainly dunes; and *loess* (windblown silt), further divided into *primary loess,* which is essentially unaltered since deposition, and *secondary loess,* which has been transported and reworked over a short distance by water or has been intensely weathered in place (8).

Extensive deposits of windblown sand and silt cover thousands of square miles in the United States (Figure 3.26). Sand dunes and related deposits are found along the coasts of the Atlantic and Pacific oceans and the Great Lakes. Inland sand is found in areas of Nebraska, southern Oregon, southern California, Nevada and northern Indiana, and along large rivers flowing through semiarid regions, as, for example, the Columbia and Snake rivers in Oregon and Washington. The majority of loess is located adjacent to the Mississippi Valley but is also found in the Pacific Northwest and Idaho.

Sand dunes are constructed from sand moving close to the ground. They have a variety of sizes and shapes and develop under a variety of conditions (Figures 3.27, 3.28, and 3.29). Regardless of where they are located, how they form, or

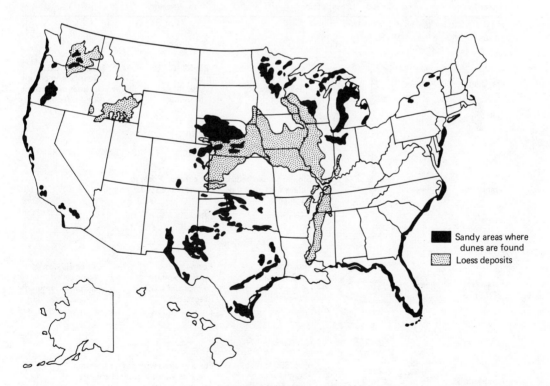

FIGURE 3.26 *Distribution of windblown landforms within the United States. [Reproduced, by permission, from Douglas S. Way,* Terrain Analysis *(Stroudsburg, Pennsylvania: Dowden, Hutchinson & Roth, 1973).]*

whether they are active or relic or otherwise stabilized, they tend to cause environmental problems. Migrating sand is particularly troublesome, and stabilization of

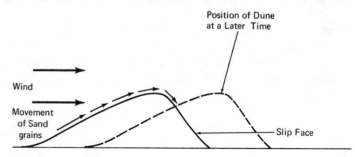

FIGURE 3.27 *Idealized diagram showing the movement of a*
sand dune. The wind moves the individual sand grains along
the surface of the gentle windward side of the dune until they
fall off the steeper leeward face. In this manner, the dune
slowly migrates in the direction of the wind. (Source: Robert J.
Foster, Physical Geology, *2nd ed. Columbus: Charles E. Merrill,*
1975.)

Dune Type	Sketch	Cross Section		Remarks
Barchan	 *Wind* →	Max. size: 100 ft high, 1000 ft point to point	25–50 ft per year movement	Most common. Generally in groups in areas of constant wind direction.
Transverse	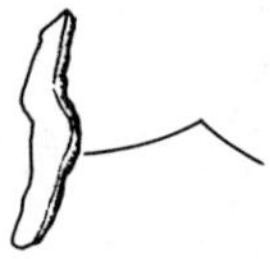 →	Similar to barchans but not curved. Form in areas with strong winds where more sand is available.		
Parabolic	 →	Max. size: 100 ft high	Form in areas with moderate winds and some vegetation.	Extremely curved types called *hairpin*. Common on seacoasts.
Longitudinal (Seif)	 →	Max. 300 ft high, 60 mi long. Avg. 10 ft high, 200 ft long	Form in areas with high, somewhat variable winds and small amount of sand available.	

FIGURE 3.28 *Types of sand dunes. (Source: Foster,* Physical
Geology, *2nd ed., Merrill, 1975.)*

FIGURE 3.29 *Barchan dunes in the Columbia River Valley, Oregon. (Photo by G. K. Gilbert, courtesy of the U.S. Geological Survey.)*

sand dunes is a major problem in construction and maintenance of highways and railroads that cross sandy areas of deserts. Building and maintenance of reservoirs in sand dune terrain are even more troublesome and tend to be extremely expensive. Such reservoirs should be constructed only if very high water loss can be tolerated. Canals in sandy areas should be lined to hold water and control erosion (8).

In contrast to sand, which seldom moves over a few feet off the ground, windblown silt and dust can be carried in huge dust clouds thousands of feet in altitude (Figure 3.30). A typical dust storm 300 to 400 miles in diameter may carry over 100 million tons of silt and dust, sufficient to form a pile 100 feet high and 2 miles in diameter (16). Terrible dust storms in the 1930s probably exceeded even this, perhaps carrying over 150 thousand tons of dust per square mile.

Loess, or windblown silt, in the United States was derived primarily during the Pleistocene Ice Ages from glacial outwash in the vicinity of major streams that carried the glacial meltwater from the ice front. Retreat of the ice left large, unvegetated areas adjacent to rivers that were very susceptible to wind erosion. We

FIGURE 3.30 *Dust storm caused by cold front at Manteer, Kansas, 1935. (Photo courtesy of Environmental Science Services Administration.)*

know this because the loess generally decreases rapidly in thickness with distance from the major rivers.

Loess is a mixture of fine sand, silt, and clay in which the grains are arranged in an open framework. Loess is porous, and its vertical permeability is greater than its horizontal. This is due in part to the presence of long, vertical tubes in the loess that are probably casts of plant roots that developed as the loess accumulated. Although loess may form nearly vertical slopes (Figure 3.31), when it is subjected to a load (such as a building) and wetted, it rapidly consolidates, a process called *hydroconsolidation,* which results as clay films or calcium carbonate cement around the silt grains is washed away. Therefore, loess may be a dangerous foundation material. Settling and cracking of a house reportedly took place overnight when water from a hose was accidentally left on. On the other hand, when loess is properly compacted and remolded, it acquires considerable strength and resistance to erosion, and, therefore, a reliable platform for building footings may be constructed (8).

Ice   This cold-climate phenomenon has been an important environmental topic for several years. As more people live and work in the higher latitudes, we will have to learn more about how to insure the best use for these sometimes fragile environments.

Only a few thousand years ago, the most recent continental glaciers retreated from northern Indiana, Michigan, and Minnesota. Figure 3.32 shows the maximum extent of the Pleistocene ice sheets. Four times in the last 2 to 3 million years, the ice advanced southward, and scientists are still speculating whether it will advance once again. We may, indeed, still be in the Ice Age. The effects of recent glacial events are easily seen in the landscape. The flat, nearly featureless ground moraine or till plains of northern Indiana are composed of material carried and

FIGURE 3.31   *Nearly vertical exposure of loess near Chester, Illinois. (Photo courtesy of James Brice.)*

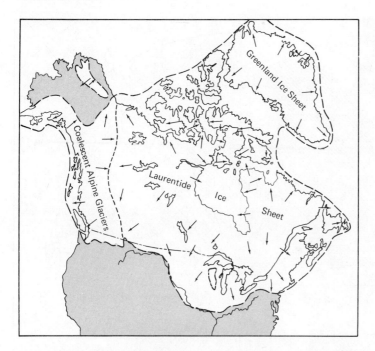

FIGURE 3.32 *Maximum extent of ice sheets during the Pleistocene glaciation. (Source: Foster,* Physical Geology, *2nd ed., Merrill, 1975.)*

deposited by continental glaciers (*till*). The till buried preglacial river valleys, and it is hard to believe that beneath the glacial deposits is a topography formed by running water much like the hills and valleys of southern Indiana where glaciers never reached.

Continental and mountain glaciers produced a variety of erosional and depositional landforms. Because of the variable types of forms and deposits, the environmental geology in a recently glaciated area may be complex. Glacial-related deposits include sands and gravels from streams in, on, under, and in front of the ice, as well as heterogeneous material (till) deposited directly by the ice. In addition, numerous lakes formed when ice blocks buried by glacial deposits melted. These lakes often contained peat and other organic material forming high organic soils. The wide variety of possibilities of earth materials in glaciated areas requires that detailed evaluation of the physical properties of surficial and subsurface materials be done before planning, designing, and constructing structures such as dams, highways, and large buildings.

In the higher latitudes, *permafrost,* permanently frozen ground, is a widespread natural phenomenon that still underlies about 20 percent of the land area in the world (Figure 3.33). Two main types of permafrost are defined by the areal extent of the phenomenon: *discontinuous* permafrost and *continuous* permafrost. At higher latitudes, discontinuous permafrost is characterized by scattered islands of thawed ground that exist in a predominantly frozen area. However, toward the southern border of the permafrost, the percentage of unfrozen ground increases until all of the ground is unfrozen. In continuous permafrost areas, the only icefree areas are beneath deep lakes or rivers. The distribution of these types of permafrost for Alaska, where about 85 percent of the state is underlain by frozen ground, is shown in Figure 3.34. Known thickness of the permafrost in Alaska varies from

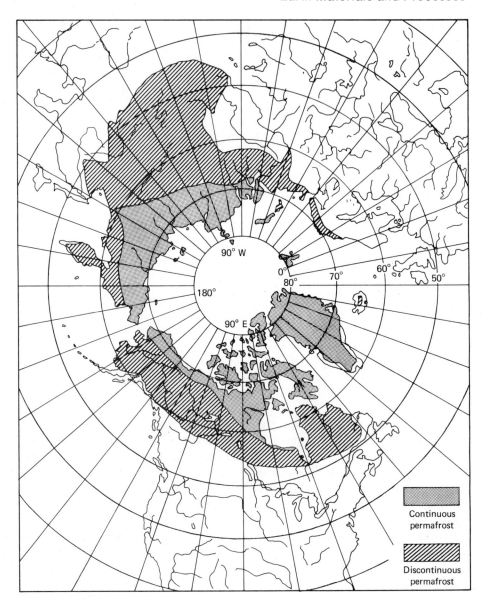

FIGURE 3.33 *Extent of permafrost zones in the Northern Hemisphere. (From Ferrians, Kachadoorian, and Greene,* Permafrost and Related Engineering Problems in Alaska, *U.S. Geological Survey Professional Paper 678, 1969.)*

about 1300 feet in northern Alaska to less than one foot at the southern margin of the frozen ground (17).

   A cross section through permafrost (Figure 3.35) shows an upper active layer that thaws during the summer. There are also unfrozen layers within the

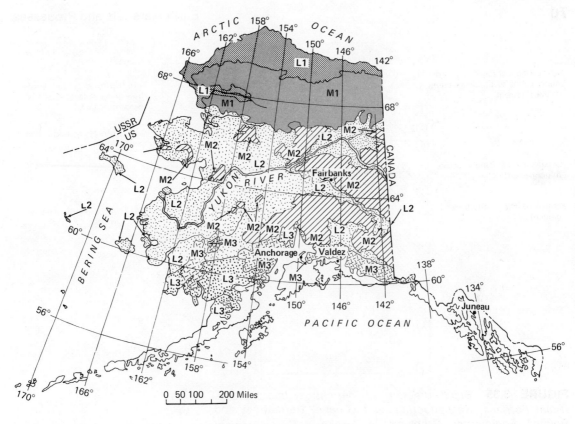

EXPLANATION

AREAS WITHIN PERMAFROST
REGION

*Mountainous areas, generally underlain
by bed rock at or near the surface.*

Underlain by continuous permafrost.

Underlain by discontinuous permafrost.

Underlain by isolated masses of permafrost.

*Lowland areas, generally underlain by
thick, unconsolidated deposits.*

Underlain by thick permafrost in areas
of either fine-grained or coarse-
grained deposits

Underlain by moderately thick to thin
permafrost in areas of fine-grained
deposits, and by discontinuous or isolated masses of permafrost in areas of
coarse-grained deposits.

Underlain by isolated masses of permafrost in areas of fine-grained deposits,
and generally free of permafrost in
areas of coarse-grained deposits.

AREAS OUTSIDE PERMAFROST
REGION

Generally free of permafrost, but a few
small isolated masses of permafrost
occur at high altitudes and in lowland
areas where ground insulation is high
and ground insolation is low, especially near the border of the permafrost region.

FIGURE 3.34 *Extent and distribution of permafrost in Alaska.
(From Ferrians, Kachadoorian, and Greene,* Permafrost and
Related Engineering Problems in Alaska, *U.S. Geological
Survey Professional Paper 678, 1969.)*

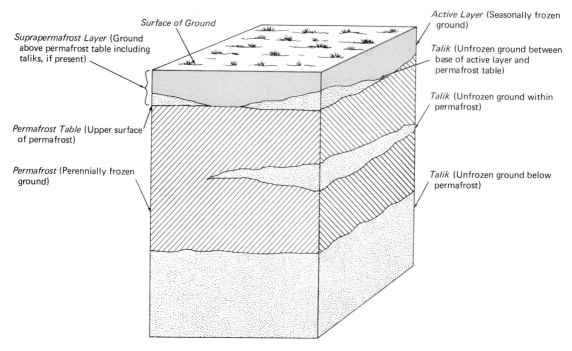

FIGURE 3.35 *Block diagram of permafrost morphology. (From Ferrians, Kachadoorian, and Greene,* Permafrost and Related Engineering Problems in Alaska, *U.S. Geological Survey Professional Paper 678, 1969.)*

permafrost below the active layer. The thickness of the active layer depends upon such factors as exposure, slope, amount of water, and particularly the presence or absence of a vegetation cover which greatly affects the thermal conductivity of the soil.

Special engineering problems are associated with the design, construction, and maintenance of structures such as roads, railroads, airfields, pipelines, and buildings in permafrost areas. Lack of knowledge about permafrost has led to very high maintenance costs and relocation or abandonment of highways, railroads, and other structures. Figure 3.36 shows a gravel road with severe differential subsidence caused by thawing of permafrost. Figure 3.37 shows a tractor trail constructed by bulldozing off the vegetation cover over permafrost on the North Slope of Alaska. The small ponds on the abandoned trail formed during the first summer following the trail construction. They will continue to grow deeper and wider as the permafrost continues to thaw (17).

The entire range of engineering problems in permafrost areas is extensive and beyond the scope of our discussion. Specific types of problems occur with different types of earth materials. In general, the major problems are associated where permafrost occurs in fine-grained, poorly drained, frost-susceptible materials. In these materials, there is generally a lot of ice which melts if the thermal regime changes. Melting produces unstable materials, resulting in settling, subsidence,

FIGURE 3.36   *Gravel road in Alaska showing severe differential subsidence due to thawing of permafrost. (Photo by O. J.
Ferrians, Jr., courtesy of the U.S. Geological Survey.)*

FIGURE 3.37   *Tractor trail on the North Slope of Alaska. The
small ponds are due to thawing of the permafrost in the roadway. (Photo by O. J. Ferrians, Jr., courtesy of the U.S. Geological Survey.)*

landslides, and lateral or downslope flowage of saturated sediment. This thawing of permafrost and subsequent frost heaving and subsidence caused by freezing and thawing of the active layer are responsible for most of the engineering problems in the arctic and subarctic regions (17).

Experience has shown that there are two basic methods of construction on permafrost: the *active* method; and the *passive* method. The active method is used where the permafrost is thin or discontinuous or contains a relatively small amount of ice. The method consists of thawing the permafrost, and if the thawed material has sufficient strength, convential construction is used. This works best in coarse-grained, well-drained soils that are not particularly frost-susceptible. The passive method is used where it is impractical to thaw the permafrost. The basic principle is to keep the permafrost frozen and not upset the natural quasi-equilibrium between environmental factors that are so sensitive that even the passage of a tracked vehicle that destroys the vegetation cover will upset the equilibrium and initiate melting which once started is nearly impossible to reverse. Special design of structures and foundations to minimize melting of the permafrost is the key to the passive method (17).

SUMMARY AND CONCLUSIONS

Processes that create, maintain, change, and destroy earth materials (water, minerals, rocks, and soil) are collectively referred to as the *geologic cycle.* More correctly, the geologic cycle is a group of subcycles, including the *tectonic, hydrologic, rock,* and *geochemical* cycles (1). In his various activities, man uses earth materials found in various parts of these cycles. The materials may be found in a concentrated or "pure" state; but after various man-induced processes, they are returned to the earth and are dispersed, contaminated, or polluted in another part of the geologic cycle (5). Once dispersed or otherwise altered, these materials cannot be concentrated or made available for human use in any reliable timetable.

Application of geologic information to environmental problems requires an understanding of both geologic and engineering properties of earth materials. Different minerals, rocks, and soils behave predictively, but they behave differently for various land uses. Thus, clay soils and compaction shales are generally poor foundation materials for large engineering structures, while granite, with few fractures, is satisfactory for most purposes. In addition, the strength of rocks is greatly influenced by active fracture systems, or old fractures which have been weathered and altered to unstable minerals.

Most continental landforms are produced by running water, and an understanding of both surface and subsurface hydrogeologic processes is necessary to work on many environmental problems. Streams and ground-water systems are basically open systems in which a dynamic equilibrium between various parts of the system is established. Any change, man-made or natural, will cause the other parts to readjust to compensate and produce a new equilibrium.

Buffalo Creek, West Virginia, 1972. Debris in channel and floodplain was deposited by a catastrophic flood that was caused by the failure of a coal-waste dam. Over 100 lives were lost. (Photo courtesy of the U.S. Army Corps of Engineers)

Buffalo Creek, West Virginia, 1972. The twisted railroad tracks were deformed by the tremendous force of the flood waters following the collapse of the coal-waste dam upstream. (Photo courtesy of the U.S. Army Corps of Engineers)

Flooding in central Pennsylvania associated with the torrential rains in the aftermath of Hurricane Agnes in 1972. (Photo courtesy of the U.S. Army Corps of Engineers)

Destruction of railroad tracks caused by flood waters in Rapid City, South Dakota, 1972. The flood claimed over 200 lives. (Photo courtesy of the U.S. Army Corps of Engineers)

Windblown sand, silt, and dust cover many thousands of square miles. Major deposits of sand are concentrated along coastal and interior areas, and windblown silt, or loess deposits, are concentrated along major rivers that carried the meltwater from continental glaciers during the Pleistocene glaciation.

Engineering problems caused by migrating and stabilized sand dunes involve expensive construction and maintenance costs for highways, buildings, and hydraulic structures.

Loess deposits, unless properly compacted, make hazardous foundations for engineering purposes because, on wetting, loess may hydroconsolidate and settle.

Geologically recent continental glaciation has produced a variety of different types of earth materials such as glacial till, highly organic soils, and water transported sediment, all of which may be located in a given area. As a result, careful evaluation of surface and subsurface deposits is necessary before planning, designing, and constructing engineering structures.

Today permafrost underlies about 20 percent of the world's land area, producing a fragile and sensitive environment. Special engineering procedures are necessary to minimize adverse effects from man-induced thawing of fine-grained, poorly drained, frozen ground.

References

1. Longwell, C. L.; Flint, R. F.; and Sanders, J. E. 1969. *Physical geology*. New York: John Wiley & Sons.
2. Le Pichon, X. 1968. Sea-floor spreading and continental drift. *Journal of Geophysical Research* 73: 3661–97.
3. Isacks, B.; Oliver, J.; and Sykes, L. 1968. Seismology and the new global tectonics. *Journal of Geophysical Research* 73: 5855–99.
4. Dewey, J. F. 1972. Plate tectonics. *Scientific American* 22: 56–68.
5. Committee on Geological Sciences. 1972. *The earth and human affairs*. San Francisco: Canfield Press.
6. Billings, M. P. 1954. *Structural geology*. 2nd ed. Englewood Cliffs, New Jersey: Prentice-Hall.
7. Jessup, W. E. 1964. Baldwin Hills dam failure. *Civil Engineering* 34: 62–64.
8. Krynine, D. P., and Judd, W. R. 1957. *Principles of engineering geology and geotechnics*. New York: McGraw-Hill.
9. Schultz, J. R., and Cleaves, A. B. 1955. *Geology in engineering*. New York: John Wiley & Sons.
10. Holden, C. 1971. Nuclear waste: Kansas riled by AEC plans for atom dump. *Science* 172:249–50.
11. Dasmann, R. F. 1972. *Environmental conservation*. 3rd ed. New York: John Wiley & Sons.
12. Hunt, C. B. 1972. *Geology of soils*. San Francisco: W. H. Freeman.
13. Flawn, P. T. 1970. *Environmental geology*, New York: Harper & Row.

14. Leopold, L. B., and Maddock, T., Jr. 1953. *The hydraulic geometry of stream channels and some physiographic implications*. U.S. Geological Survey Professional Paper 252.
15. Mackin, J. H. 1948. Concept of the graded river. *Geological Society of America Bulletin* 59: 463–512.
16. Way, D. S. 1973. *Terrain analysis*. Stroudsburg, Pennsylvania: Dowden Hutchinson & Ross.
17. Ferrians, O. J., Jr.; Kachadoorian, R.; and Greene, G. W. 1969. *Permafrost and related engineering problems in Alaska*. U.S. Geological Survey Professional Paper 678.

Part Two

HAZARDOUS EARTH PROCESSES

The earth is a dynamic, evolving system with complex interactions of internal and external processes. Internal processes are responsible for moving giant lithospheric plates and thus the continents. Interactions at plate junctions generate internal stress that causes rock deformation, resulting in earthquakes, volcanic activity, and tectonic creep (slow surface movement due to movement along a fault zone). These processes trigger numerous external events such as landslides, mudflows, and tsunamis (giant sea waves). Other external activities, such as running water or moving waves, are the result of the interaction among the hydrosphere, atmosphere, and lithosphere. For example, tremendous flooding may result from hurricane activity. However, the nature and extent of an external event is affected by internal processes which produce the uplifted land surface that, for example, running water and waves erode.

Man generally does not perceive the full significance of processes which periodically cause loss of life and property. The casual attitude toward natural hazards is a complex manifestation of an internal optimism people have about their life and their home, combined with a lack of understanding of physical relations that control the magnitude (severity) and frequency of hazardous processes.

Chapters 4 through 8 will discuss man's interaction with natural geologic events that have taken (and will continue to take) human life and that have damaged and destroyed (and will continue to damage and destroy) property. These events include river floods, landslides, earthquakes, volcanic eruptions, and coastal processes. Especially important are the natural and man-induced aspects of the magnitude and frequency of hazardous earth processes and how they relate to human use and interest. These will be discussed in detail when appropriate for each hazard.

4

RIVER FLOODING

More than any other surface process, running water is responsible for sculpturing the landscape. Rivers erode their valleys and, in the process, provide a system of drainage that is delicately balanced with the climate, topography, rocks, and soil. Rivers are also basic transportation systems that carry rock and soil material delivered to the channel by the tributaries and the slope processes on the valley sides. Much of the sediment transported in rivers is periodically stored by deposition in the channel and on the adjacent floodplain (Figure 4.1). These areas, collectively called the *riverine environment,* are the natural domain of the river. Lateral migration of bends of rivers and over-bank flow combine to produce the floodplain, which is periodically inundated by water and sediment. This natural process of over-bank flow is termed *flooding.* Channel discharge at the point where water overflows the channel is called the *flood stage* and may or may not coincide with damage to man-made structures. The term *flood stage* frequently connotes that the stream or river has reached a high water condition likely to cause damage to personal property on the floodplain. This is obviously based on a human perception of the event and, therefore, will change or vary as human use on the floodplain changes (1).

FIGURE 4.1 *Block diagram show-ing floodplain and river channel.*

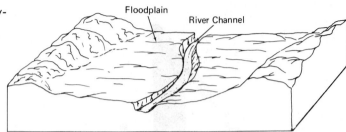

A summary of flood damage in the United States is shown in Figure 4.2, and a few of the more severe floods are listed in Table 4.1. These data empha-size the significant impact of this most universally experienced natural hazard. In the United States, the lives lost due to flooding number about 80 persons per year, with property damage averaging nearly $225 million per year. This loss of life is rela-tively low compared to loss in preindustrial societies that lack monitoring and warning systems before the flood and disaster relief after. During one 20-year period (1947–67), Asia, excluding Russia, reported 154,000 lives lost to floods, compared with about 1300 lives lost in the United States during the same period. Although the preindustrial societies with dense world populations on floodplains lose a larger proportion of lives, they have a relatively lower amount of property damage than industrial societies (1, 2).

TABLE 4.1 *Severe river floods in the United States.*

Year	Month	Location	Lives Lost	Property Damage[a]
1937	Jan.–Feb.	Ohio and lower Mississippi river basins	137	417.7
1938	March	Southern California	79	24.5
1940	Aug.	Southern Virginia and Carolinas, and Eastern Tennessee	40	12.0
1947	May–July	Lower Missouri and middle Mississippi river basins	29	235.0
1951	June–July	Kansas and Missouri	28	923.2
1955	Dec.	West Coast	61	154.5
1963	March	Ohio River basin	26	97.6
1964	June	Montana	31	54.3
1964	Dec.	California and Oregon	40	415.8
1965	June	Sanderson, Texas (flash flood)	26	2.7
1969	Jan.–Feb.	California	60	399.2
1969	Aug.	James River basin, Virginia	154	116.0
1971	Aug.	New Jersey	3	138.5
1972	June	Black Hills, South Dakota	242	163.0
1972	June	Eastern United States	113	3000.0
1973	March–June	Mississippi River	—	1200.0

Source: NOAA, *Climatological Data, National Summary,* 1970, 1972, and 1973.
[a]In millions of dollars.

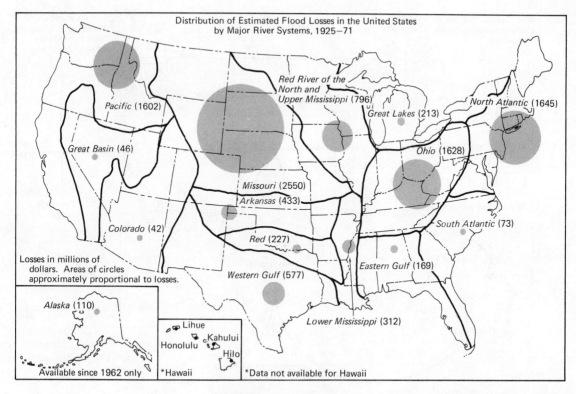

FIGURE 4.2 *Distribution of estimated flood losses in the United States by major river systems from 1925 to 1971. (U.S. Department of Commerce,* Climatogical Data, National Summary, *1972.)*

Most river flooding is a function of the total amount and distribution of precipitation and the rate at which it infiltrates into the rock or soil and the topography. However, some floods result from rapid melting of ice and snow in the spring and, on rare occasions, from the failure of a dam.

The Magnitude and Frequency of Floods

Flooding is intimately related to the amount and intensity of precipitation. Catastrophic floods reported on television and in newspapers are produced by infrequent, large, intense storms. Smaller floods or *flows* are produced by less intense storms, which occur more frequently. All flow events that can be measured or estimated from stream-gauging stations (Figure 4.3) can be arranged in order of their magnitude of discharge, generally measured in cubic feet per second (cfs). The list of flows so arranged is called an *array* and can be plotted on a discharge-frequency curve (Figure 4.4) by deriving the recurrence interval R for each flow from the relationship

$$R = \frac{N+1}{M}$$

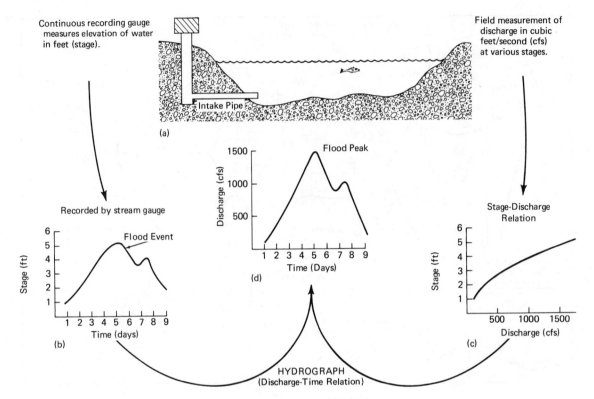

FIGURE 4.3 *Field data (a) consist of a continuous recording of the water level (stage), which is used to produce a stage-time graph. Field measurements (b) at various flows also produce a stage-discharge graph (c). Then graphs (b) and (c) are combined to produce the final hydrograph (d).*

where R is a recurrence interval in years, N is the number of years of record, and M is the rank of the individual flow in the array (3). The highest flow for nine years of data for the stream shown on Figure 4.4 is almost 10,000 cfs and so has a rank M equal to one (4). Therefore, the recurrence interval of this flood is

$$R = \frac{N+1}{M} = \frac{9+1}{1} = 10$$

which means that a flood with a magnitude equal to or exceeding 10,000 cfs can be expected about every ten years. Studies of many streams and rivers show that channels are formed and maintained by bank-full discharge with a recurrence interval of 1.5 to 2 years (1000 cfs on Figure 4.4). Therefore, we can expect a stream to emerge from its banks and cover part of the floodplain with water and sediment once every year or so.

The longer flow records are collected, the more accurate the prediction of floods is likely to be. However, designing structures for a 10-year, 25-year, 50-year, or even 100-year flood, or in fact any flow below possible maximum, is itself a calculated risk because prediction of a flood of a certain magnitude is based on probability, and, therefore, there is an element of chance. Theoretically, a 25-year

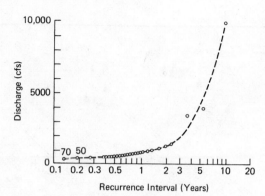

FIGURE 4.4 *Example of a flood frequency curve. (L. B. Leopold, U.S. Geological Survey Circular 559, 1968.)*

flood should happen on the average of every 25 years, but two 25-year floods could occur in any one year (5). Therefore, we conclude that as long as we continue to build dams, highways, bridges, homes, and other facilities on flood-prone areas, we can expect continued loss of lives and property.

In discussing the concepts of magnitude and frequency of floods, we should distinguish between *upstream* floods and *downstream* floods (Figure 4.5). Upstream floods are in the upper parts of drainage areas and are generally produced by intense rainfall of short duration over a relatively small area. Although these floods can be severe over a relatively small area, they generally do not cause floods in the larger streams they join downstream. Downstream floods, on the other hand, cover a wide area. They are usually produced by storms of long duration that saturate the soil and produce increased runoff. Flooding on small tributary basins is limited, but the contribution of increased runoff from thousands of tributary basins may cause a large flood downstream, which can be recognized by the downstream migration of an ever increasing flood wave with large rise and fall of discharge (6). The example of the Chattanooga-Savannah River system in Figure 4.6(a) shows the downstream migration with time of a flood crest. It illustrates that a progressively longer time is necessary for the rise and fall of water as the flood wave proceeds downstream; and it dramatically illustrates the tremendous increase in discharge from low-flow conditions to over 60,000 cfs in five days and 160 miles of downstream flow (7). It is the large downstream floods that usually make television and newspaper headlines. Figure 4.6(b) illustrates the same flood in terms of discharge per unit area, thus eliminating the effect of downstream increase in discharge. This better illustrates the shape and form (sharpness of peaking) of the flood wave as it moves downstream (7).

A few upstream floods of very high magnitude have been caused directly by failure of man-made structures. For example, the most destructive flood in West Virginia's history was caused by the failure of a coal-waste dam on the middle fork of Buffalo Creek. On the morning of February 26, 1972, in that flood, which lasted only three hours, at least 118 lives were lost; 500 homes were destroyed, leaving 4000 persons homeless; and over $50 million of property damage occurred. A wall of water 10 to 20 feet high swept through the Buffalo Creek Valley at an average rate of about 7 feet per second, or 5 miles per hour. For three days before the disaster, nearly 4 inches of rain fell in the area, producing a 10-year flood in local streams similar to Buffalo Creek. The volume of water released by the

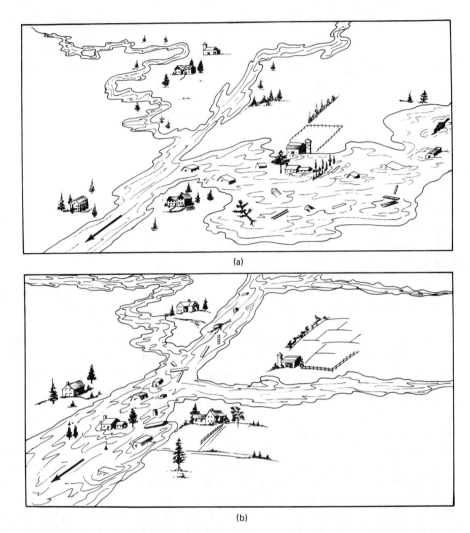

FIGURE 4.5 *Idealized diagram comparing upstream floods (a) to downstream floods (b). Upstream floods generally cover small areas and occur during the growing season, whereas downstream floods cover wide areas and occur during the dormant season. (U.S. Department of Agriculture.)*

collapse of the coal-waste dam was approximately 17.6 million cubic feet, producing a flood about 40 times greater than a naturally occurring 50-year flood. The U.S. Geological Survey concluded that the failure of the dam contributed almost all of the peak flood flow and that direct runoff from other sources was not significant. It was further concluded that there were several causes of the dam failure. First, the dam was neither designed nor constructed to withstand the amount and depth of water it impounded. It was, in fact, a waste pile that grew as more and more material was

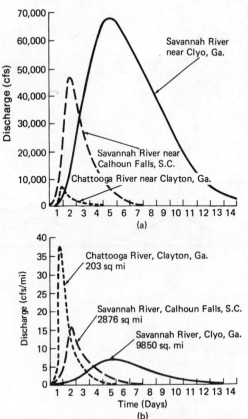

FIGURE 4.6 *Downstream movement of a flood wave on the Savannah River, South Carolina and Georgia. [From William G. Hoyt and Walter B. Langbein,* Floods *(copyright 1955 by Princeton University Press) Fig. 8, p. 39. Reprinted by permission of Princeton University Press.]*

dumped. Second there was no spillway or other adequate water level control in the dam to provide a means of removing water. Third, sludge waste from the coal mining operation was an inadequate foundation material for the dam, and seepage through the foundation caused openings to form through which water flowed, greatly reducing the stability of the dam. Fourth, stability of the dam was further lowered because of its great thickness, and the dam became saturated and somewhat buoyant. Fifth, the dam itself was constructed of coal waste which included fine coal, shale, clay, and mine rubbish, all material which disintegrates rapidly and is very unstable. A safe, economical dam cannot be constructed of this material alone (8).

After the failure of the dam, the impounded water emptied into Buffalo Creek in 15 minutes. At the time, the flow in the creek was well below bank-full stage. The flood wave traveled three hours before reaching the mouth of Buffalo Creek at the Town of Man. Along the way, the peak flattened and the flood took longer to pass a particular point (Figure 4.7). When the wave hit the Guyandotte River at 11:00 A.M., it produced a sudden high peak in the already rising river (Figure 4.8).

Because the Buffalo Creek flood was not naturally occurring, it is not strictly valid to compare it with other natural floods. Nevertheless, this example is a serious warning of possible effects of man's activities on even small streams (8).

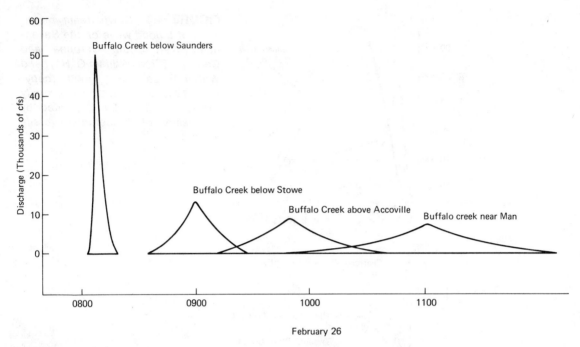

FIGURE 4.7 *Estimated flood hydrographs for Buffalo Creek on February 26, 1972. (Davies, Bailey, and Donovan, U.S. Geological Survey Circular 667, 1972.)*

Urbanization and Flooding

Human use of land in the urban environment has increased both the magnitude and the frequency of floods in small drainage basins of a few square miles. The rate of increase is a function of the percentage of the land that is covered with rough pavement and cement (which is referred to as *impervious cover*) and the percentage of area served by storm sewers. These two perimeters are collectively a measure of the degree of urbanization. The graph in Figure 4.9 shows that an urban area with 40 percent impervious surface and 40 percent served by storm sewers can expect to have about three times as many overbank flows as before urbanization. This relation holds for floods of small and intermediate frequency, but as the size of the drainage basin increases, floods of high magnitude with frequencies of 50 years or so are not much affected by urbanization (Figure 4.10).

Floods are a function of rainfall-runoff relations, and urbanization causes a tremendous number of changes in these relations. One study showed that urban runoff is 1.1 to 4.6 times preurban runoff (5). (See Figure 4.11.) The largest differences correspond to larger storms. Estimates of discharge for different recurrence intervals at different degrees of urbanization are shown in Figure 4.12. They dramatically indicate the tremendous increase of runoff with increasing impervious areas and storm sewer coverage.

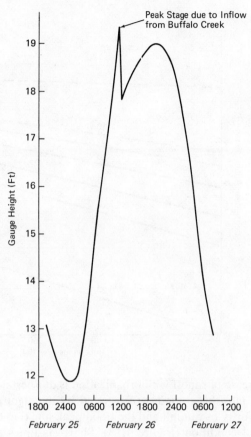

FIGURE 4.8 *Hydrograph of the Guyandotte River at Man during the period February 25 to February 27, 1972. The flood inflow from Buffalo Creek produced a sudden high peak of 19.34 feet at the gauging station. Peak discharge was 31,600 cfs. (Davies, Bailey, and Donovan, U.S. Geological Survey Circular 667, 1972.)*

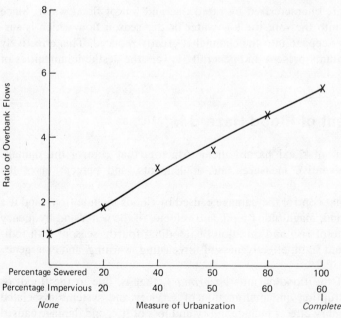

FIGURE 4.9 *Relationship between the ratio of overbank flows and measure of urbanization. This figure shows that as the degree of urbanization increases, the number of over-bank flows per year also increases. (L. B. Leopold, U.S. Geological Survey Circular 559, 1968.)*

FIGURE 4.10 *Graph showing the variation of flood frequency ratio with percentage of impervious area. Note that the smaller floods with recurrence intervals of just a few years are much more affected by urbanization than the larger floods. The 50-year flood is little affected by the amount of area that is rendered impervious. (L. A. Martens, U.S. Geological Survey Water Supply Paper 1591C, 1968.)*

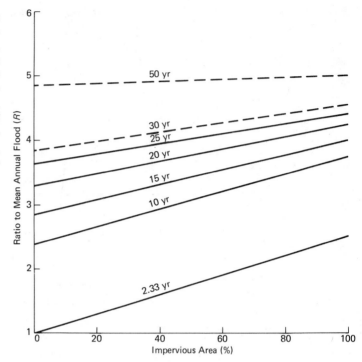

The increase of runoff with urbanization is due certainly to the fact that less water infiltrates into the ground, as suggested by the significant reduction in time between the majority of rainfall and the flood peak hits (lag-time) for an urban condition and a rural condition (Figure 4.13). Short lag-times are referred to as *flashy discharge* and are characterized by rapid rise and fall of flood water. Since little water infiltrates into the soil, the low water or dry season flow which is sustained by groundwater seepage into the channel is greatly reduced. This effectively concentrates any pollutants present and generally lowers the aesthetic amenities of the stream (4).

Nature and Extent of Flood Hazard

The nature and extent of flood hazard include factors that control the damage caused by floods, preventive measures and adjustments, and perception of the hazard.

Factors that control the damage caused by floods include first, land use on the floodplain; second, magnitude (depth and velocity of the water and frequency of flooding); third, rate of rise and duration of flooding; fourth, season; fifth sediment load deposited; and sixth, effectiveness of forecasting, warning, and emergency systems.

Effects of the flooding may be *primary*, that is, due to the flood, or *secondary*, due to disruption and malfunction of services and systems associated with the flood (9). Primary effects include injury and loss of life, and damage caused

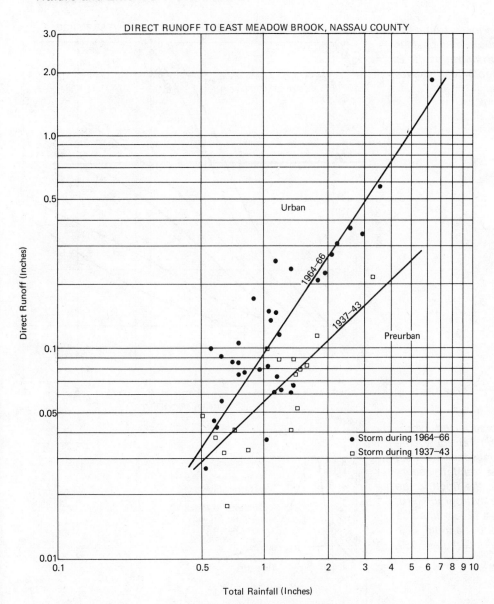

DIRECT RUNOFF TO EAST MEADOW BROOK, NASSAU COUNTY

Direct Runoff (Inches)

Total Rainfall (Inches)

FIGURE 4.11 *Comparison of the rainfall-runoff relationships for preurban and urban conditions. The data are for individual storms in one subarea of Nassau County, New York. (G. E. Seaburn, U.S. Geological Survey Professional Paper 627B, 1969.)*

by swift currents, debris, and sediment to farms, homes (Figures 4.14 and 4.15), buildings, railroads, bridges (Figure 4.16), roads, and communication systems. In addition, erosion and deposition of sediment in the rural and urban landscape may involve a loss of considerable soil and vegetation (Figure 4.17). Secondary effects may include short-term pollution of rivers, hunger and disease, and displacement of

FIGURE 4.12 *Flood frequency curve for a one-square-mile basin in various states of urbanization. (L. B. Leopold, U.S. Geological Survey Circular 559, 1968.)*

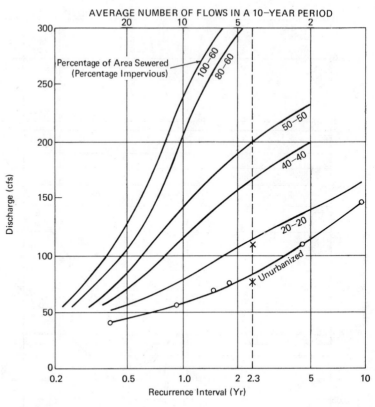

FIGURE 4.13 *Generalized hydrographs. Hydrograph (a) shows the typical lag-time between the time when most of the rainfall occurs and the time when the flood in the stream occurs. Hydrograph (b) shows the decrease in lag-time due to urbanization. (L. B. Leopold, U.S. Geological Survey Circular 559, 1968.)*

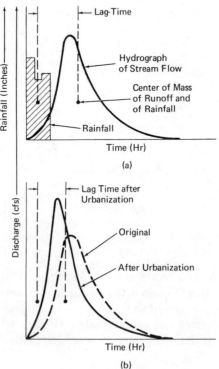

FIGURE 4.14 *Damage to farm and buildings during the Missouri River floods of 1952. (Photograph by Forsythe, courtesy of the U.S. Department of Agriculture.)*

persons who have lost their homes. In addition, fires may be caused by short circuits or broken gas mains (9).

Preventive and adjustment measures for floods involve different approaches. *Prevention* includes engineering structures and projects such as levees and flood walls to serve as physical barriers against high water; reservoirs to store water which is later released at safe rates; channel improvements *(channelization)* to increase the channel size and to move the water quickly off the land; and channel diversions to route flood waters around areas requiring protection. *Adjustment* measures include floodplain regulation and flood insurance (9).

From an environmental view, the best approach to minimize flood damage in urban areas is floodplain regulation. This is not to say that physical barriers, reservoirs, and channel works are not desirable. In areas developed on floodplains, they will be necessary to protect lives and property (Figure 4.18). However, we need to recognize that the floodplain belongs to the river system and encroachments increase flooding (Figure 4.19). Therefore, an ideal solution would be to discontinue development of areas subject to flooding that necessitates new physical barriers. In other words, we must "design with nature." Realistically, in most cases, the most effective and practical solution is a combination of physical barriers and floodplain regulations that will result in less physical modification of the river system. For example, reasonable floodplain zoning in conjunction with a diversion channel project or upstream reservoir may result in a smaller diversion channel or reservoir than if no floodplain regulations were used.

The purpose of floodplain regulation is to obtain the most beneficial use of floodplains while minimizing flood damage and cost of flood protection (10). It is a compromise between indiscriminate use of floodplains, which results in loss of life and tremendous property damage, and complete abandonment of floodplains, which gives up a valuable natural resource. A preliminary step to floodplain regulation is flood-hazard mapping, which is a means of providing floodplain information

FIGURE 4.15 *Damage caused by flooding of the Missouri River in western Iowa, 1952. The upper photograph shows disruption of transportation systems (railroad) and the lower photograph shows a closeup of the flood damage to government-stored grain. (Photograph by Forsythe, courtesy of the U.S. Department of Agriculture.)*

for land-use planning (11). These maps may delineate past floods or floods of a particular frequency, say, the 100-year flood, and are useful in deriving regulations for private development, purchasing of land for public use such as parks and recreational facilities, and providing guidelines for future land-use of floodplains.

In urbanizing areas, the accuracy of flood-hazard mapping based entirely on stream flow data is very questionable. A better map may be produced by assuming projected urban conditions with estimated percentage of impervious areas. A theoretical 100-year flood map can then be produced. The U.S. Geological Survey in Charlotte, North Carolina, is producing such maps to be used with the Mecklenburg County Floodway Regulations (Figure 4.20). Two districts are mapped in the flood hazard area: the Floodway District, and the Floodway Fringe District (Figure 4.21).

FIGURE 4.16 *Flood damage to state highway in Ohio. (Photograph by N. Richard Mowrey, courtesy of the U.S. Department of Agriculture, Soil Conservation Service.)*

The *Floodway District* is the portion of the channel and floodplain of a stream designated to provide passage of the 100-year regulatory flood without increasing the elevation of the flood by more than one foot. On this land, permitted uses include farming, pasture, outdoor plant nurseries, wildlife sanctuaries and

FIGURE 4.17 *Aerial view of flood damage to farmlands during the Missouri River floods of 1952. (Photograph by Forsythe, courtesy of the U.S. Department of Agriculture.)*

FIGURE 4.18 *Dike protecting commercial property from flooding of the Missouri River in 1973. (Photo courtesy of the Department of the Army, U.S. Corps of Engineers.)*

game farms; loading areas, parking areas, rotary aircraft ports and similar uses provided they are further than 25 feet from the stream channel; golf courses, tennis courts, archery ranges and hiking and riding trails; streets, bridges, overhead utility lines, and storm drainage facilities; temporary facilities for certain specified activities such as circuses, carnivals, and plays; boat docks, ramps, and piers; and dams, if they are constructed in accordance with approved specifications. Other uses of the Floodway District, such as open storage of any equipment or material, any structures designed for human habitation or underground storage of fuel or flammable liquids, require special permits.

The second district is the *Floodway Fringe* District, which consists of the land located between the Floodway District and the maximum elevation subject to flooding by the 100-year regulatory flood. Permitted uses in this area include any uses permitted in the Floodway District; residential accessory structures, provided they are firmly anchored to prevent flotation; fill material that is graded to a minimum of 1 percent grade and protected against erosion; and structural foundation if

FIGURE 4.19 *Idealized diagram showing the effect of floodplain encroachment on increasing flood heights. (Water Resources Council,* Regulation of Flood Hazard Areas, *vol. 1, 1971.)*

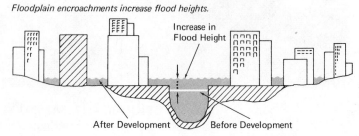

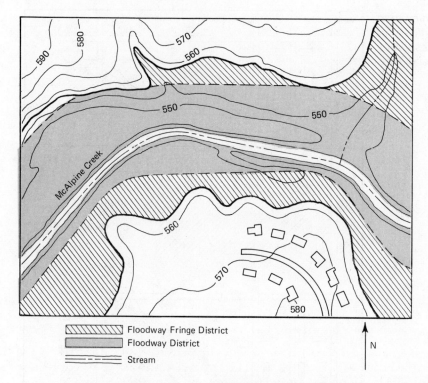

FIGURE 4.20 *Example of a flood hazard map showing the Floodway Fringe District and the Floodway District. (County Engineer, Mecklenburg County, North Carolina.)*

firmly anchored to prevent flotation. Above-ground storage of processing or any material that is flammable or explosive or that could cause injury to human, animal, or plant life in time of flooding is prohibited on the Floodway Fringe District.

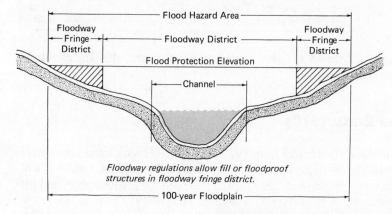

FIGURE 4.21 *Idealized diagram showing a valley cross section and its flood hazard area, Floodway District, and Floodway Fringe District. (Water Resources Council, Regulation of Flood Hazard Areas, vol. 1, 1971.)*

FIGURE 4.22 *Typical zoning map before and after the addition of flood regulations. (Water Resources Council,* Regulation of Flood Hazard Areas, *vol. 1, 1971.)*

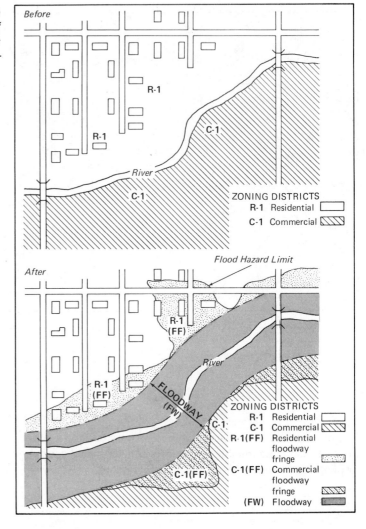

Figure 4.22 shows a typical zoning map before and after the establishment of floodplain regulations.

Floods of the Early 1970s

Floods recorded in the United States from 1972 through 1973 took hundreds of lives and caused several billion dollars of property damage. Floodplain residents will certainly long remember these years both for the high and dangerous waters and for the incalculable suffering of the survivors. Floods in June of 1972 rank as two of the worst flood disasters in the history of the United States (2). The first occurred on the ninth and tenth of June in the Black Hills area of South Dakota, and the second occurred in the eastern United States on the twenty-first through the twenty-fourth of June.

The South Dakota event claimed 242 lives in a catastrophic flash flood. Damages were estimated at $163 million. Most of the deaths and property losses were along Rapid Creek near Rapid City, South Dakota. The flood resulted from torrential rains that struck the Black Hills, delivering at places 15 inches of rain in five hours. Total rainfall from the six-hour storm nearly equaled the yearly average. A storm of this magnitude is likely to strike only every several thousand years (2).

The floods in the eastern United States were produced by tremendous rainstorms associated with Hurricane Agnes. In terms of property damage (exceeding $3,000,000,000), this was the most devastating flood ever to strike this country. One hundred thirteen persons lost their lives. Torrential rains and floods were recorded essentially everywhere east of the Appalachians from North Carolina to New York State (2).

Total loss of property from flood damage in 1973 was second only to that in 1972. The largest flood of the year was the spring flooding of the Mississippi River, which resulted in approximately $1.2 billion in property damage. Although tens of thousands of persons had to be evacuated and suffered appreciably as millions of acres of farmland were inundated throughout the Mississippi River Valley, fortunately, there were few deaths caused directly by the flooding (12).

Perception of Flooding

At the institutional level, there appears to be adequate sensitivity in the perception of flooding. However, on the individual level, the situation is not as clear. Several conclusions have been drawn from perception studies. Accurate knowledge of flood hazard does not inhibit all persons from moving onto the floodplain. Flood hazard maps are not effective as a form of communication. Upstream development is frequently the scapegoat for downstream floodplain residents threatened by floods. People are tremendously variable in their knowledge of flooding, anticipation of future flooding, and willingness to accept adjustments caused by the hazard (1).

SUMMARY AND CONCLUSIONS

River flooding is the most universally experienced natural hazard. The loss of life is relatively low in developed countries with adequate monitoring and warning systems compared with that in preindustrial societies. However, property damage is much greater in industrial societies where floodplains are often extensively developed.

The magnitude and frequency of flooding are a function of the intensity and distribution of precipitation and the rate of infiltration of water into the soil, rock, and topography. Human use and interest in urban areas has increased the flood hazard in small drainage basins by increasing the amount of ground covered with buildings, roads, parking lots, etc., which forces storm water to run off the land rather than infiltrate it.

Two types of floods are recognized: *upstream* floods produced by intense rainfall of short duration over a relatively small area; and *downstream* floods produced by storms of long duration over a large area which saturate the soil, causing increased runoff from thousands of tributary basins and resulting in large floods in the major rivers.

Factors that control the damage caused by flooding include land use on the floodplain; magnitude and frequency of the flooding; amount of sediment deposited; and effectiveness of forecasting, warning, and emergency systems. From an environmental view, the best solution to minimizing flood damage is floodplain regulation. However, in highly urban areas, it will remain necessary to utilize physical barriers, reservoirs, and channel works to protect existing development. The realistic solution to minimizing flood damage involves a combination of floodplain regulation and engineering techniques.

There is an adequate sensitivity in the perception of flooding at the institutional level. However, on the individual level more public-awareness programs are needed to help people perceive the hazard of living in flood-prone areas.

References

1. Beyer, J. L. 1974. Global response to natural hazards: floods. In *Natural hazards,* ed. G. F. White, pp. 265–74. New York: Oxford University Press.
2. U.S. Department of Commerce, 1972. *Climatological data, national summary* 23, no. 13.
3. Linsley, R. K., Jr.; Kohler, M. A.; and Paulhus, J. L. 1958. *Hydrology for engineers.* New York: McGraw-Hill.
4. Leopold, L. B. 1968. *Hydrology for urban land planning.* U.S. Geological Survey Circular 559.
5. Seaburn, G. E. 1969. *Effects of urban development on direct runoff to East Meadow Brook, Nassau County, Long Island, New York.* U.S. Geological Survey Professional Paper 627B.
6. Agricultural Research Service. 1969. *Water intake by soils.* Miscellaneous Publication No. 925.
7. Strahler, A. N., and Strahler, A. H. 1973. *Environmental geoscience.* Santa Barbara, California: Hamilton Publishing.
8. Davies, W. E.; Bailey, J. F.; and Kelly, D. B. 1972. *West Virginia's Buffalo Creek flood: a study of the hydrology and engineering geology.* U.S. Geological Survey Circular 667.
9. Office of Emergency Preparedness. 1972. *Disaster preparedness* 1, 3.
10. Bue, C. D. 1967. *Flood information for flood plain planning.* U.S. Geological Survey Circular 539.
11. Shaeffer, J. R.; Ellis, D. W.; and Spieker, A. M. 1970. *Flood-hazard mapping in metropolitan Chicago.* U.S. Geological Survey Circular 601C.
12. U.S. Department of Commerce. 1973. *Climatological data, national summary* 24, no. 13.

5

LANDSLIDES AND RELATED PHENOMENA

Slopes are the most common landforms, and although most slopes appear stable and static, they are actually dynamic, evolving systems. Slope processes are one significant reason why stream valleys are much wider than the stream channel and adjacent floodplain. Material on most slopes is constantly moving down the slope at rates which vary from imperceptible creep of soil and rock to thundering avalanches and rockfalls which move at tremendous speeds. As with floods, it may not be the largest (rare and infrequent) nor the smallest (most common) event that moves the greatest amount of material down slopes as valleys develop. Rather, events of moderate magnitude and frequency may play the most important role.

Landslides and related phenomena such as mudflows, earthflows, rockfalls, snow or debris avalanches, and subsidence are natural events that would occur with or without human activity. However, human use and interest has led to both an *increase* in some of these events such as landslides on hillside developments due to oversteeping of slopes and a *reduction* of landslides by stabilizing structures or techniques on naturally sensitive slopes likely to experience landslides.

Important aspects of landslides and related phenomena are the type of earth material involved, the slope, the climate, and water. Depending on these variables, resulting movements will involve flowage, sliding, falling, or subsidence of earth materials at various rates. A brief classification of these events is given in Figure 5.1. It reflects the terminology used in this discussion and in other chapters. Table 5.1 is a historical summary of landslide information in the United States. Of particular interest is the estimate of the frequency of the hazard; the frequency is much greater in mountain areas, thus reflecting the importance of topography.

The causes of many landslides can be examined by studying relations between forces which tend to make earth material slide (*driving* forces) and forces which tend to oppose such movement (*resisting* forces). The most common driving force is the weight of the slope material, including anything superimposed on the slope, such as trees, buildings, etc. The most common resisting force is the shear strength of the slope material. If the slope material is homogeneous, the type of slide movement is likely to be *rotational* along a potential slide plane. On the other hand, if the slope material has directions of weaknesses such as fractures or clay layers along bedding planes, the resulting slide is called a *translation* or *slab* slide. Figure 5.2 shows these two basic types of landslides. If the shear resistance is greater than the driving force, then the slope is stable. However, stability can change rapidly. Anything that increases the weight, such as snow, infiltration of water, or structures such as buildings, walls, etc., can lower the stability of slopes. In addition, anything that lowers the shear resistance increases the possibility of landslides. The most likely thing to lower shear resistance is an increase in water pressure (excess moisture) which has a buoyancy effect. In other cases, stream erosion along banks and wave erosion along coasts tend to cut the toe (lower area) of slopes, and this tends to have the same effect as adding weight to the slope. Man's activities in building roads and excavating pads for homes in hillside areas have the same effect as streams and waves and have resulted in many landslides.

In many cases, the *real* causes of landslides, increase in driving force or decrease in resisting force discussed above, are marked by *immediate* causes such as

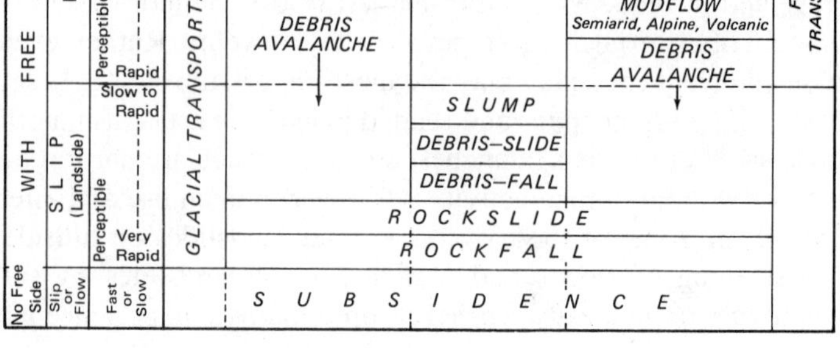

FIGURE **5.1** *Classification of landslides. [Reprinted, by permission, from C. F. Sharpe,* Landslides and Related Phenomena *(New York: Columbia University Press), 1938.]*

TABLE 5.1 *Frequency of several types of landslides in the United States.*

Type of Slide	Major Areas	Number of Historical Slides	Approx. Frequency per 100 sq. mi.	Approx. Frequency per 40,000 sq. mi.	Est. Prop. Damage (Million $ Adjusted to 1971 Values)	Recorded Deaths
Rockslide and rockfall	White, Blue Ridge, Great Smoky, Rocky Mtns., and Appalachian Plateau	Several hundred	—	1 per 10 yr	30	42
Rockslump and rockfall	Widespread in central and western U.S.; prevalent in Colo. Plateau, Wyo., Mont., southern Calif., Oreg., and Wash.	Several thousand	Av. 10 per yr in hill areas; 1 per yr in plateaus	100 per yr in hill areas; 10 per yr in plateau areas	325	188
	Appalachian Plateau	Several thousand	1 per 10 yr	70 per yr	350 (mainly in highway and railroad damage)	20
	Calif. Coast Ranges, Northern Rockies	Several hundred	1 per 10 yr	10 per yr	30	—
Slump	Maine, Conn. Riv. Valley, Hudson Valley, Chicago, Red Riv., Puget Sound, Mont. glac. lakes, Alaska	About 70	1 per 100 yr	1 per yr	140	103
	Long Island, Md., Va., Ala., S. Dak., Wyo., Mont., Colo.	Several hundred	1 per 50 yr	1 per yr	30 (mainly to highways and foundations)	—
	Mississippi and Missouri Valleys, eastern Washington and southern Idaho	Several hundred	1 per 10 yr	1 per yr	2	—
	Appalachian Piedmont	About 100	—	1 per yr	Less than 1	—
Debris flow and mudflow	White, Adirondack and Appalachian Mtns.	Several hundred	1 group slides, 10+ per group, per 100 yr in White Mtns. and North Carolina	1 group slides, 10+ per group, per 15 yr	100	89

SOURCE: *Disaster Preparedness,* Office of Emergency Preparedness, 1972.

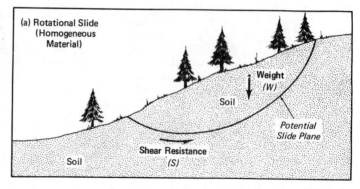

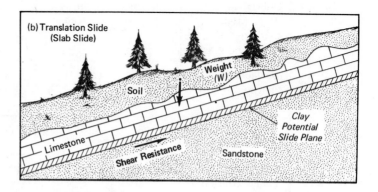

earthquake shocks, vibrations, or a sudden increase in the amount of moisture in slope material. The distinction between real and immediate causes is very important in court cases where a landslide case is heard and a distinct statement by an earth scientist concerning the cause of a landslide is expected (1). For example, a translation slide may have as an *immediate* cause heavy rains which saturated the earth material; but, the *real* cause is the potential to slide upon long, weak, clay layers. A similar hypothetical example might be given for a man-made slope in a housing development in which the immediate cause is an earthquake shock, but the real cause is a poorly designed slope.

Water is almost always directly or indirectly involved with landslides, and thus its role in landslides is of particular importance. Water is the universal solvent, and much weathering of rocks, which slowly reduces their shear strength, takes place near the surface of the earth. The weathering contributes to the instability of rocks and is especially significant in areas with limestone, which is very susceptible to chemical weathering. It has been argued that water has a lubricating effect on individual soil grains and potential slide planes. This notion is probably incorrect because water may actually act as an antilubricant, and very little water is needed to produce nearly saturated conditions, implying that all slopes in a humid climate are fully lubricated (2). The amount of water in a soil is very significant. For example, dry sand will form a cone when dumped; moist sand will stand nearly vertical because surface tension in the water holds the grains together; and saturated sand with all of its pore spaces filled will flow as mud. Probably the most important effects of water in producing landslides are as follows: First, wave or stream erosion may erode the toe area slopes, increasing the driving force. Second, excess water may cause a rise in water pressure, decreasing the resistance or shear along a potential slide plane. Third, rapid draw down, such as when the water level in a reservoir or river is lowered quickly at a rate of at least several feet per day, may increase the driving force due to the unsupported weight of water in the banks and decreasing shear resistance by the high water pressure. Fourth, water contributes to the spontaneous liquefaction of clay-rich sediment, or "quick clay." Some clays, when disturbed, lose their shear strength, behave as a liquid, and flow. The shaking of clay below Anchorage, Alaska, during the 1964 earthquake produced this effect and was very destructive (Figures 5.3 and 5.4). Fifth, seepage from artificial sources such as reservoirs or unlined canals into adjacent slopes may lower the cohesion and thus the shear strength by removing the cementing materials, as happened in the St. Francis Dam failure. Seepage may also cause an increase of water pressure in adjacent slopes, causing a reduction in the resisting force. A rise in water pressure occurs with many landslides, and it is safe to say that most landslides are, in fact, due to an abnormal increase of the water pressure in the slope-forming materials (2).

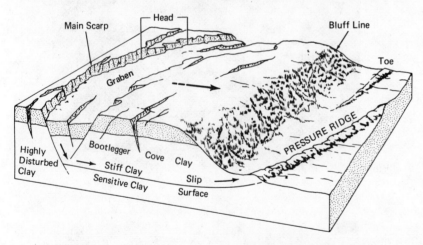

FIGURE 5.3 *Block diagram showing how landslides developed in response to the 1964 Alaskan earthquake in Anchorage. The graben is a down-drop block of earth material. (Source: W. R. Hansen, U.S. Geological Survey Professional Paper 542A, 1965.)*

Human Use and Landslides

Effects of human use and interest on the magnitude and frequency of landslides vary from insignificant or nearly so to very significant. In cases where man's activities have little to do with the magnitude and frequency of landslides, we need to learn all we can about where, when, and why they occur to avoid hazardous areas and minimize damage.

FIGURE 5.4 *Landslides at Anchorage, Alaska, caused by the 1964 earthquake. The slides are similar to those diagrammed in Figure 5.2. (a) Government Hill School slide. (b) Aerial view of slide near the hospital on Fourth Avenue. (c) Slides that caused the collapse of Fourth Avenue near C Street. [Photos (a) and (b) by W. R. Hansen, courtesy of U.S. Geological Survey. Photo (c) courtesy of U.S. Army.]*

(a)

(b)

(c)

In cases where human use has increased the number and severity of landslides, we need to learn how to recognize, control, and minimize their occurrence wherever possible.

Mixtures of adverse geologic conditions such as weak soil or rock and potential slip plains on steep slopes with torrential rains, heavy snowfall, or seasonably frozen ground (permafrost) will continue to produce landslides, mudflows, and avalanches regardless of man's activities. These are natural processes reacting to natural conditions. For example, the night of January 22, 1967, produced a landslide disaster of tremendous magnitude in Brazil. Following an electrical storm and cloudburst of 3 and 1/2 hours, an area of about 75 square miles was devastated by landslides and erosion (floods) which claimed the lives of about 1700 people. Damage to property and effects on industry were inestimable as tens of thousands of landslides turned the green hills into wastelands and valleys into a mud bath (Figure 5.5). Virtually all the slides were debris avalanches and mudflows (3). The slopes in disaster areas characteristically have only a thin veneer of residual soil over hard rock, resulting in a contact of relatively easy parting (Figure 5.6). It is interesting to note that more avalanches cut areas of high vegetation than grass-covered slopes, and many destroyed forests untouched by man for many years. Combination of high-frequency vibrations of the electrical storm with the pressure of the falling water is considered as primary in producing the avalanches. However, rising water pressure, groundwater movement, and erosion probably play a part also. This magnitude of landslide activity, while unique in recorded history, has probably been relatively frequent along the mid-southern coast of Brazil during the geologic evaluation of the slopes (3).

The above example is similar to a widespread episode of debris avalanches that occurred in August of 1969. The remnants of hurricane Camille moving eastward from Kentucky and the Appalachian Mountain ridges mixed with a mass of saturated air to provide thunderstorms of catastrophic proportions. These storms locally produced 28 inches of rain in 8 hours and triggered a great many debris avalanches in central Virginia (Figures 5.7 and 5.8). The storms claimed 150 lives,

FIGURE 5.5 *Floresta Creek Valley mudflow at the base of the Sierra Das Araras Escarpment. Before the mudflow, the valley bottom contained a village and a highway construction camp. Several hundred people lost their lives when the mudflow, which was about 12 feet in depth, moved through the area. (Photo by F. O. Jones, courtesy of U.S. Geological Survey.)*

FIGURE 5.6 *Debris avalanche in thin, residual soils. (Photo by F. O. Jones, courtesy of U.S. Geological Survey.)*

most to flooding. However, most people died from broken bones and blunt-force injuries rather than drowning. It is impossible to estimate how many died as a result of the avalanches, but the debris they delivered to channels in floods certainly was significant. The avalanches generally followed preexisting depressions, moved a layer of soil and vegetation 1 to 3 feet thick, and left a pronounced linear scar of exposed bedrock. The average amount of rock and soil debris moved was 88,000 cubic feet, or about 40,000 tons (4). Although the loss of life in Virginia was terrible, it was relatively low because of the sparce population. Inhabitants of Yungay and Ranrahirca, Peru, were not so fortunate when, in 1970, a debris avalanche triggered by an earthquake roared 12,000 feet down Mt. Huascaran at a speed of 200 miles per hour, killing about 40,000 persons while depositing many feet of mud and boulders and leaving only scars where the villages had been (5). (See Figure 5.9.)

Many landslides have been caused by interactions involving adverse geologic conditions, excess moisture, and man-initiated changes in the landscape and slope material. Examples include the Vaiont Reservoir slide of 1963, the Handlova slide in 1960, and numerous slides in urban areas such as Rio de Janeiro and Los Angeles.

The world's worst dam disaster occurred on October 9, 1963, when approximately 2600 lives were lost at the Vaiont Dam in Italy. As reported by George Kiersch, the disaster involved the world's highest thin-arch dam (875 feet at the crest), and, strangely, no damage was sustained by the main shell of the dam or the abutments (6). The tragedy was caused by a huge landslide in which more than 312,000,000 cubic yards of rock and other debris moved at speeds of 60 miles per

FIGURE **5.7** *Debris avalanche scar. (Photo by G. P. Williams, courtesy of U.S. Geological Survey.)*

hour down the north face of Mt. Toc, located above the reservoir, and completely filled it with slide material for 1.1 miles along the axis of the valley to heights of nearly 500 feet above the reservoir level (Figure 5.10). The rapid movement created a tremendous updraft of air and propelled rocks and water up the south side of the valley, higher than 800 feet above the reservoir level. The slide in accompanying blasts of air and water and rock produced strong earthquakes recorded many miles away. It literally blew the roof off one man's house well over 800 feet above the reservoir and pelted the man with rocks and debris. The filling of the reservoir produced waves of water over 300 feet high that swept over the abutments of the dam. One mile downstream, the waves were still over 230 feet high, and everything for miles downstream was completely destroyed. The entire event, slide and flood, was over in less than seven minutes. The landslide followed a three-year period of monitoring the rate of creep on the slope which varied from less than 1 to as many as 12 inches per week, until September, 1963, when it increased to 10 inches per day. Finally, on the day before the slide, it was about 40 inches per day. Engineers expected a landslide, but one of much smaller magnitude, and they did not realize until October 8, one day before the slide, that a large area was moving as a uniform, unstable mass. Animals grazing on the slope had sensed danger and moved away on October 1.

The slide was caused by a combination of factors. First, there were adverse geologic conditions, including *weak rocks* and *limestone* with open fractures, sinkholes, and clay partings which were inclined toward the reservoir, producing unstable blocks (Figure 5.11), and *very steep topography* that created a strong driv-

FIGURE 5.8   *Debris avalanche deposits of logs, boulders, and other particles along the east branch of Hat Creek in Virginia. (Photo by G. P. Williams, courtesy of U.S. Geological Survey.)*

ing (gravitational) force. Second, there was increased water pressure in the valley rocks due to the impounded water in the reservoir. Groundwater migration into bank storage raised the water pressure and reduced the resistance to shearing (resisting force) in the slope. The rate of creep before the slide increased as the water table rose in response to higher reservoir levels. Third, heavy rains from late September until the day of the disaster further increased the weight of the slope materials, raised the water pressure in the rocks, and produced runoff that continued to fill the reservoir even after engineers tried to lower the reservoir level. It was concluded that the immediate cause of the disaster was an increase in the driving force accompanied by a great decrease in the resisting force; but the real cause was due to excess groundwater which raised the groundwater pressure, which in turn changed the weight of the slope material and produced a buoyancy effect along plains of weakness in the rock (6).*

The Handlova Slide in Czechoslovakia is an interesting example of an extremely hazardous landslide caused by a mixture of human use and adverse geologic conditions (Figure 5.12). A large coal-burning power plant in Handlova used a soft coal which emitted a large amount of fly ash that was carried and deposited to the south by prevailing winds. The land was changed by the accumulation of ash to such an extent that some land formerly used for grazing purposes had to be plowed. This allowed rains to infiltrate at a greater rate and disturb an already delicate groundwater situation. After a high rainfall in 1960, which further raised the water table, a large landslide involving approximately 26,000,000 cubic yards of

*G. A. Kiersch, *Civil Engineering* 34 (1964), pp. 32–39.

FIGURE 5.9 *Oblique aerial view of the debris avalanche that destroyed Yungay and Ranrahirca, Peru, killing about 40,000 people. (Photo by George Plafker, courtesy of U.S. Geological Survey.)*

earth material developed. The slide moved nearly 500 feet in one month and threatened to severely damage the town. A well-organized program to control the slide and stop the movement by draining the slide material was successful within 60 days. Even so, 150 homes were destroyed (7).

Human use and interest in the landscape are most likely to cause landslides in urban areas where high densities of people and supporting structures such as roads, homes, and industries exist. Examples from Rio de Janeiro, Brazil, and Los Angeles, California, illustrate.

Rio de Janeiro, with a population exceeding 4 million people, may have more slope-stability problems than any other city its size (3). Combinations of steep slopes and fractured rock mantled with surficial deposits contribute to the problem. In earlier times, many such slopes were logged for lumber and fuel and to clear space for agriculture. This early activity was followed by landslides associated with heavy rainfall. Recently, lack of room on flat ground has led to increased urban development on slopes. Vegetation cover has been removed, and roads leading to development sites at progressively higher areas are being built. Excavations have cut the toe (lower part) of many slopes and severed the soil mantle at critical points (Figure 5.13). In addition, the placing of slope fill material below excavation areas has increased the load (driving force) on slopes already unstable before the fill.

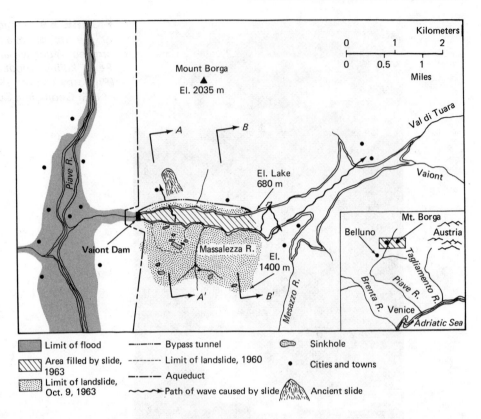

FIGURE 5.10 *Sketch map of the Vaiont Reservoir showing the 1963 landslide which overtopped the dam and caused severe flooding and destruction over large areas downstream. (Modified from Kiersh, 1964,* Civil Engineering *34.)*

Furthermore, this area periodically experiences tremendous rainstorms. Thus, it is easy to understand why Rio de Janeiro has a serious problem. In 1966, following heavy rains, numerous landslides occurred. The terrible storms of 1967 discussed earlier missed the city. If they had not, the results would have been catastrophic (3).

Los Angeles, and more generally Southern California, has experienced a remarkable frequency of landslides associated with hillside development. For example, in one storm period during 1962, landslides and mudflows in the Los Angeles area claimed two lives and forced the evacuation of more than 100 homes. Millions of dollars in property damage to structures such as homes, swimming pools, patios, and utilities occurred (8). (See Figure 5.14.) Landslides in Southern California result from complex physical conditions, including great contrasts in topography, rock and soil types, climate, and vegetation. Interactions between the natural environment and human activity are complex and notoriously unpredictable. For this reason, the area has the sometimes dubious honor of showing the ever-increasing value of urban geology (8). As a result, more consulting geologists in Southern California are employed in the analysis of slope stability than in any other field, and Los Angeles has led the nation in developing codes concerning grading (artificial excavation and filling) for development.

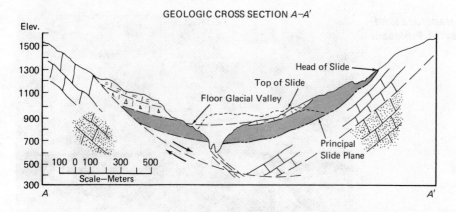

GEOLOGIC CROSS SECTION A–A'

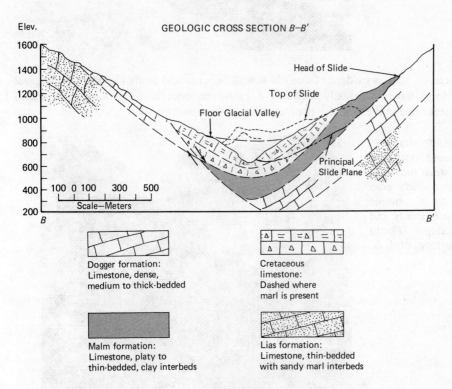

GEOLOGIC CROSS SECTION B–B'

Dogger formation:
Limestone, dense,
medium to thick-bedded

Cretaceous
limestone:
Dashed where
marl is present

Malm formation:
Limestone, platy to
thin-bedded, clay interbeds

Lias formation:
Limestone, thin-bedded
with sandy marl interbeds

FIGURE 5.11 *Generalized geologic cross sections through the slide area of the Vaiont River Valley. (Modified from Kiersh, 1964, Civil Engineering 34.)*

Even if no people were present in Southern California, landslides would be fairly common. The bulk of material that moves from valley sides and sea cliffs in Southern California is probably moved by landslides. Figure 5.15 shows one of the larger recent slides, the Portuguese Bend slide of 1956. As computed from geologic mapping, slides effect 60 percent of the length of sea cliffs, and the retreat of the cliff is probably controlled by landslides (8). (See Figures 5.16 and 5.17.) Similar estimates for slopes are not available, but the complex geology and terrain

FIGURE 5.12 *The Handlova land-slide. (Photo courtesy of Professor M. Matula.)*

features, as well as evidence from old landslide scars and landslide deposits, suggest that slopes have historically been active. However, man has tremendously increased the magnitude and especially the frequency of landslides.

FIGURE 5.13 *Landslide that de-molished several houses and two apartment buildings. More than 132 people died. The large slide was evidently facilitated by a smaller landslide caused by a highway cut that overloaded the slope. (Photo by F. O. Jones, courtesy of U.S. Geological Survey.)*

FIGURE 5.14 *Damage to property in Los Angeles caused by a mudflow. Note the enormous size of material that mudflows can carry. (Photo courtesy of the Los Angeles Times.)*

The grading process in Southern California has been responsible for many landslides. It took natural processes many thousands, if not millions, of years to produce valleys, ridges, and hills. In this century, man has developed the machines to grade them. Leighton writes: "With modern engineering and grading practices and appropriate financial incentive, no hillside appears too rugged for future development" (8). No earth material can withstand the serious assault of modern technology. Therefore, man is a geological agent capable of carving the landscape as do glaciers and rivers, but at a tremendously faster pace. For example, he can convert steep hills almost overnight into a series of flat lots and roads. Such conversions have led to numerous man-made landslides. As shown in Figure 5.18, oversteepened slopes mixed with increased water from sprinkled lawns or septic systems, as well as additional weight of fill material and a house, make formerly stable slopes unstable. Any man-made project that steepens or saturates a slope, or increases the height of or places an extra load on it, may cause a landslide (8). (See Figure 5.19.)

Identification, Prevention, and Correction of Landslides

In order to minimize the landslide hazard, it is necessary to *identify* areas in which landslides are likely to occur, *design* slopes or engineering structures to *prevent* landslides, and *control* and *stop* slides after they have started moving.

———— Approx. limits of ancient landslide   —·—·— Approx. limits of active landslide

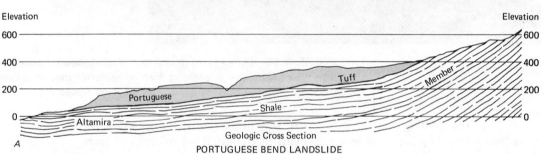

FIGURE 5.15 *Aerial view of Palos Verdes showing the approximate known limits of a large, ancient landslide and the outline of the Portuguese Bend landslide. Arrows show the direction of movement. Geologic cross section shows that volcanic tuft (consolidated ash) is sliding over shale. (Photo and cross section courtesy of County of Los Angeles, Department of County Engineer.)*

Identifying areas with a high potential for landslides is a first step in developing a plan to avoid development in areas with a likely landslide hazard. Slide tendency can be recognized by examining geologic conditions, slope, and precipitation. This information can then be used to produce slope stability maps as shown in Figure 5.20 for an area near San Clemente, California. That these maps can be useful in predicting a possible hazard is well illustrated in the before-and-after sequence of photographs in Figure 5.21. The house in the upper left corner of the top photograph (a) was mapped as being on unstable earth material (see the arrow), and, as shown in the lower oblique air photo (b), a landslide eventually destroyed the home. Of course, this type of mapping is needed before homes are constructed, but this example certainly verifies the validity and worth of such mapping.

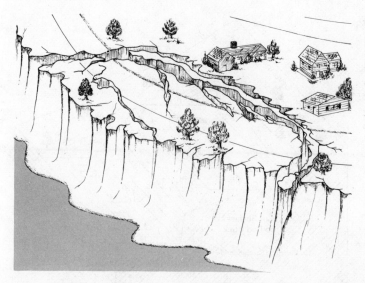

FIGURE 5.16 *How wave erosion of sea cliffs facilitates landslides. Notice the cracking, an early sign that a slide may be imminent.*

Prevention of large, natural landslides is nearly impossible, but good engineering practice can do much to minimize the hazard when it cannot be avoided. Methods of prevention include removal of unstable slope materials, selection of a safe slope factor in excavation, provisions for surface and subsurface drainage, and construction of retaining walls or other structures (5). A less common method involves the insertion of heavy bolts (rock bolts) through holes drilled through potentially unstable rocks into stable rocks. This technique was used to secure the slopes at the Glen Canyon Dam on the Colorado River and the Hanson Dam on the Green River in Washington (9).

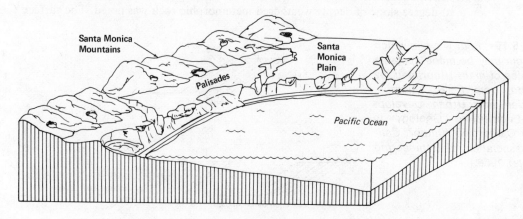

FIGURE 5.17 *Block diagram of a portion of Santa Monica Mountains, Santa Monica Plain, and Pacific Palisades in Southern California. Landslides are common along the sea cliffs, especially in the Pacific Palisades area. [Reprinted, by permission, from F. B. Leighton, "Landslides and Urban Development," Engineering Geology in Southern California (Whittier, California: Association of Engineering Geologists), 1966.]*

FIGURE **5.18** *Development of man-made translation landslides. Stable slopes may be made un-stable by removing support from the bedding plane surfaces. The cracks shown in the upper part of the diagram are one of the early signs that a landslide is likely to occur soon. [Reprinted, by per-mission, from F. B. Leighton, "Land-slides and Urban Development,"* Engineering Geology in Southern California *(Whittier, California: As-sociation of Engineering Geolo-gists), 1966.]*

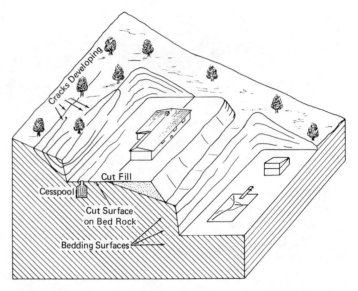

Once movement of a slide has started, the best way to stop it is to attack the process that started the slide. In most cases, the cause of the slide is an increase in water pressure, and in such cases, an effective drainage program must be initiated. This may include surface drains at the head of the slide to keep additional surface water from infiltrating and subsurface drainpipes or wells to remove water and lower the water pressure. The draining will tend to increase the resisting force of the slope material and, therefore, stabilize the slope.

The tremendous success of drainage was demonstrated by the following observation of Karl Terzaghi (1950). After a high-magnitude rainstorm, movement on a 30-degree slope of deeply weathered metamorphic rock was noted. The surface

FIGURE 5.19 *Four ways in which a stable cut may be made unstable by activities of man. [Reprinted, by permission, from F. B. Leighton, "Landslides and Urban Develop-ment,"* Engineering Geology in Southern California *(Whittier, Cali-fornia: Association of Engineering Geologists), 1966.]*

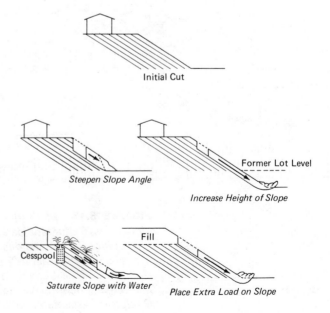

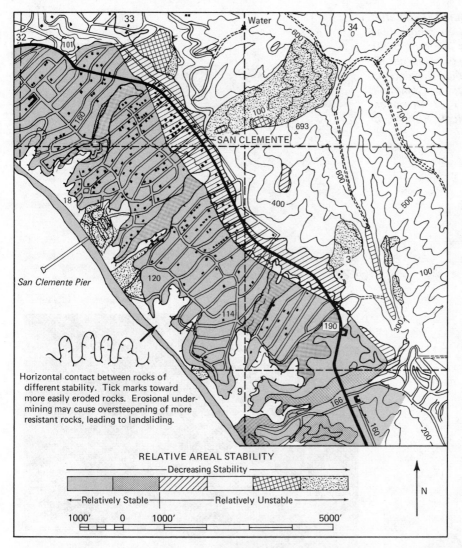

FIGURE 5.20 *Relative slope stability map of a portion of the San Clemente area. (Modified after Ian Campbell,* Environmental Planning and Geology, *U.S. Department of Housing and Urban Development, 1971.)*

of sliding was approximately 130 feet below the surface, and the slide area was about 500 feet wide by 1000 feet long and involved one-half million cubic yards of material. The slide was close to a hydroelectric power station, and thus immediate action was deemed necessary. Field work established that if the water level could be lowered approximately 15 feet, then the increase in resisting force would be sufficient to stabilize the slide. Drainage was accomplished by trenches and horizontal drill holes extending into the water-bearing zones of the rock. After drainage, the movement stopped, and even though the next rainy season brought record rainfall, no new movement was observed (2).

(a)

(b)

When slope materials are fine-grained and relatively impermeable, drainage is difficult at best and will probably be be ineffective. In this case, common correction methods include reducing the slope angle and constructing artificial barriers such as retaining walls. Unusual methods have also been used. For example, tunnels have been excavated in the slope and hot air circulated to dry out the unstable material. In other cases, unstable soil material has been frozen during construction to stabilize the slope. Drying the material with hot air was used with apparent success in the Pacific Palisades area of Southern California, and the freezing technique was used when a slide threatened to impede construction of the Grand Coulee Dam. Freezing was done by using 377 freezing points. The refrigeration system provided 160,000 pounds of ice per day, and a frozen dam 40 feet high, 20 feet thick, and 100 feet long formed to stabilize the slope material (9).

Subsidence

Various interactions between geologic conditions and human activity have also been involved with numerous incidents of subsidence. Most subsidence is caused by withdrawal of fluids from subsurface reservoirs or collapse of surface and near-surface soil and rocks over subterranean voids.

Withdrawal of fluids such as oil with associated gas and water, ground-water, and mixtures of steam and water for geothermal power have caused subsidence (10). In all cases, the general principles are the same. Fluids in earth materials below the earth's surface have a high fluid pressure that tends to support the material above. This is why a large rock at the bottom of a swimming pool seems lighter. Buoyancy is produced by the liquid that tends to lift it. If that support or buoyancy is removed by pumping out the fluid, the support is reduced. Subsidence at the surface may be the result. The actual subsidence mechanism involves compaction of individual grains of the earth material as the grain-to-grain load increases due to a lowering of the fluid pressure. Subsidence of oil fields generally involves considerable reduction of fluid pressure, up to 4000 pounds per square inch, at great depth (thousands of feet), and over a relatively small area, less than 50 square miles. On the other hand, subsidence due to withdrawal of groundwater generally involves a relatively low reduction of fluid pressure, often less than 200 pounds per square inch, at relatively shallow depths (less than 2000 feet) and over a large area, sometimes many hundreds of square miles (10).

The Wilmington oil field in the harbor area of Long Beach in Los Angeles, California, is a spectacular incidence of land subsidence that has caused over $100 million in damage to structures and several hundred ruptured oil wells and has caused localized flooding (Figure 5.22). Subsidence was first noted there in 1940 and by 1974 had increased to 29 feet in the central area (Figure 5.23). Predictions of ultimate subsidence as great as 45 feet promoted a joint effort to stop it. A repressuring program was started in 1958. Tremendous quantities of water were injected into the rocks from which the oil was pumped out, raising the fluid pressure. This actually reversed the subsidence, and by 1963 there had been appreciable rebound amounting to as much as 15 percent of the initial subsidence in some points of the oil field (10). Figure 5.24 shows the vertical movement at Pier A from 1952 to 1974 where the rebound was significant.

Thousands of square miles of the central valley of California have subsided. More than 2000 square miles in the Los Banos-Kettleman City area alone have subsided more than 1 foot, and within this area, one 70-mile stretch has subsided an average of over 10 feet with a maximum of 28 feet (Figure 5.25). The cause was overpumping of groundwater from a deep confined aquifer. As the water was literally mined, the fluid pressure was reduced and the grains were compacted (11). The effect at the surface was subsidence.

Another example of subsidence caused by overpumping is found in Mexico City. Population in Mexico City increased from less than one-half million in

FIGURE 5.22   *Flooding at Long Beach, California, caused by subsidence resulting from withdrawal of oil. (Photo courtesy of the city of Long Beach.)*

1895 to 1 million in 1920 and 5 million in 1960. As population grew, so did the demand for groundwater, which is pumped from sand and gravel aquifers separated by clay and silt deposits. These conditions are continuous at depths of about 60 meters to over 500 meters. The discharge of thousands of private and municipal wells far exceeds the natural recharge of the aquifers (10). The reduction of fluid pressure as the water table lowers has resulted in as much as 7 meters of subsidence, most of which has occurred since 1940. The continuous sinking has produced many problems in drainage and building construction (10). One building, the Palace of Fine Arts, constructed in 1934, has subsided approximately 3 meters, and, reportedly, steps that once went up to the first floor now go down to that floor.

Subsidence is also caused when subterranean earth materials are removed by natural or man-made processes. Rock such as limestone and dolomite are soluble, and subterranean voids often form. Lack of support for overlying rock may lead to collapse and the formation of large sinkholes, some of which are over 100 feet across and 50 feet deep. One near Tampa, Florida, collapsed suddenly in 1973, swallowing part of an orange orchard.

What might be the largest sinkhole in the United States formed in 1972 near Montevallo, Alabama (12). A massive hole 400 feet wide and 150 feet deep,

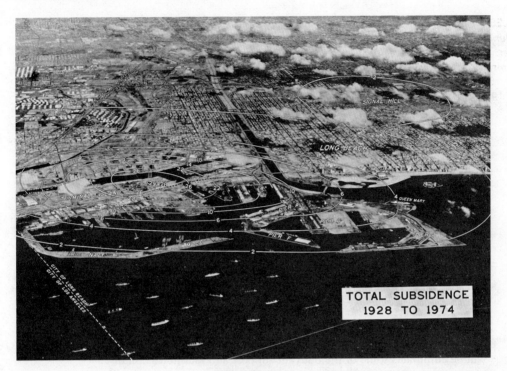

FIGURE 5.23 *Subsidence in the Long Beach Harbor area, California, from 1928 to 1974. (Photo courtesy of the Port of Long Beach.)*

named the *December Giant* by the press, developed suddenly when topsoil and subsurface clay collapsed into an underlying limestone cavern (Figure 5.26). The collapse was caused by a loss of support to the clay and soil over the cavern. A nearby resident reported hearing a roaring noise accompanied by breaking timber and earth tremors that shook his house. Sinkholes of this type have caused considerable damage to highways, homes, sewage facilities, and other structures (Figure 5.27). Fluctuations in the water table, natural or man-induced, are probably the triggering

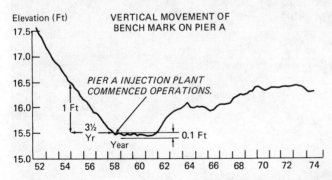

FIGURE 5.24 *Vertical movement of bench mark (point of known elevation) on Pier A from 1952 to 1974. (Courtesy of the Port of Long Beach.)*

FIGURE 5.25 *Principal areas of land subsidence in California due to groundwater withdrawal. (Reprinted, by permission, from W. B. Bull, Geological Society of America Bulletin 84, 1973.)*

mechanism. High water table conditions favor solutional enlargening of the cavern, and the buoyancy of the water helps support the overburden. Lowering of the water table eliminates some of the buoyant support and facilitates collapse.

Serious subsidence events have been associated with mining of salt, coal, and other minerals. Salt is often mined by solution methods in which water is injected through wells into salt deposits. The salt dissolves, and water supersaturated in salt is pumped out. The removal of salt leaves a cavity in the rock and weakens the support for the overlying rock. This may lead to large-scale subsidence. In 1970, one event near Detroit produced a subsidence pit 400 feet across and 300 feet deep. Another near Saltville, Virginia, produced 250 feet of subsidence relatively quickly (Figure 5.28). Two homes went down with the Saltville subsidence. According to local residents, one family moved out the day before the event because one member

FIGURE 5.26 *A very large sinkhole known as the* December Giant. *(Photo courtesy of Bill Warren and the Geological Survey of Alabama.)*

FIGURE 5.27 *Home destroyed by large sinkhole. (Photo courtesy of the U.S. Department of Housing and Urban Development.)*

FIGURE 5.28 *Large subsidence pit resulting from subsurface solutional mining of salts near Saltville, Virginia.*

of the family dreamed the mountain was falling. As of 1975 large and still actively growing fractures in the rocks surrounding the subsidence area indicated that the area involved with the subsidence was considerably larger than what actually subsided into the subterranean void below.

Full recovery where all the coal is removed from subsurface mines has produced subsidence problems. The Pittsburgh area is a good example. Mining has been going on there over 100 years. In early years, companies purchased mining rights permitting removal of the coal with no responsibility for surface damage. The results were not so serious when mining was conducted under farmland, but with recent rapid urbanization progressing faster than coal can be extracted, problems have resulted. If all of the coal is removed, the chance of subsidence and damage to homes is high. However, if about 50 percent of the coal is left, then this amount is usually sufficient support (Figure 5.29). The Bituminous Mine Subsidence and Land Conservation Act of 1966 provided for protection of public health, welfare, and safety by regulating coal mining. However, this act will cause hundreds of millions of tons of coal to remain in the ground, attesting to the nature of trade-offs when there is a conflict in surface and subsurface human use of the land (13).

Subsidence incidents have also been reported over coal mines which have not been worked for over 50 years (12). On a January morning in 1973, a few residents of Wales were driving when a section of the road suddenly collapsed into a pit 10 meters deep. Their car tottered on the brink while they scrambled to safety. The collapse was over an air shaft of a lost mine. The subsidence disrupted some

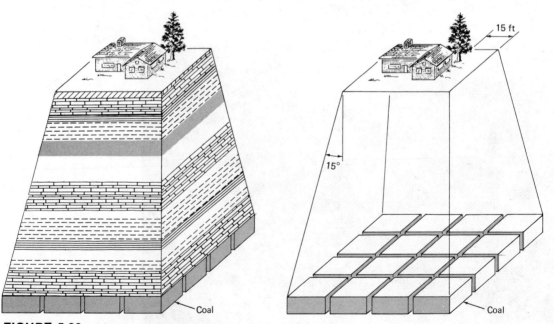

FIGURE 5.29 *Idealized diagram showing necessary support for a home over a coal mine. Support based on a rock fracture of 15 degrees on all four sides of the house foundation with 15-foot-wide safety area around the foundation. (After A. E. Van-dale,* Mine Engineering *19, 1967.)*

utility service. Other similar subsidences have happened in the past and will likely occur in the future.

Perception of the Landslide Hazard

The common reaction of homeowners in Southern California is, "It could happen on other hillsides, but never this one" (8). As with flooding, landslide hazard maps will probably not prevent persons from moving into hazardous areas, and prospective hillside occupants who are initially unaware of the hazards probably cannot be swayed by technical information. Furthermore, the infrequency of large slides may tend to reduce the awareness of the hazard. This is especially true where evidence of past events is not easily viewed.

SUMMARY AND CONCLUSIONS

The most common landforms are slopes, which are dynamic, evolving systems in which surficial material is constantly moving down slope at rates varying from imperceptible creep to thundering avalanches.

Important aspects of landslides are the type of earth material on the slope, the climate, and water. The cause of most landslides can be determined by examining the relations between the forces that tend to make earth materials slide (*driving* forces) and forces that tend to oppose movement (*resisting* forces). The most common driving force is the weight of the slope materials, and the most common resisting force is the shear strength of the slope materials.

The role of water in landslides is especially significant and is nearly always directly or indirectly involved in landslides. Water in streams, lakes, or oceans erodes the toe area of slopes, increasing the driving forces. Excess water increases the weight of the slope materials while raising the water pressure, which in turn decreases the resisting forces in slope materials. A rise in water pressure occurs before many landslides, and, in fact, most landslides are due to an abnormal increase in water pressure in the slope-forming materials.

Effects of human use on the magnitude and frequency of landslides varies from insignificant to very significant. Where activities of people have little effect on landslides, we need to learn all we can about where, when, and why landslides occur so that we can avoid development in hazardous areas or, if necessary, provide protective measures. In cases where human use has increased the number and severity of landslides, we need to learn how to recognize, control, and minimize these occurrences.

Withdrawal of fluids such as oil and water and subsurface mining of salt, coal, and other minerals has led to a subsidence hazard. In the case of fluid withdrawal, the cause of subsidence is a reduction of fluid pressures that tend to support overlying earth materials. In the case of solid material removal, the subsidence may result due to loss of support of the overlying material.

To minimize landslide hazard, it is necessary to have identification, prevention, and correction procedures. Monitoring and mapping techniques facilitate identification. Prevention of large natural slides is nearly impossible, but good engineering practices can do much to minimize the hazard when it cannot be avoided. Correction of landslides must be designed to attack the processes that started the slide.

Perception of the landslide hazard by most people, unless they have prior experience, is negligible. Furthermore, hillside residents, like floodplain occupants, are not easily swayed by technical information.

References

1. Krynine, D. P., and Judd, W. R. 1957. *Principles of engineering geology and geotechnics.* New York: McGraw-Hill.
2. Terzaghi, K. 1950. Mechanisms of landslides. The Geological Society of America: *Application of Geology to Engineering Practice,* Berken vol.: 83–123.
3. Jones, F. O. 1973. *Landslides of Rio de Janeiro and the Sierra das Araras Escarpment, Brazil.* U.S. Geological Survey Professional Paper 697.
4. Williams, G. P., and Guy, H. P. 1973. *Erosional and depositional aspects of Hurricane Camille in Virginia, 1969.* U.S. Geological Survey Professional Paper 804.
5. Office of Emergency Preparedness. 1972. *Disaster preparedness* 1, 3.
6. Kiersch, G. A. 1964. Vaiont Reservoir disaster. *Civil Engineering* 34: 32–39.
7. Legget, R. F. 1973. *Cities and geology.* New York: McGraw-Hill.
8. Leighton, F. B. 1966. Landslides and urban development. In *Engineering geology in Southern California*, eds. R. Lung and R. Proctor, pp. 149–97. Special Publication of the Los Angeles Section of the Association of Engineering Geology.
9. Morton, D. M., and Streitz, R. 1975. Mass movement. In *Man and his physical environment*, eds. G. D. McKenzie and R. O. Utgard, pp. 61–70. Minneapolis: Burgess.
10. Poland, J. F., and Davis, G. H. 1969. Land subsidence due to withdrawal of fluids. In *Reviews in Engineering Geology*, eds. D. J. Varnes and G. Kiersch, pp. 187–269. The Geological Society of America.
11. Bull, W. B. 1974. Geologic factors affecting compaction of deposits in a land-subsidence area. *Geological Society of America Bulletin* 84: 3783–802.
12. Cornell, J., ed. 1974. *It happened last year—earth events—1973.* New York: Macmillan.
13. Vandale, A. E. 1967. Subsidence: a real or imaginary problem. *Mining Engineering* v. 19, no. 9, 86–88.

6

EARTHQUAKES AND
RELATED PHENOMENA

The earth is a dynamic, evolving system. Its outer layer of lithosphere is broken into several large and numerous small *plates* that move relative to one another. As new lithosphere is produced at oceanic ridge systems and older lithosphere either is consumed at subduction zones or else slides past another plate, tremendous stress builds up in the rocks. When the stress exceeds the strength of the rocks, the rocks break and an earthquake is produced. This action can be compared to the pushing together and sliding of two rough boards past one another. Friction along the boundary slows their motion, but rough splinters break off and motion occurs at various places along the boundary. This is similar to what happens to plate boundaries where one plate slides past another or one overrides another. The rocks take up the stress and deform, and if the stress continues, they eventually break along weak zones called *faults*. This breaking of rocks and resulting movement along faults produces seismic waves that are recorded as earthquakes. Some of the waves travel within the earth (primary and secondary waves), while other, more complex waves travel along the surface of the earth. It is the surface waves that produce most of the damage to buildings and other structures.

Because most natural earthquakes are initiated near plate boundaries (Figure 3.2), there tend to be linear to curvilinear continuous zones along which most activity will take place (Figure 6.1). However, large earthquakes have also occurred far from plate boundaries, as, for example, the 1886 earthquake which devastated Charleston, South Carolina, claiming 60 lives and causing $23 million in property damage.

The point or area within the earth where the first motion along a fault takes place is called the *focus* of the earthquake. The *epicenter* is generally the point on the surface of the earth directly above the focus. However, in some cases, as for an earthquake originating deep in a subduction zone, the epicenter may not be directly over the focus. The location of an earthquake reported by the news media is the epicenter.

A large earthquake ranks as one of nature's most catastrophic and devastating events. Earthquakes have destroyed large cities and taken thousands of lives in a few seconds. One sixteenth century earthquake in China reportedly claimed 850,000 lives, and as recently as 1923, an earthquake near Tokyo, Japan,

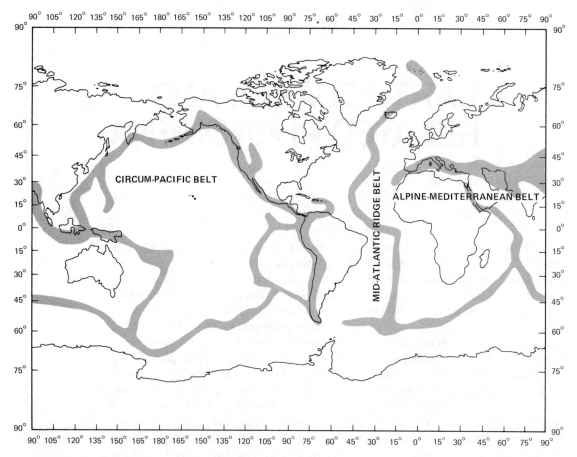

FIGURE 6.1 *Map of the world showing the three major earthquake belts. (Base map from NOAA.)*

claimed 143,000 lives. It has been estimated that if a large earthquake were to hit San Francisco again, as it did in 1906, as many as 100,000 people might die in that one event. Table 6.1 lists some of the major earthquakes that have recently (geologically) occurred in the United States. Based upon past magnitude and frequency of earthquakes, a hazard map has been produced (Figure 6.2). Although this map is valuable, considerable more data are necessary to identify more precisely hazardous areas to assist in developing building codes and determining insurance rates.

Year	Locality	Damage $ million	Lives Lost
1886	Charleston, South Carolina	23.0	60
1906	San Francisco, California	524.0	700
1925	Santa Barbara, California	8.0	13
1933	Long Beach, California	40.0	115
1940	Imperial Valley, California	6.0	9
1952	Kern County, California	60.0	14
1959	Hebgen Lake, Mont. (damage to timber and roads)	11.0	28
1964	Alaska and U.S. West Coast (includes tsunami damage from earthquake near Anchorage)	500.0	131
1965	Puget Sound, Washington	12.5	7
1971	San Fernando, California	553.0	65

TABLE 6.1 *Selected major recorded earthquakes in the United States*

SOURCE: U.S. Department of Commerce, OEP data, 1971.

Like natural hazards previously discussed, earthquakes often produce primary effects and secondary effects. *Primary* effects are caused directly by the phenomena. They include violent ground motion accompanied by fracturing that may cause permanent displacement of several feet, for example, as in the 1906 earthquake in San Francisco which produced 26 feet of horizontal displacement. These violent motions produce sudden surface accelerations of the ground which can snap and uproot large trees and knock people to the ground. This motion may shear or collapse large buildings, bridges, dams, tunnels, pipelines, and other rigid structures (1). Damage due to the 1964 Alaskan earthquake, excluding personal property and income, is shown on Figure 6.3. Extensive damage to transportation systems such as railroads, airports, and buildings occurred (Figures 6.4 and 6.5). Fortunately, Alaska has a relatively low population density. What would be the effect of a large earthquake in a highly populated area? A hint was given by the 1971 earthquake that occurred on the fringe of a highly populated area of the San Fernando Valley in California. The earthquake released less than 1 percent of the energy of the Alaskan earthquake, and yet it claimed one-half the lives and did more property damage than the Alaskan event of much larger magnitude. Damage to homes and larger buildings was extensive in the city of San Fernando (Figures 6.6 and 6.7). In addition, the lower Van Norman Dam above the highly populated valley was severely damaged (Figure 6.8) and on the brink of catastrophic failure, threatening the lives of 80,000 people who evacuated their homes (2, 3). They returned safely to their homes four days later after the water in the reservoir had been lowered to a safe level.

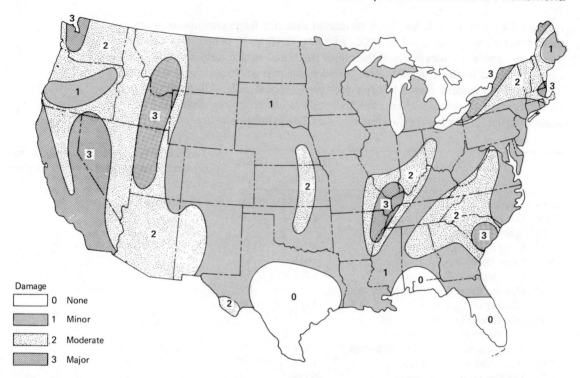

Damage

☐ 0 None

▨ 1 Minor

▨ 2 Moderate

▨ 3 Major

FIGURE 6.2 *Earthquake risk map for the conterminous United States. This map shows very generally those areas that are most likely to experience a major destructive earthquake. However, care must be used in interpreting this map. For example, although parts of both California and South Carolina have experienced a large earthquake and, therefore, are assigned the same value on the map, there is not necessarily an equal probability of earthquake occurrence in these two areas. (Environmental Science Service Administration, 1969.)*

        Secondary effects of earthquakes include a variety of *short-range* events, such as landslides, fires, tsunamis, and floods, and *long-range* effects, including phenomena such as regional subsidence or emergence of landmasses and regional changes in groundwater levels.

        Landslides of all varieties are triggered or directly caused by earthquakes. They can be very destructive and can cause the loss of many lives, as demonstrated by the 1970 Peru earthquake. In that earthquake, more than 70,000 people died, and of these, 40,000 were killed by a gigantic avalanche. The 1964 Alaskan earthquake produced literally thousands of landslides and avalanches, some of which devastated Anchorage (4, 5).

        Damage and even catastrophic destruction have resulted from *fires* indirectly caused by earthquakes. Disruption of electrical power lines and broken gas lines may start fires that are difficult to control because fire-fighting equipment may be damaged and essential water mains may be broken. In addition, access to fires is hampered by blocked and damaged roads. Earthquakes in Japan and the

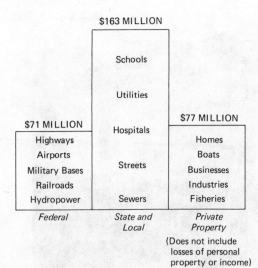

$163 MILLION

Schools

Utilities

$71 MILLION   Hospitals $77 MILLION

Highways   Homes
Airports   Boats
Military Bases   Streets   Businesses
Railroads   Industries
Hydropower   Sewers   Fisheries

Federal   *State and*   *Private*
                 Local   *Property*

(Does not include
losses of personal
property or income)

FIGURE 6.3 *Earthquake damages in Alaska as estimated by the Office of Emergency Planning. (Source: Hansen and Eckel, U.S. Geological Survey Professional Paper 541, 1966.)*

United States have produced terrible fires. The San Francisco earthquake of 1906 has repeatedly been referred to as the "San Francisco fire," and, in fact, 80 percent of the damage was caused by terrible fires which ravaged the city for several days. The 1923 earthquake in Japan killed 143,000 people, 40 percent of whom died in a fire storm that engulfed an open space where people had gathered in an unsuccessful attempt to reach safety (1).

Tsunamis or seismic sea waves (giant waves) are secondary effects generated by coastal or submarine earthquakes, as well as other submarine distur-

FIGURE 6.4 *Wreckage of control tower at Anchorage International Airport, Alaska. Six-story tower failed under sustained seismic shaking. (Photo by George Plafker, courtesy of U.S. Geological Survey.)*

FIGURE 6.5 *Damage to Penney's department store in Anchorage, Alaska. Building failed after sustained seismic shaking. At the time of the photograph, most of the rubble had been cleared from the streets. (Photo by George Plafker, courtesy of U.S. Geological Survey.)*

bances such as volcanic activity. Tsunamis are extremely destructive and perhaps the most serious of secondary effects caused by earthquakes. These giant waves claimed the vast majority of lives lost in the 1964 Alaskan earthquake.

      Regional changes in land elevations are another secondary effect of some large earthquakes. The Alaskan earthquake of 1964 caused vertical deformation (uplift and subsidence) of more than 100,000 square miles (5). The deformation includes two major zones of warping. Each is about 600 miles long and over 130 miles wide (Figure 6.9). The seaward zone is one of uplift as great as 30 feet. The landward zone is characterized by subsidence as great as 8 feet centered near the vicinity of Portage. Effects of this nature vary from severely disturbed coastal marine life to changes in groundwater levels. In addition, there was flood damage to some communities that experienced subsidence (Figure 6.10), whereas in areas of uplift, canneries and fishermen's homes were placed above the reach of most high tides. Before the earthquake, these structures and their docks were on the water (Figure 6.11).

Tectonic Creep

Slow, nearly continuous movement along a fault zone not accompanied by felt earthquakes is known as *tectonic creep*. This process can slowly damage roads, sidewalks, building foundations, and other structures (Figure 6.12). It has damaged culverts under the football stadium at the University of California at Berkeley (Figure 6.13).

(a)

(b)

FIGURE 6.6 *Damage to homes in San Fernando, California, caused by the 1971 earthquake. [Photo (a) by R. Castle, courtesy of U.S. Geological Survey. Photo (b) courtesy of Los Angeles City Department of Building and Safety.]*

Periodic repairs have been necessary as the cracks develop, and movement of 1.25 inches in eleven years was measured (6). Faster rates of tectonic creep have been recorded on the Calaveras fault zone near Hollister, California. There, a winery located on the fault is slowly being pulled apart at about 1/2 inch per year (7).

FIGURE 6.7   *Rescue operations at Veterans Hospital after the 1971 earthquake. (Photo courtesy of Los Angeles City Department of Building and Safety.)*

FIGURE 6.8   *Severely damaged as a result of the 1971 earthquake, the lower Van Norman Dam threatened the lives of thousands of people. (Photo by R. E. Wallace, courtesy of U.S. Geological Survey.)*

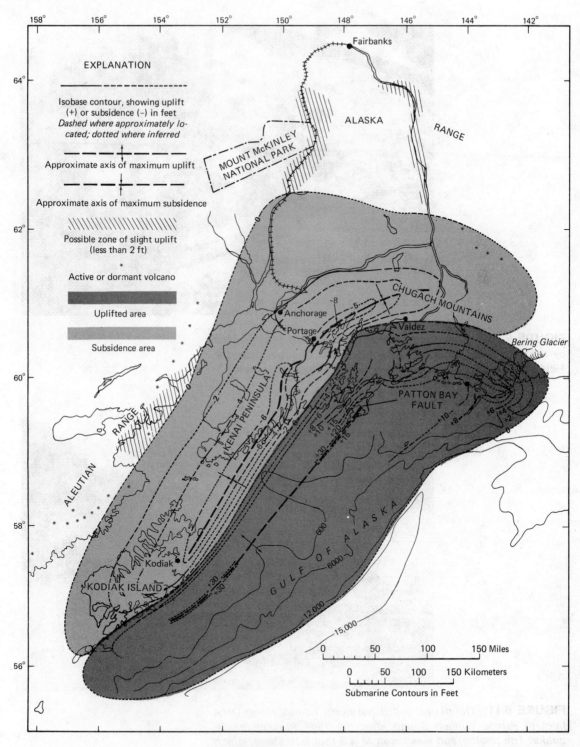

FIGURE 6.9 *Map showing the distribution of tectonic uplift and subsidence in south central Alaska due to the Alaskan earthquake of 1964. (From E. B. Eckel, U.S. Geological Survey Professional Paper 546, 1970.)*

FIGURE 6.10   *Flooding of the Portage area of Alaska caused by regional subsidence due to tectonic land changes from the 1964 earthquake. (Photo courtesy of U.S. Geological Survey.)*

FIGURE 6.11   *Canneries and fishermen's homes along Orca Inlet in Prince William Sound after the 1964 Alaskan earthquake. The photograph was taken at a 9-foot tide stage, which would have reached beneath the docks prior to the earthquake. (Photo courtesy of U.S. Geological Survey.)*

FIGURE 6.12 *Curb displaced by fault creep on the Hayward fault. This type of movement is generally very slow and not accompanied by felt earthquakes.*

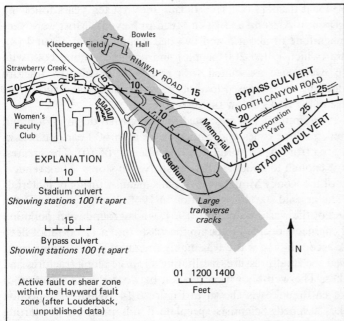

FIGURE 6.13 *Map showing the location of the University of California Memorial Stadium, Berkeley, California; the active fault of shear zone within the Hayward fault zone; and the stadium culvert where major cracking has taken place. (From Radbruch et al., U.S. Geological Survey Circular 525, 1966.)*

Damage due to tectonic creep generally occurs along narrow zones that delineate a fault in which movement is slow and continuous.

Magnitude and Frequency of Earthquakes

The *magnitude* of an earthquake is a measure of the amount of energy released and is useful in making quantitative comparisons of earthquakes. It is determined by the amplitude of the largest horizontal trace recorded on a *seismograph* (an instrument for measuring earthquake vibrations). The scale of magnitude is logarithmic, so a magnitude of 6, for example, produces a displacement 10 times larger than does a magnitude of 5. However, the energy released may be 30 to 60 times greater. Thus, between 27,000 and 216,000 shocks of magnitude 5 are required to release as much energy as a single earthquake of magnitude 8. There are probably about one million earthquakes a year that can be felt; however, only a few are strong enough to be recorded any considerable distance from their source. It has been estimated that every 100 years there are about 200 strong earthquakes with a magnitude exceeding 7.75.

Earthquakes Caused by Human Use of the Land

Activities of man have increased and caused earthquake activity. The damage caused is regrettable, but the lessons learned may be used in the future to control or stop large catastrophic earthquakes. Three ways man has caused earthquakes are by loading the earth's crust, disposing of waste into deep wells, and setting off underground nuclear explosions (8).

About 600 local tremors occurred during the first ten years following completion of Hoover Dam in Arizona and Lake Mead in Nevada. Most were very small, but one had a magnitude of about 5 and two had magnitudes of about 4 (8). An earthquake in India (which killed 200 people) and one in Zambia, both with magnitudes of about 6, were also associated with reservoirs. Evidently, fracture zones (faults) are activated by the increased load of the water on the land and by the increased water pressure in the rocks below the reservoir.

Events in 1966 dramatically suggested that deep-well disposal of liquid chemical waste activated fracture zones and caused more than 600 earthquakes in the Denver, Colorado, area from April, 1962, to November, 1965 (9). The greatest magnitude was 4.3, large enough to knock bottles off shelves in stores. The chemical waste was a by-product of the Rocky Mountain Arsenal's manufacturing of materials for chemical warfare. The arsenal started production in 1942, but evaporation from earth reservoirs disposed of the waste water until 1961, when groundwater pollution became a problem and engineers decided to pump the waste into a 12,000-foot-deep disposal well. The rock accepting the waste was highly fractured metamorphic rock. The new liquid increased the fluid pressure, facilitating slippage along fractures and precipitating earthquakes. The well-known correlation between rate of injection of waste and frequency of earthquakes is shown in Figure 6.14. When the injection stopped, the earthquakes stopped. Scientists speculate if this principle of starting and stopping earthquakes by varying the fluid pressure in rocks can be used to prevent large earthquakes.

Numerous earthquakes have been triggered by nuclear explosions at the Nevada test site (8). After the underground explosion of a 1.1-megaton nuclear device, thousands of aftershocks were recorded. The depth of the shocks ranged from the surface down to 7 kilometers. Magnitudes did not exceed 5.0, which is considerably less than the magnitude of 6.3, due to the initial explosion. Analysis of the aftershocks suggested they were triggered primarily by a release of natural

Homes in Anchorage, Alaska, destroyed by landslides associated with the 1964 earthquake. (Ward's Natural Science Establishment, Inc. photo)

San Francisco Bay area. The active faults, including the San Andreas, in this area, combined with poor land-use planning, result in a serious earthquake hazard. (ERTS photo)

Regardless of their mental distress after earthquakes, people generally do not really understand the earthquake hazard in areas of potential disaster. In addition, there appears to be a considerable hiatus between what people say and what they do. For example, near Tokyo, Japan, six new earthquake antidisaster centers and a steel mill are being constructed. While the centers are to aid victims of earthquakes, the steel mill is being placed near sea level on landfill which is part of a reclamation project for Tokyo Bay. The site is subject to severe earthquake and tsunami hazard (13). At the same time, across the Pacific Ocean, people in San Francisco continue their casual attitude toward their possible earthquake, which promises to be a tremendous disaster. People in hazardous areas could take down dangerous structures on old buildings that hang out over the streets. They could practice better planning for siting of buildings, and they could stop filling in bays and other sensitive areas. But they probably will not because a hazardous event, even a tremendously catastrophic one that comes but once every few generations, is simply not perceived by the general population as a real threat. However, it is encouraging that some communities are developing ordinances to limit development along active faults (Figure 6.25) and are considering land-use alternatives (Figure 6.26).

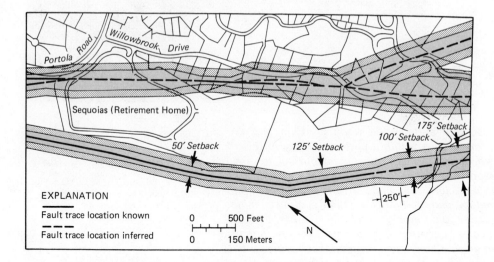

FIGURE 6.25 *Example of a fault hazard's easement required for building setbacks from active fault traces by town ordinances in Portola Valley, California. Where the location of the fault trace is well known, no new buildings are allowed to be constructed within 50 feet of each side of the fault. Structures with occupancies greater than single family dwellings are required to be 125 feet from the fault. Where the precise location of the fault trace is less well known, more conservative setbacks of 100 feet for single family residents and 175 feet for higher occupancies are required. (From Mader et al., and U.S. Geological Survey Circular 690, 1972.)*

FIGURE 6.26 *Oblique aerial photograph showing approximate location of fault traces within a part of the Hayward fault zone, some of which experienced movement during the 1868 earthquake. Present development compatible with a possible earthquake hazard include the undeveloped land, the plant nursery, the cemetery, and the highway. Other such uses might include golf courses, riding stables, drive-in theaters, or other recreational activities. Locating schools, police departments, hospitals, and other public facilities such as these along active fault traces is unwise land-use planning. (Photo by R. E. Wallace, courtesy of U.S. Geological Survey.)*

SUMMARY AND CONCLUSIONS

Large earthquakes rank as one of nature's most catastrophic and devastating events. Most earthquakes are located on tectonically active areas where lithospheric plates, on which the continents and ocean basins are superposed, interact.

Primary effects of earthquakes are violent ground motion accompanied by fracturing which may shear or collapse large buildings, bridges, dams, tunnels, and other rigid structures. Secondary effects include short-range events, such as fires, tsunamis, and floods, and long-range effects, such as regional subsidence and uplift of landmasses and regional changes in groundwater levels.

Tectonic creep resulting from slow movement along fault zones is a less hazardous process associated with earthquake activity. Nevertheless, it can cause considerable damage to roads, sidewalks, building foundations, and other structures.

The magnitude of an earthquake is a measure of the amount of energy released. It is determined by the amplitude of the largest horizontal trace recorded on a standard seismograph. It has been estimated that there are about one million earthquakes every year, two of which, on the average, have sufficiently high magnitude (greater than 7.75) to cause catastrophic damage.

Man has caused increasing earthquake activity in three ways: first, by loading the earth's crust by constructing large reservoirs; second, by disposing of liquid waste in deep disposal wells which raises fluid pressures in rocks and facilitates movement along fractures; and third, by setting off underground nuclear explosions. The accidental damage caused by the first two ways is regrettable, but lessons learned from all the ways we have caused earthquakes may be used to eventually control or stop large earthquakes.

Prediction and control of earthquakes are now subjects of serious research. Optimistic scientists believe that eventually we will be able to make long-range and short-range predictions of earthquakes. Many believe such predictions may be based on previous patterns and frequency of earthquakes; anomalous uplift and subsidence; and continuous monitoring of ground tilt, strain in rocks, microearthquake activity, magnetic activity, radon concentration, and electrical resistivity changes in earth materials. Scientists hope to control earthquakes by manipulating fluid pressure in rocks along faults and thereby cause a series of small earthquakes rather than let a natural catastrophic event occur. It has also been suggested that nuclear explosions, in conjunction with manipulation of fluid pressure, might be used to release natural tectonic stress slowly before a large earthquake.

Warning systems and earthquake prevention are not yet reliable alternatives. However, some steps can be taken, for example, structural protection, land-use planning, and insurance and relief measures.

The earthquake hazard in areas of potential disaster remains poorly perceived by inhabitants of the areas. Their lack of concern results probably because a process that produces even catastrophic damage may not be perceived as a real threat if it strikes the same area only every few generations.

References

1. Office of Emergency Preparedness. 1972. *Disaster preparedness* 1, 3.
2. U.S. Geological Survey and the National Oceanic and Atmospheric Administration. 1971. *The San Fernando, California, earthquake of February 9, 1971.* U.S. Geological Survey Professional Paper 733.

3. National Academy of Sciences and National Academy of Engineering. 1973. The San Fernando earthquake of February 9, 1971: lessons from a moderate earthquake on the fringe of a densely populated region. In *Focus on environmental geology*, ed. R. W. Tank, pp. 66–76. New York: Oxford University Press.

4. Hansen, W. R. 1965. *The Alaskan earthquake, March 27, 1964: effects on communities*. U.S. Geological Survey Professional Paper 542A.

5. Eckel, E. B. 1970. *The Alaskan earthquake, March 27, 1964: lessons and conclusions*. U.S. Geological Survey Professional Paper 546.

6. Radbruch, D. H. et al. 1966. *Tectonic creep in the Hayward fault zone, California*. U.S. Geological Survey Circular 525.

7. Steinbrugge, K. V., and Zacher, E. G. 1960. Creep on the San Andreas fault. In *Focus on environmental geology*, ed. R. W. Tank, pp. 132–37. New York: Oxford University Press.

8. Pakiser, L. C.; Eaton, J. P.; Healy, J. H.; and Raleigh, C. B. 1969. Earthquake prediction and control. *Science* 166: 1467–74.

9. Evans, D. M. 1966. Man-made earthquakes in Denver. *Geotimes* 10: 11–18.

10. Press, F. 1975. Earthquake prediction. *Scientific American* 232: 14–23.

11. Wallace, R. E. 1974. *Goals, strategy, and tasks of the earthquake hazard reduction program*. U.S. Geological Survey Circular 701.

12. Emiliani, C.; Harrison, C.; and Swanson, M. 1969. Underground nuclear explosions and the control of earthquakes. *Science* 165: 1255–56.

13. Nicholas, T. C., Jr. 1974. Global summary of human response to natural hazards: earthquakes. In *Natural hazards*, ed. G. F. White, pp. 274–84. New York: Oxford University Press.

7

VOLCANIC ACTIVITY

Volcanic activity is, worldwide, a very rare process that usually affects sparsely populated areas. However, volcanic eruptions can be tremendously destructive, and if such an event occurs near a densely populated area, a catastrophe of unheard proportions will be recorded (1).

Causes of volcanic activity are generally related to plate tectonics, and most active volcanoes are located near plate junctions where *magma* (molten rock) is produced as spreading or sinking lithospheric plates interact with other earth materials. Approximately 80 percent of all active volcanoes are located in the "ring of fire" which circumscribes the Pacific Ocean. This area is essentially the Pacific plate (Table 7.1). In the United States, several volcanoes in Hawaii, about 25 in Alaska, and several in the Cascade Range of the Pacific Northwest can be considered active.

Volcanoes in the Cascade Range of Washington, Oregon, and California continue to pose potential danger to the population and agricultural centers of the Pacific Northwest (Figure 7.1). Table 7.2 summarizes recent activity of major volcanoes in the Cascade Range. Table 7.3 lists the types and frequency of eruptions

TABLE 7.1 *Distribution of the world's active volcanoes.*

Area	Percentage of Active Volcanoes
Pacific	79
Western Pacific Islands	45
North and South America	17
Indonesian Islands	14
Central Pacific Islands (Hawaii, Samoa)	3
Indian Ocean Islands	1
Atlantic	13
Mediterranean and Asia Minor	4
Other Areas	3
	100

SOURCE: A. Rittman, *Disaster Preparedness,* Office of Emergency Preparedness, 1962.

TABLE 7.2 *Summary of recent activity of major volcanoes in the Cascade Range.*

	Mt. Baker	Glacier Peak	Mt. Rainier	Mt. St. Helens	Mt. Adams	Mt. Hood	Mt. Jefferson	Three Sisters	Mt. Mazama (Crater Lake)	Newberry Volcano (Newberry Crater)	Mt. Shasta	Glass Mountain area	Lassen Peak—Chaos Crags area
Active in historic time	X		X	X		X					?	?	X
Known products of eruptions, last 12,000 years													
Lava flow	X		X	X	X	X	X	X	X	X	X	X	X
Tephra (airborne rock debris)	X	X	X	X		?		X	X	X	X	X	X
Pyroclastic flow		X	X	X					X		X		X
Mudflow	X	X	X	X	X	X					X		X
Estimated population at risk													
More than 1000	X		X	X		?			X		X		?
Less than 1000		?			X		X	X		X		X	

SOURCE: Crandell and Mullineaux, "Technique and Rationale of Volcanic Hazards." *Environmental Geology* 1 (New York: Springer-Verlag, 1975).

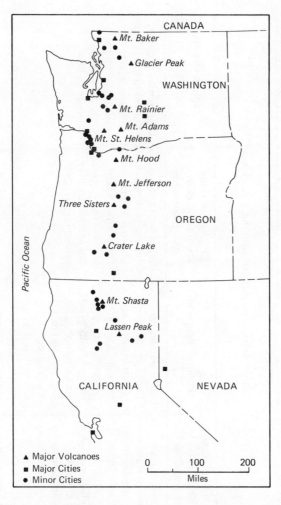

FIGURE 7.1 *Map of the Cascade Range showing the locations of some of the major volcanoes and cities in their vicinity. (Source: Crandell and Waldron,* Disaster Preparedness, *Office of Emergency Preparedness, 1969.)*

that have occurred at Mt. Rainier in the last few thousand years, the various effects that might be expected, and possible warning signs for future eruptions.

Three common types of volcanoes are the shield and composite volcanoes and the volcanic dome (Figure 7.2). Each type of volcano has a characteristic style of activity that is a result primarily of the silica content of the magma.

Shield volcanoes, by far the largest of the volcanoes, are characterized by relatively nonexplosive activity, resulting from the low silica content (about 50 percent) of the magma. Shield volcanoes are built up almost entirely from numerous lava flows. They are common in the Hawaiian Islands and are also found in some areas of the Pacific Northwest and Iceland.

Composite volcanoes, known for their beautiful cone shape, are associated with a magma of intermediate silica content (about 60 percent) which is more viscous than the low-silica magma of shield volcanoes. Volcanic activity characteristic of composite volcanoes is a mixture of explosive activity and lava flows. Examples in the United States include Mt. St. Helens and Mt. Rainier (Figure 7.1). These volcanoes can be dangerous. A study of volcanic activity on and near Mt. Rainier established that in the last 10,000 years, there have been one eruption in-

TABLE 7.3 *Frequency and type of eruptions that have occurred at Mt. Rainier in the last few thousand years, various effects that might be expected, and possible warning signs for future eruptions.*

| | Volcanic Activity Not Necessarily Associated with the Rise of New Magma | Types of Volcanic Activity Associated with the Rise of New Magma | | | |
	Steam Explosion (Generally on Small Scale)	Steam Explosion (May Be on Small to Very Large Scale)	Pumice[b] Eruption	Eruption of Bombs and (or) Block-and-Ash Avalanches	Eruption of Lava Flows
Direct effects	Formation of small-scale rockfalls and avalanches. Effects confined chiefly to flanks[a] of volcano and valley floors immediately adjacent.	Formation of rockfalls and avalanches that grade downvalley into debris flows.[c] Effects confined chiefly to flanks of volcano and valley floors. Formation of air-laid rubble deposits whose distribution would be limited to flanks of volcano and areas closely adjacent.	Fall of pumice on flanks of volcano; widest distribution beyond flanks in a downwind direction, thus probably greatest to the northeast, east, or southeast. Thickness probably will be less than a foot beyond 5-mile radius of summit. Anticipate thin ashfall over broad area without regard for topography.	Fall of hot to incandescent bombs, ash, and rock fragments. Distribution chiefly on flanks of volcano, but avalanches may extend into valleys.	Distribution probably limited to flanks of volcano. Major effects expectable only in immediate vicinity of flow.

	1 in 10,000 years	1 in 5000 years	1 in 2500 years	1 in 2000 years	1 in 10 to 100 years
(Direct effects)	Extensive melting of glaciers possible, caused by internal volcanic heat and by steam moving toward outside of volcano. Expected result: floods and debris flows on valley floors. Ejection of water from craters if flows occur at summit of volcano. Extrusion of lava flow beneath or onto glacier might cause catastrophic floods.	Ejection of water from summit craters in early stage of eruption. Possible result: floods and debris flows on valley floors. Possible floods and debris flows caused by eruption of hot debris onto snow.	Debris flows may result from extensive downslope movements of pumice onto valley floors.	Very large avalanches of moist, altered rock on flanks of volcano grading directly into debris flows on valley floors. Debris flows may be very long, extending tens of miles beyond park boundaries.	
Indirect effects				Possible floods and (or) debris flows caused by damming of rivers by avalanche deposits. Principal effects probably would be limited to valley floors within and closely adjacent to park.	
Possible warning signs of impending activity	Increase in frequency and magnitude of local earthquakes. Large increase in stream discharge unrelated to meteorological conditions. Appearance of clouds of water vapor associated with steam jets, possibly accompanied by small steam explosions and rockfalls. Increase in fumarolic activity and increased melting of snow at summit cone. Abnormal glacier melting at hot spots; appearance of melt pits in glaciers. Increase in temperature of fumaroles. Increase of sulfur and chlorine in fumarolic gases.			Appearance of steam jets and clouds of water vapor, possibly accompanied by explosions and rockfalls. Abnormal melting of glaciers at hot spots; appearance of melt pits in glaciers.	
Indicated possible frequency	1 in 10,000 years	1 in 5000 years	1 in 2500 years	1 in 2000 years	1 in 10 to 100 years

SOURCE: U.S. Geological Survey Bulletin 1238, 1967.
a Flanks of the volcano include an area within 4 or 5 miles from the summit.
b The term *pumice* is synonymous with the term *volcanic ash eruption* in the text.
c The term *debris flow* is synonymous with the term *mudflow* in the text.

(a) Kilauea, Hawaii

(b) Mt. St. Helens, Washington

(c) Mt. Lassen, California

FIGURE 7.2 *(a)* Shield volcano *has low-silica (approximately 50 percent) magma and relatively nonviolent activity. The largest of all volcanoes, it may measure 200 kilometers in width and is constructed of many lava flows. (b)* Composite volcano *has magma of intermediate silica content (approximately 60 percent) and activity alternating from explosive to lava flows. The cone is constructed of alternating layers of pyroclastic material (mostly ash) and lava flows. (c)* Volcanic dome *has high-silica magma (approximately 70 percent). It is the most violent and explosive of all volcanoes.* [Photos (a) and (b) courtesy of U.S. Geological Survey. Photo (c) by Mary R. Hill, courtesy of the California Division of Mines and Geology.]

volving lava flows and several explosive eruptions of volcanic ash (1). Mt. St. Helens has formed primarily in the last 2500 years, and the upper part has developed in the last few hundred years. Geologists believe that there is a high probability that Mt.

St. Helens will erupt again before the end of this century. Therefore, the potential hazard must be considered when developing future land use in the Pacific Northwest.

   Volcanic domes are characterized by viscous magma with a high silica content (about 70 percent). Activity is generally explosive, making volcanic domes very dangerous. Mt. Lassen in northeastern California, which last erupted from 1915 to 1917 (Figure 7.3), is a good example of a volcanic dome. One eruption included a tremendous horizontal blast that destroyed a large area. Fear of another such eruption recently resulted in the closing of the Manzanita Lake campground and lodge in the Mt. Lassen National Park which were judged to be too close to the volcano.

   Although relatively rare, volcanic activity has, from a historical standpoint, destroyed considerable property and taken many human lives. Table 7.4 lists some of the more catastrophic events. Unfortunately, volcanic activity cannot be prevented, and, therefore, programs to minimize this hazard involve prediction, warning, preparation, and experience gained from past events.

Effects of Volcanic Activity

Primary effects of volcanic activity include lava flows, airborne clouds of volcanic debris, and pyroclastic flows. Secondary effects include mudflows and fires (1).

Lava Flows Lava flows are the most familiar product of volcanic activity (Figure 7.4). They result when magma reaches the surface and overflows the crater or volcanic vent along the flanks of the volcano. Lava flows that are low in silica are generally nonexplosive or effusive, while flows with high silica content have eruptive or explosive activity. Volcanoes on islands or continents that have intermediate to high-silica-content magma are generally eruptive, whereas submarine activity in-

FIGURE 7.3 *Eruption of Lassen Peak, June, 1914. (Photo by B. F. Loomis, courtesy of the Loomis Museum Association.)*

TABLE 7.4 *Selected Volcanic Events.*

Volcano or City	Effect
Vesuvius A.D. 79	Destroyed Pompeii and killed 16,000 people. City was buried and rediscovered in 1595.
Skaptar Jokull, Iceland, 1783	Killed 10,000 people (many died from famine) and most of the island's livestock. Also killed some crops as far away as Scotland.
Krakatoa, Indonesia, 1883	Tremendous explosion; 36,000 deaths from tsunami.
Mt. Pelee, Martinique, 1902	Ash flow killed 30,000 people in a matter of minutes.
La Soufrière, St. Vincent, 1902	Killed 2000 people and caused the extinction of the Carib Indians.
Mt. Lamington, Papua, 1951	Killed 6,000 people.
Villarica, Chile, 1963–64	Forced 30,000 people to evacuate their homes.
Mt. Helgafell, Vestmannaeyjar, Iceland, 1973	Forced 5200 people to evacuate their homes.

SOURCE: Data partially derived from C. Ollier, *Volcanoes* (Cambridge, Massachusetts: MIT Press, 1969).

volves lava with a low silica content and is usually effusive. Lava flows can be quite fluid and move very rapidly or be relatively viscous and move slowly. Most lava flows are slow, and people can easily move out of the way as they approach (2).

Several methods, such as bombing, hydraulic chilling, and constructing walls, have been employed to deflect lava flows away from populated or otherwise valuable areas. These methods have had little real success. They cannot be expected to block large flows, and they need to be further evaluated.

Observation of lava movement in Japan suggests that there are two types of pressure: *hydrostatic pressure* related to the vertical height of the lava; and *pressure from the momentum of the flow,* which must be considered when building walls to deflect lava. Although even walls of rubble and loosely laid stone have fortuitously withstood lava flows in Italy and Hawaii, experience suggests that proper planning is necessary for best results (3). The following criteria are suggested by Mason and Foster: First, walls about 10 feet high are usually sufficient. Second, the upslope side should be steeper to prevent overriding of the wall by the lava. Third, the wall should be located diagonally to the slope so that the lava will be deflected to an area where little damage will occur. Fourth, guide channels should be located upslope to facilitate the direction of lava to the proper location. Fifth, walls should be located below likely vents and as far upslope as possible.

Bombing of the lava flows has been attempted to stop the advance of lava flows. It is most effective against lava flows in which fluid lava is confined to a channel by congealing lava on the margin of the flow. The objective is to partly block the channel by bombing, thereby causing the lava to pile up and facilitating an upstream break through which the lava escapes to a less damaging route. Successive bombing at higher and higher points as necessary to control the threat is the

FIGURE 7.4 *Lava flows moving slowly through a papaya orchard (a) and a village (b) during the 1959–60 eruption of Kilauea volcano, Hawaii. The lava fountain in (a) is approximately 400 feet high. (Photos by J. P. Eaton, courtesy of U.S. Geological Survey.)*

(a)

(b)

established procedure. Bombing has some merit for future research, but it probably provides little protection, is unpredictable, and, in any event, cannot be expected to affect large flows. Furthermore, poor weather conditions and the abundance of smoke and falling ash may reduce the effectiveness of bombing (3).

      The world's most ambitious program to control lava flows was initiated in 1973 on the Icelandic island of Heimaey, when lava flows with a low silica content nearly closed the harbor to the island's main town and thereby threatened the continued use of the island as Iceland's main fishing port. The situation prompted immediate action. Experiments on the island of Surtsey in 1967 showed that water could be used to stop the advance of lava. Favorable conditions existed on Heimaey to try it on a large scale. First, the main flows were viscous and slow moving, allowing the time necessary to initiate a control program. Second, transportation by sea and the local road system was adequate to move the necessary pumps, pipes, and heavy equipment. Third, water was readily available. The procedure was to first cool the margin and surface of the flow with numerous fire hoses fed from the 5-inch pipe. Then bulldozers were moved up on the slowly advancing (less than 3 feet per hour) flow, making a track or road on which the plastic pipe was placed. The pipe did not melt as long as water was flowing in it, and small holes in the pipe also helped cool the particularly hot spots. This was done along various parts of the flow, and at each place, several cubic feet of water per second was delivered at a distance of about 150 feet in back of the edge of the flow (Figures 7.5 and 7.6). This was done in conjunction with diversion barriers constructed by bulldozers mounding up loosened material in front of the advancing flow. Lava tended to pile up against these barriers. The watering at each location lasted about two weeks, or until the steam stopped coming out of the lava in that particular area. Watering had little effect the first day, but then that part of the flow began to slow down. The program undoubtedly had an important effect on lava flows. It tended to restrict their movement and thus reduced property damage. After the outpouring of lava stopped in June, 1973, the harbor was still usable (4).

FIGURE 7.5 *Volcanic activity threatening homes on the island of Heimaey in 1973. Watering of the lava can be seen in the background. (Photo courtesy of Icelandic Airlines.)*

FIGURE 7.6 *Close-up of an attempt to control the movement of a lava flow with water on the island of Heimaey. (Photo courtesy of Icelandic Airlines.)*

Pyroclastic Activity Pyroclastic activity is characteristic of high-silica magma. It is the eruptive or explosive volcanism in which all types of volcanic debris from ash to very large particles (tephra) are physically blown from a volcanic vent into the atmosphere. Two main types of activity are recognized: *volcanic ash eruptions,* in which a tremendous quantity of rock fragments, natural glass fragments, and gas are blown high into the air by explosions from the volcano; and *volcanic ash flows,* which are hot avalanches of ash, rock, and glass fragments that are mixed with gas and are blown out a vent and move very rapidly down the sides of the volcano.

Volcanic ash eruptions can cover hundreds and even thousands of square miles with a carpet of volcanic ash. An eruption at Crater Lake about 7000 years ago blanketed an area of several hundreds of thousands of square miles in the northwestern United States with ash. Approximately 800 square miles was covered with more than 6 inches of ash (Figure 7.7). Several volcanoes in the Cascade Range could have similar eruptions in the future (2).

There are several hazards from volcanic ash eruptions. First, destruction of vegetation, including crops and trees, may result (Figure 7.8). Second, surface water may be contaminated by sediment, and temporary increase in acidity of the water may result. The increase in acidity generally lasts only a few hours after the eruption ceases. Third, there may be structural damage to buildings caused by the increased load on roofs. One inch of ash will place an extra 7 tons of weight on a roof with a surface area of about 1500 square feet. This was a major problem during the 1973 Icelandic eruption. Many houses, public buildings, and businesses

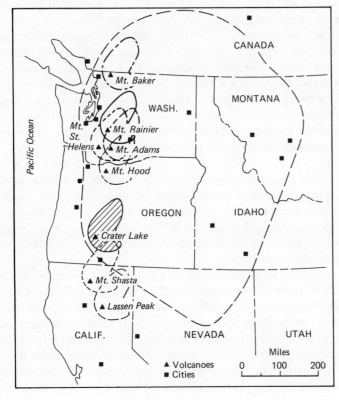

FIGURE 7.7 *Map showing area covered by a volcanic ash eruption at Crater Lake about 7000 years ago. The outer line shows the maximum limits of the ash fall. The shaded pattern shows the area covered by 6 inches or more of volcanic ash. This same area is superposed on other major volcanoes of the range. (Source: Crandell and Waldron, Disaster Preparedness, Office of Emergency Preparedness, 1969.)*

FIGURE 7.8 *Volcanic ash damaged or killed the trees shown here during the 1959–60 Kilauea volcano, Hawaii. (Photo by J. P. Eaton, courtesy of U.S. Geological Survey.)*

were literally buried in tephra (Figure 7.9). Many buildings collapsed from the excess load, but inhabitants did save many by diligently shoveling the debris from the roofs (4). Fourth, health hazards such as irritation of the respiratory system and eyes are caused by contact with the ash and associated caustic fumes (2).

Volcanic ash flows, which may move up to 70 miles per hour down the sides of a volcano, can be catastrophic if a populated area is in the path of the flow. It was such a flow of hot, incandescent ash, steam, and other gases that on the morning of May 8, 1902, roared through the West Indies town of St. Pierre, killing 30,000 people. Only one man, a prisoner in jail, survived, and he was severely burned and horribly scarred. Rumor has it that he spent the rest of his life touring circus sideshows as the "Prisoner of St. Pierre." Ash flows may be as hot as hundreds of degrees Celsius and may incinerate everything in their path. Such flows have occurred on volcanoes of the Pacific Northwest in the past and probably can be expected in the future. Fortunately, they seldom occur in populated areas.

Mudflows and Fires Two serious secondary effects of volcanic activity are *mudflows,* produced when a large volume of loose volcanic ash and other ejecta become saturated and unstable and move suddenly downslope, and *fires.* A small mudflow resulting from the 1915 eruption of Mt. Lassen in California is shown in Figure 7.10. Figure 7.11 shows the effects of mudflows and hot fiery blasts to forested land. What if there had been a town there?

Gigantic mudflows have originated on the flanks of volcanoes in the Pacific Northwest (1). The paths of two mudflows, the Osceola and Electron, that originated on Mt. Rainier are shown in Figure 7.12. Deposits of the Osceola mudflow are 5000 years old. This mudflow moved over 50 miles from the volcano and involved over 2½ billion cubic yards of debris, equivalent to 5 square miles piled to a depth exceeding 500 feet. Deposits of the younger, 500-year-old Electron mudflow reached about 35 miles from the volcano and involved in excess of 200 million cubic yards of mud. Over 30,000 people now live on the area covered by these old

(a)

(b)

FIGURE 7.9 *(a) Volcanic eruption (1973) on the island of Heimaey, about 10 miles southwest of Iceland. This eruption forced the evacuation of 5200 residents and covered more than one-half of the buildings with ash and lava. (b) House nearly buried in ash. (Photos courtesy of Icelandic Airlines.)*

FIGURE 7.10 *Several small mudflows on the west side of Lassen Peak. Four such mudflows reached Manzita Lake at the base of the mountain. Notice the new lava at the top of the peak. (Photo by B. F. Loomis, courtesy of the Loomis Museum Association.)*

flows, and there is no guarantee that similar flows will not occur again. Figure 7.13 shows the potential risk of mudflows and tephra accumulation for Mt. Rainier. A person in the valley in view of such a flow would describe it as a wall of mud several tens of feet high moving at about 20 miles per hour. He would see it at a distance of perhaps one mile and would need a car headed in the right direction toward high ground to escape being buried alive (2).

Crandell and Waldron emphasize the potential hazard of volcanic mud-flows as compared to flood hazards. Floods are usually preceded by heavy rains that cause a gradual rise in water level. Persons in flood-prone areas generally have time to escape, and when the flood recedes, the water and danger are essentially gone. However, catastrophic mudflows can occur with little or no warning. They are likely to start when the volcano is hidden by clouds of smoke, and after the event, the mud, perhaps tens of feet thick, remains. Since mudflows are confined to valleys, there is another possible hazard when the valley is artificially dammed to produce hydro-electric power. A large mudflow could fill a reservoir, pushing the water over the spillway and causing a severe flood downstream. On the other hand, if used wisely, reservoirs might be a safety factor for all except the really large mudflows. The water level in reservoirs might be drawn down during an upstream volcanic event, and the storage basin behind the dam could be used to contain a possible mudflow. This is not the designed function of the dam, but it is a fortunate safety mechanism (2).

(a)

(b)

FIGURE 7.11 *Jessen Meadow with Mt. Lassen in the background. Photograph (a) was taken in August of 1910. Photograph (b) was taken in 1925, from the same viewpoint but after a mudflow and hot blast from the volcano. Notice that all the timber in the 1910 photograph has been destroyed. (Photo by B. F. Loomis, courtesy of the Loomis Museum Association.)*

FIGURE 7.12 *Map of Mt. Rainier and vicinity showing the extent of the Osceola mudflow in the White River Valley (shaded) and the Electron mudflow (dot pattern) in the Puyallup River Valley. (Source: Crandell and Mullineaux, U.S. Geological Survey Bulletin 1238, 1967.)*

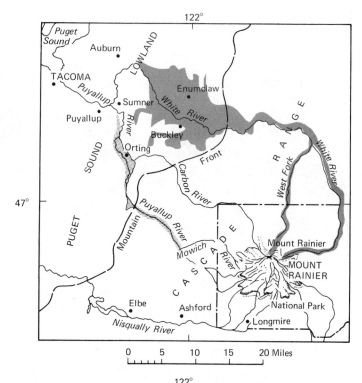

FIGURE 7.13 *Sketch map of Mt. Rainier and vicinity showing relative degrees of potential hazards from ash falls (tephra), mudflows, and floods that might result from an eruption. [Source: Crandell and Mullineaux, "Techniques and Rationale of Volcanic Hazards." Environmental Geology 1 (New York: Springer-Verlag, 1975).]*

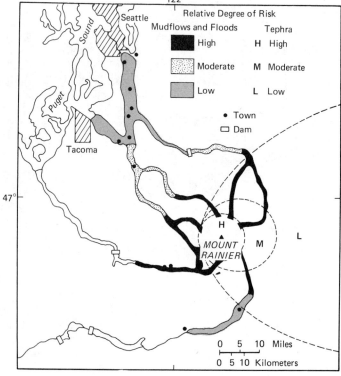

Prediction of Volcanic Activity

It is unlikely that we will be able to accurately predict all volcanic activity in the near future. However, valuable information is being gathered about events that occur before eruptions. The Hawaiian volcanoes, especially Kilauea, have supplied most of the data. The summit of the volcano actually tilts and swells prior to an eruption, and it subsides during the actual outbreak (Figure 7.14). This movement,

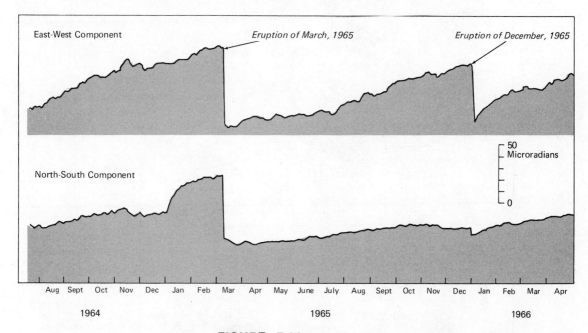

FIGURE 7.14 *Graph showing east-west component and north-south component of ground tilt recorded from 1964 to 1966. Notice the abrupt change in ground tilt before eruption. (Source: R. S. Fiske and R. Y. Koyanagi, U.S. Geological Survey Professional Paper 607.)*

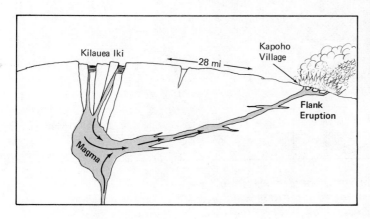

FIGURE 7.15 *Idealized diagram showing the 1960 flank eruption on Kilauea which destroyed the village of Kapoho. (After Eaton and Schmidt,* Atlas of Volcanic Phenomena, *U.S. Geological Survey.)*

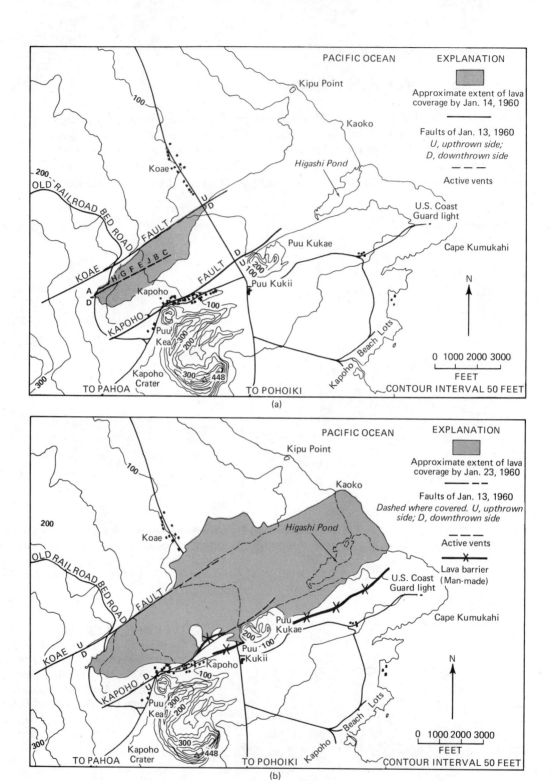

FIGURE 7.16 *Volcanic eruption on flank of Kilauea from January 14 to February 5, 1960. Notice the lava barriers which failed. (Source: Richter et al., U.S. Geological Survey Professional Paper 537E, 1970.)*

Eruption of a Hawaiian volcano. (Photo courtesy of Hawaii Visitors Bureau)

Rift zone on the flank of Kilauea Volcano, Hawaii. The rifts, delineated by large cracks caused by tensional stress and accompanied by earthquakes, form prior to a flank eruption. (Ward's Natural Science Establishment Inc. photo)

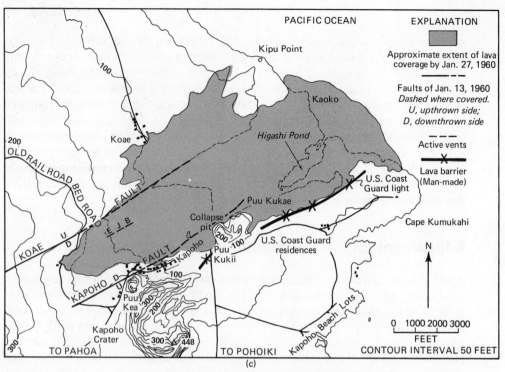

PACIFIC OCEAN

Kipu Point

Kaoko

Higashi Pond

Koae

OLD RAILROAD BED ROAD

200

FAULT

KOAE
U
D

E J B

Collapse
pit

Puu Kukae

Kapoho

FAULT

Puu
Kukii

200

100

U.S. Coast Guard
residences

U.S. Coast
Guard light

Cape Kumukahi

KAPOHO
D
U

FAULT

100

Puu
Kea

300

200

Kapoho
Crater

300   448

300

TO PAHOA

TO POHOIKI

Kapoho Beach Lots

EXPLANATION

Approximate extent of lava
coverage by Jan. 27, 1960

Faults of Jan. 13, 1960
*Dashed where covered.
U, upthrown side;
D, downthrown side*

— X — Active vents
Lava barrier
(Man-made)

N

0   1000 2000 3000
FEET

CONTOUR INTERVAL 50 FEET

(c)

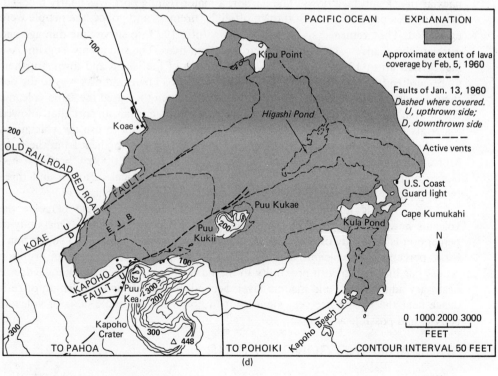

PACIFIC OCEAN

Kipu Point

Higashi Pond

Koae

200

OLD RAILROAD BED ROAD

FAULT

KOAE
U
D

E J B

Puu Kukae

Puu
Kukii

200

U.S. Coast
Guard light

Cape Kumukahi

Kula Pond

KAPOHO
D
U

FAULT

100

Puu
Kea

300

200

Kapoho
Crater

300

△ 448

TO PAHOA

TO POHOIKI

Kapoho Beach Lots

EXPLANATION

Approximate extent of lava
coverage by Feb. 5, 1960

Faults of Jan. 13, 1960
*Dashed where covered.
U, upthrown side;
D, downthrown side*

Active vents

N

0   1000 2000 3000
FEET

CONTOUR INTERVAL 50 FEET

(d)

in conjunction with earthquake swarms that reflect the moving subsurface magma and announce a coming eruption, was used to predict volcanic activity in the vicinity of the farming community of Kapoho on the flank of the volcano (Figure 7.15), 29 miles from the summit. The inhabitants were evacuated before the activity which overran, thrust aside, and floated away lava barriers (walls) and eventually destroyed most of the village (5). (See Figure 7.16.) It is expected that because of their characteristic swelling and earthquake activity before eruptions, the Hawaiian volcanoes will be predictable. However, applying this to less active volcanoes such as in the Cascade Range is much more difficult. Nevertheless, close monitoring of earthquakes and tilting or swelling of volcanoes offer the most promising lines of research that may lead to better prediction of volcanic eruptions.

Adjustment to and Perception of the Volcanic Hazard

Apart from the psychological adjustment to losses, the only major human adjustment to volcanic activity is evacuation (6). The small number of adjustments probably reflects the fact that few eruptions are near populated areas. The experience of about 5000 Icelandic people on the island of Heimaey offers an example of a hardy people responding to a serious destructive hazard. In January, 1973, the dormant Mt. Helgafell came alive, and subsequent eruptions nearly buried the town of Vestmannaeyjar in ash and lava flows. The harbor, a major fishing port, was nearly blocked. With the exception of about 300 town officials, firemen, and police, the people were evacuated. They returned six months later in July, 1973, to survey the damage and estimate the chances of rebuilding their town and lives. The situation was grim. Ash had drifted up to 12 feet thick and covered much of the island and town. Furthermore, molten lava was still steaming near the volcano. Their first task was to dig out their homes and shops, salvaging what they could. Then they used the same volcanic debris that had buried their town to pave new roads and an airport that allowed materials to be moved in and distributed. They also decided to use the volcano to advantage, and by January, 1974, the first home was heated by heat from the lava. More homes are expected to be similarly heated, including 200 new prefabricated homes and a new hospital. More significantly, two fish-meal factories and three freezing plants were reopened, revitalizing the island's fishing industry (7).

Little information is available concerning how people perceive the volcanic hazard because eruptions seldom occur near populated areas. One study of perception is the study of volcanic activity in Hawaii which makes two suggestions. First, perception of volcanic activity and what people will do in case an eruption occurs has little to do with proximity to the hazard, home ownership, knowledge of necessary adjustments, and income level. Second, a person's age and length of residence near the hazard are very significant in how knowledgeable he is of volcanic activity and possible adjustments (6).

SUMMARY AND CONCLUSIONS

Volcanic activity is, worldwide, a rare process generally related to plate tectonics. Most volcanoes are located at plate junctions where magma is produced as spreading or sinking lithospheric plates interact with other earth material. Regardless of how rare volcanoes are, they have destroyed considerable property and have taken many human lives.

Primary effects of volcanic activity include lava flows and pyroclastic activity. Secondary effects include mudflows and fires. All of these effects have occurred in the recent history of the Cascade Range of the Pacific Northwest, and there is no reason to believe that this activity will not occur there in the future.

Several methods, including construction of walls, bombing, and hydraulic chilling, have been used in attempting to control lava flows. These methods have had a varying degree of success and are in need of further evaluation.

Pyroclastic activity includes volcanic ash eruptions, which may cover large areas with a carpet of ash, and ash flows, which move up to 70 miles per hour down the side of a volcano.

Mudflows are generated when melting snow and ice mix with volcanic ash. Such flows are a serious hazard and may devastate an area many miles from the volcano.

With sufficient monitoring of ground tilting and earthquake activity, prediction of volcanic activity is possible. Prediction is being done with some success in Hawaii, and applying these techniques to other areas may lead to better prediction of volcanic activity. On a worldwide scale, however, it is unlikely that we will be able to accurately predict all volcanic activity in the near future.

Apart from the psychological adjustments to losses, the only major human adjustment to volcanic activity is evacuation.

Perception of the volcanic hazard is apparently a function of how old a person is and how long he has lived near the hazard. Direct proximity to a volcano, home ownership, knowledge of necessary adjustments, and income level have little effect on the perception.

References

1. Office of Emergency Preparedness. 1972. *Disaster preparedness* 1, 3.
2. Crandell, D. R., and Waldron, H. H. 1969. Volcanic hazards in the Cascade Range. In *Geologic hazards and public problems, conference proceedings,* eds. R. Olsen and M. Wallace, pp. 5–18. Office of Emergency Preparedness Region 7.
3. Mason, A. C., and Foster, H. L. 1953. Diversion of lava flows at Oshima, Japan. *American Journal of Science* 251: 249–58.
4. Williams, R. S., Jr., and Moore, J. G. 1973. Iceland chills a lava flow. *Geotimes* 18: 14–18.
5. Richter, D. H.; Eaton, J. P.; Murata, K. J.; Ault, W. U.; and Krivoy, H. L. 1970. *Chronological narrative of the 1959–60 eruption of Kilauea Volcano, Hawaii.* U.S. Geological Survey Professional Paper 537E.
6. Murton, Brian J., and Shimabukuro, Shinzo. 1974. Human response to volcanic hazard in Puna District Hawaii. In *Natural hazards,* ed. G. F. White, pp. 151–59. New York: Oxford University Press.
7. Cornell, James, ed. 1974. *It happened last year—earth events—1973.* New York: Macmillan.

8

COASTAL HAZARDS

Although coastal areas are varied in topography, climate, and vegetation, they are generally dynamic environments. It is there that continental processes and oceanic processes converge to produce a landscape that is characteristically capable of rapid change. The impact of hazardous coastal processes is considerable because many populated areas are located near the coast. This is especially true in the United States where it is expected that most of the population will eventually be concentrated along the nation's 94,000 miles of shoreline (including the Great Lakes). Today, all but one of the nation's 13 largest cities occur in the coastal zone, and approximately 75 percent of the population live in coastal states (1).

The coastal hazards that are most serious are *tropical cyclones,* which claim many lives and cause enormous amounts of property damage every year; *tsunamis,* or seismic sea waves, which are particularly hazardous to coastal areas in the Pacific Ocean; and *coastal erosion,* which continues to produce considerable property damage that requires human adjustment.

173

Tropical Cyclones

The most serious coastal hazards are tropical cyclones which historically have taken hundreds of thousands of lives in a single storm. A tropical cyclone that struck the northern Bay of Bengal in Bangladesh in November of 1970 produced a 20-foot rise in sea. Flooding killed approximately 300,000 people, caused $63 million in crop losses, and destroyed 65 percent of the total fishing capacity of the coastal region (2).

Tropical cyclones, known as *typhoons* in most of the Pacific Ocean and *hurricanes* in the Western Hemisphere, cause *damage* and *destruction* from high winds; *river flooding* which results from intense precipitation and which usually causes more deaths and destruction than the wind; and *storm surges* (wind-driven oceanic waters) which are the most lethal aspect of tropical cyclones (Figure 8.1). The great majority of lives in these storms are lost from drowning (3). Property damage from hurricanes can be staggering. Two hurricanes in the United States in the last ten years each caused nearly $1.5 billion in property damage. Despite the increasing population along the Atlantic and Gulf coasts, the loss of lives from hurricanes has decreased significantly. This decrease is due to more effective detection and warning. However, the amount of property damage has greatly increased (Figure 8.2), and there is growing concern that continued increase of population accompanied by unsatisfactory evacuation routes, building codes, and refuge sites may contribute to hurricane catastrophies along the Atlantic and Gulf coasts (3).

A secondary effect of tropical cyclones is flash flooding caused by intense rainfall as the storm moves inland. Hurricane Camille in 1969 first damaged the Louisiana and Mississippi coastlines and, two days later, caused record amounts of precipitation in the mountains of western Virginia, causing more than $100 million in damage (3).

FIGURE 8.1 *Storm surge from Hurricane Camille, 1969. (Photo courtesy of NOAA.)*

Trends of Losses from Hurricanes in
the United States

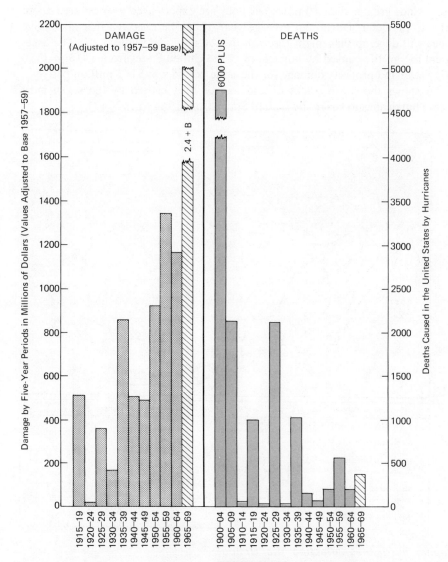

FIGURE 8.2 *Trends of damage and death from hurricanes in the United States from 1915 to 1969. (Source:* Project Stormfury *—1970, U.S. Department of Commerce, 1970.)*

The magnitude and frequency of tropical cyclones are not affected by human activity. Most hurricanes form in a belt between 8 degrees and 15 degrees north and south of the equator, and the areas most likely to experience cyclones in this zone are those with warm surface-water temperatures. The storms are generated as tropical disturbances, and they dissipate as they move over the land. Wind speeds in these storms are greater than 70 miles per hour and blow in a large spiral around a relatively calm center called the "eye" of the hurricane. Winds of 74 miles per hour or greater are generally recorded over an area about 100 miles in diameter, while gale-force winds greater than 40 miles per hour are experienced over an area about 400 miles in diameter. During an average year, it can be expected that about five hurricanes will develop that might threaten the Atlantic and Gulf coasts. The average annual loss of life caused by hurricanes in the United States from 1915 to 1970 was 107. Estimated property damage for the same period was $142 million per year. Figure 8.3 shows the storm tracks of and loss of lives caused by the seven most devastating hurricanes to strike the United States from 1964 to 1970 (3).

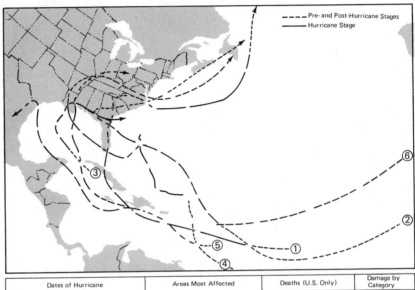

Dates of Hurricane	Areas Most Affected	Deaths (U.S. Only)	Damage by Category
1. August 20-September 5, 1964 CLEO	Southern Florida, eastern Virginia		8
2. August 28-September 16, 1964 DORA	Northeastern Florida, southern Georgia	5	8
3. September 28-October 5, 1964 HILDA	Louisiana	38	8
4. August 27-September 12, 1965 BETSY	Southern Florida, Louisiana	75	9
5. September 5-22, 1967 BEULAH	Southern Texas	15	8
6. August 14-22, 1969 CAMILLE	Mississippi, Louisiana, Alabama, Virginia, West Virginia	255 (68 missing 11-10-69)	9
7. July 30-August 5, 1970 CELIA	Texas, New Mexico	11	8

FIGURE 8.3 *Catastrophic hurricanes affecting the United States from 1964 to 1970. The track for Celia is not shown. Category-eight damage ranges from $50 million to $500 million; category-nine, from $500 million to $5 billion. (Source:* Some Devastating North Atlantic Hurricanes of the 20th Century, *U.S. Department of Commerce, 1970. Updated by NOAA.)*

Tsunamis

Tsunamis (seismic sea waves) are often incorrectly called *tidal waves.* They are extremely destructive and a very serious natural hazard. Fortunately, they are relatively rare and are usually confined to the Pacific basin. The frequency of these events in the United States is about one every eight years (3). Table 8.1 lists the casualties and damage in the United States from tsunamis during the period from 1900 to 1971.

Tsunamis apparently originate when ocean water is vertically displaced by large earthquakes or other phenomena. In open water, the waves may travel at speeds as great as 600 miles per hour, and the distance between successive crests may exceed 100 miles. Wave heights in deep water may be only a foot or so; but on entering shallow coastal waters, the waves slow to less than 40 miles per hour, and the wave heights may increase to more than 50 feet. Figures 8.4 and 8.5 show a tsunami that struck Hawaii.

Tsunamis claimed most of the lives in the 1964 Alaskan earthquake and can cause catastrophic events thousands of miles from where they were generated, as exemplified by the 1960 seismic waves that killed 61 people in Hawaii. The waves were caused by an earthquake in Chile and took only 15 hours to reach Hawaii. Although there is no way to predict tsunamis, it is possible to detect them and warn coastal communities in their path. Once an earthquake or another tsunami-generating disturbance is recognized, the arrival time of a sea wave can be predicted

TABLE 8.1 *Casualties and damage in the United States caused by tsumanis from 1900 to 1971.*

Year	Dead	Injured	Estimated Damage ($000)	Area
1906	—	—	5	Hawaii
1917	—	—	*	American Samoa
1918	—	—	100	Hawaii
1918	40	—	250	Puerto Rico
1922	—	—	50	Hawaii, California, American Samoa
1923	1	—	4000	Hawaii
1933	—	—	200	Hawaii
1946	173	163	25,000	Hawaii, Alaska, West Coast
1952	—	—	1200	Midway Island, Hawaii
1957	—	—	4000	Hawaii, West Coast
1960	61	282	25,500	Hawaii, West Coast, American Samoa
1964	122	200	104,000	Alaska, West Coast, Hawaii
1965	—	—	10	Alaska

SOURCE: Eskite, "Analysis of ESSA Activities Related to Tsunami Warning," NOAA.

*Damage reported but no estimates available.

(a)

(b)

FIGURE 8.4 *Sandy Beach on the Island of Oahu (a) moments before a tsunami generated by an earthquake in the vicinity of the Aleutian Trench struck the beach. Note: The arrow on photograph (a) is the exact location of photograph (b), which shows Sandy Beach a few minutes later after the tsunami struck. Notice the man circled on the lower photograph fleeing from the wave and the several automobiles, also circled, that were swept off the highway by the surging wave. (Photo by Y. Ishii, courtesy of* Honolulu Advertiser.*)*

(a)

FIGURE 8.5 *Kuhio Beach (a) a few moments before the tsunami of May, 1960, struck the beach. (b) The same beach moments after the wave arrived. Notice the ship on the horizon in both photographs. (Photo by Y. Ishii, courtesy of* Honolulu Advertiser.*)*

(b)

to within plus or minus 1.5 minutes per hour of travel time. This information can be used to produce a tsunami-warning-system map, as shown for Hawaii in Figure 8.6.

Damage caused by tsunamis is most severe at the water's edge, where boats, harbor, buildings, transportation systems, and utilities may be destroyed (Figures 8.7 and 8.8). In addition, the waves may be disastrous to aquatic life, including fish, mollusks (clams, etc.), and plants in the nearshore environment (3).

Coastal Erosion

When compared to other natural hazards such as earthquakes, tropical cyclones, or floods, erosion of coasts is generally a continuous, predictable process that causes a

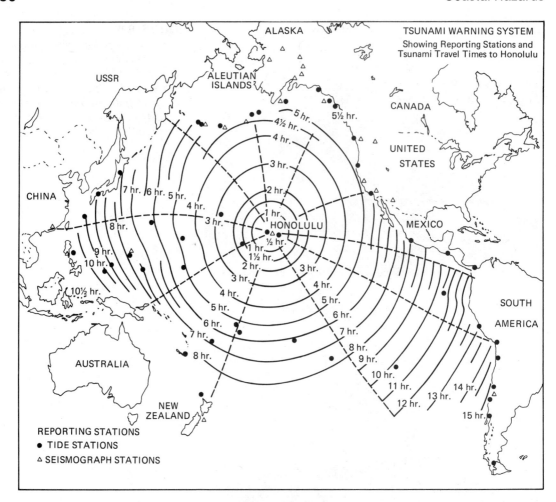

FIGURE 8.6 *Tsunami warning system. Map shows reporting stations and tsunami travel times to Honolulu, Hawaii. (Source: NOAA.)*

relatively low amount of damage in a restricted area. Nevertheless, coastal erosion has caused and continues to cause property damage, and large sums of money are expended to control it. As extensive development of coastal areas for vacation and recreational living continues, problems of coastal erosion are certain to become a more serious threat to human use.

Coastal conditions for the East Coast and West Coast are strikingly different. This difference is probably related to plate tectonics. The *West Coast* is on the leading edge of the North American plate and is an active area experiencing uplift and deformation. As a result, the coastline is characterized by rugged sea cliffs and a very irregular coastline. Such a coast where the rocks project out into the ocean (headlands) is conducive to coastal erosion (Figure 8.9). On striking a shore, waves will expend more energy on the protruding areas than the embayed areas

FIGURE 8.7   *Tsunami damage to railyards in Seward, Alaska, caused by the 1964 earthquake. (Photo courtesy of NOAA.)*

(Figure 8.10). Therefore, the effect of wave erosion is to straighten the shoreline—eroding the headlands and depositing the eroded material in the intervening bays. The *East Coast,* on the other hand, is on the trailing edge of the North American plate. It is a tectonically quiet coast. For this reason and others, the major coastline features are chains of barrier islands separated from the mainland by bays, sounds,

FIGURE 8.8   *Tsunami damage to fishing boats at Kodiak, Alaska, caused by the 1964 earthquake. (Photo courtesy of NOAA.)*

FIGURE 8.9 *Breaking waves as agents of erosion. (Photo courtesy of California Division of Beaches and Parks.)*

and lagoons. Exchange of water between the lagoons and the open sea is through inlets that form when storm waves breach the barrier islands (4).

FIGURE 8.10 *Oblique aerial photograph showing breaking waves and surf zone. Notice that most of the wave energy is expended on the headland protruding out in the water. This area erodes considerably faster than the bay areas. (Photo courtesy of California Division of Beaches and Parks.)*

   Beach erosion along the West Coast has been accelerated by human activity. It is well known that the sand on beaches is not static. Wave action constantly keeps the sand moving in the surf and swash zones (Figure 8.11), and the net result is longshore drift and beach drift, which collectively move sand along a coast (littoral transport; see Figure 8.12). The sand is supplied to the coastal area by rivers that have transported the sand from areas upstream where it was produced by the weathering of quartz-rich rocks. Man has interfered with this natural flow of sand from mountains to beach by building dams which effectively trap the sand. As a result, West Coast beaches are deprived of sediment they would naturally receive otherwise. A result of this entrapment is beach erosion. Special structures, known as groins, have been designed to trap sand and thus build a wider beach. They have been built along both the East Coast and the West Coast (Figure 8.13). Most groins are linear and are constructed perpendicular to the shore line, and most have a length approximately 50 percent greater than the width of the desired beach. They are usually spaced from one to three lengths of the structure, with greater spacing along a continuous, straight shoreline (5). Groins may be solid and trap all the sand moving along the shore, or they may be permeable and simply modify the longshore transport of sand as desired but not completely obstruct it. Construction of groins, especially the solid type, has a disadvantage: The beaches' down-drift from the groins will be impoverished of sand and suffer from erosion. Lawsuits have resulted when groins have retarded longshore drift and caused beaches to erode at valuable recreation sites.

   Beach erosion along the East Coast is due to first, tropical cyclones and severe storms; second, a rise in sea level; and third, human use and interest interfering with natural shore processes (4).

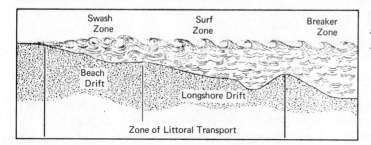

FIGURE 8.11 *Idealized diagram showing the swash, surf, and breaker zones typical of beach profiles.*

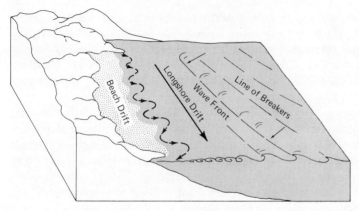

FIGURE 8.12 *Block diagram showing beach drift and longshore drift, which collectively move sand along the coast (littoral transport).*

FIGURE 8.13 *Miami Beach showing system of beach groins constructed to minimize beach erosion and trap sand. (Photo courtesy of U.S. Army Corps of Engineers.)*

Tropical cyclones and *severe storms* can greatly alter a coastline by beach erosion, erosion of the foreshore sand dune system, and opening of new inlets through the barrier islands. In addition, storm surges cut channels through the foreshore dunes, and eroded sediment is deposited in back of the beach as washover deltas (4).

Recent *rise in sea level* is eustatic (worldwide) and independent of continental movement. The rate of rise is not agreed upon but is probably something less than one centimeter per year. Evidence for submergence of the coast is present in many areas along the East Coast. Along the Outer Banks of North Carolina (Figure 8.14), the Shackleford Banks that were low-lying forests in 1850 are today salt marshes (6).

Man's interference with natural shore processes has caused considerable coastal erosion. Most problems are in areas that are highly populated and developed. For example, in the area between Palm Beach and Miami Beach on the east coast of Florida, there has been considerable erosion. The predominant direction of longshore drift is from north to south, and in some areas artificial barriers have retarded the movement of sand, which has caused some beaches to grow while the beaches to the south have eroded, resulting in damage to valuable property.

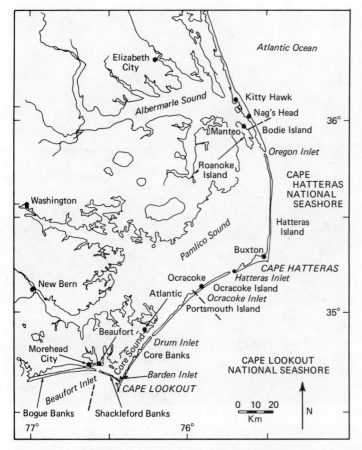

FIGURE 8.14 *The Outer Banks of North Carolina showing the locations of Cape Hatteras and Cape Lookout National Seashores. (Source: Godfrey and Godfrey,* Coastal Geomorphology, *edited by D. R. Coates. Binghamton, New York: Publications in Geomorphology, State University of New York, 1973.)*

A good example of how man has influenced coastal processes is found in the Outer Banks of North Carolina (Figure 8.14). The northern section near Cape Hatteras (Figure 8.15) has been progressively stabilized and developed (particularly near the beach), whereas the southern section near Cape Lookout is nearly unaltered by human activity.

The northern section is characterized by a continuous artificial dune system that was built to control erosion. The desire to protect roads and buildings behind the dunes led to an attempt at total stabilization. The idea was to hold everything in place and prevent the sea from washing over the island. In contrast, the southern section is characterized by a natural zone of dunes rather than a line. The zone is frequently broken by overwash passes that allow water to flood across the island (6). (See Figure 8.16.)

Several studies have shown that the building of an artificial dune line has resulted in erosion and narrowing of the beach and has necessitated spending millions of dollars to stabilize artificial dunes that tend to erode rapidly (6, 7). Even with this, the structures and roads behind the dune line will eventually have to be relocated as the shoreline continues to erode and recede. In contrast, where the barrier islands are left in a natural state, there is little erosion caused by severe storms. As a result, the dunes are not washing away but are moving back by natural processes responsible for the origin and maintenance of the barrier island system. It

FIGURE 8.15 *Cape Hatteras show-*
ing the dune line and the beach.
(Photo by R. Anderson, courtesy of
National Park Service.)

is emphasized, however, that there is considerable controversy as to whether the Core Banks are "typical" barrier islands in that overgrazing with subsequent wind erosion of the frontal dune line may have facilitated the overwash. If this is true, then a reevaluation of the benefits of frequent overwash is necessary.

Coastal erosion is a periodic problem along the coasts of the Great Lakes, and it has been particularly troublesome along the Lake Michigan shoreline. Damage is most severe during prolonged periods of high lake levels which occur following extended periods of above-normal precipitation. The relationship between precipitation and lake level has been documented by the United States Corps of Engineers since 1860. They have shown that the lake level has fluctuated about 6.5 feet during this time. Figure 8.17 shows the mean monthly level of Lake Michigan from 1947 to 1973. Lake level has been rising in recent years from a low stage in 1964, and in 1973 it reached the high level of 1952. During this highwater stage there has been considerable coastal erosion in which many buildings, roads, retaining walls, and other structures have been destroyed by wave erosion (8). (See Figure 8.18.)

During periods of below-average lake level, wide beaches that dissipate energy from storm waves and protect the shore develop. However, with rising lake-level conditions, the beaches become narrow, and storm waves exert considerable energy against coastal areas. Even a small lake-level rise of several feet on a gently sloping shore may inundate about 100 feet of beach (8).

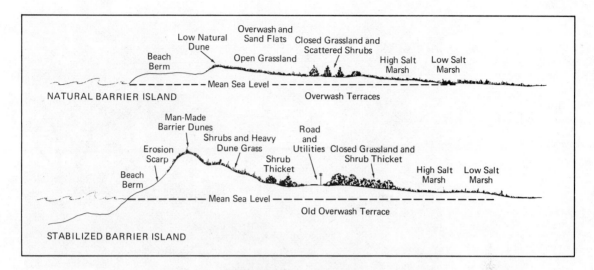

FIGURE 8.16 *Cross sections of the Barrier Islands of North Carolina. The upper diagram is typical of the more natural systems and the lower typical of the stabilized Barrier Islands.* (*From Dolan,* Coastal Geomorphology, *edited by D. R. Coates. Binghamton, New York: Publications in Geomorphology, State University of New York, 1973.*)

The severity of shore erosion of the Great Lakes at a particular site depends on first, the orientation of the coastline; second, the offshore water depths; and third, the resistance of the shoreline to wave erosion. Any protective structure constructed to control beach and shoreline erosion must effectively reduce the energy of the waves striking the shore. This may be done by constructing an offshore breakwater or a variety of onshore structures. Of course, another alternative is to relocate buildings and other structures to safer sites. A few of the construction options and relocation alternatives are summarized with advantages and disadvantages in Figure 8.19.

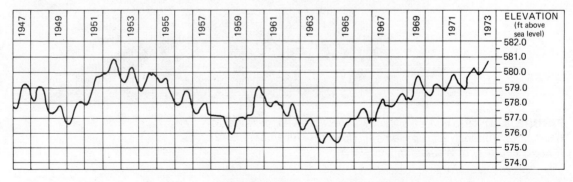

FIGURE 8.17 *Hydrograph of monthly mean level of Lakes Michigan and Huron since 1947, measured at Beach Harbor, Michigan. (After U.S. Department of Commerce, Lake Survey Center, NOAA).*

FIGURE 8.18 *Erosion of a sandy shoreline with damage to residences (upper) and pavement (lower), November 1972. (Photos courtesy of Waukegan* News-Sun.*)*

Perception of and Adjustment to Coastal Hazards

People adjust to the tropical cyclone hazard either by doing nothing and bearing the loss or by taking some kind of action to modify the potential loss. Bearing the loss is

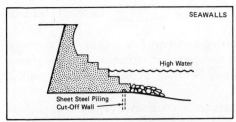

SEAWALLS

ADVANTAGES

1. Provides protection from wave action and stabilizes the backshore.
2. Low maintenance cost.
3. Readily lends itself to concrete steps to beach.
4. Stabilizes the backshore.

DISADVANTAGES

1. Extremely high first cost.

2. Subject to full wave forces; fail from scout; flanking of foundation.
3. Not easily repaired.
4. Complex design and construction problem. Qualified engineer is essential.
5. Slope design is most important.
6. More subject to catastrophic failure unless positive toe protection is provided.

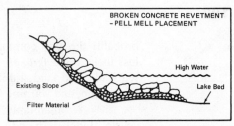

BROKEN CONCRETE REVETMENT – PELL MELL PLACEMENT

ADVANTAGES

1. Inexpensive.
2. Easy construction.

DISADVANTAGES

1. Large concrete pieces are difficult to obtain.

2. Large pieces required for underlying filter layer because of large void.
3. Extremely unattractive appearance, unless special care is taken in construction.

COST/LIN FT — $25 to $85

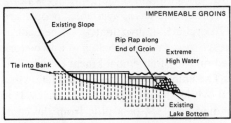

IMPERMEABLE GROINS

ADVANTAGES

1. Resulting beach protects upland areas and provides recreational benefit.
2. Moderate first cost and low maintenance cost.

DISADVANTAGES

1. Extremely complex coastal engineering design problem. Quali-

fied coastal engineering services are essential. Groins rarely function as intended.
2. Areas downdrift will probably experience rapid erosion.
3. Unsuitable in areas of low littoral drift.
4. Subject to flanking, must be securely tied into bluff.

COST/LIN FT — $125 to $150

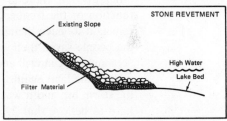

STONE REVETMENT

ADVANTAGES

1. Most effective structure for absorbing wave energy.
2. Flexible — not weakened by light movements.
3. Natural rough surface reduces wave runup.
4. Lends itself to stage construction.
5. Easily repaired — low maintenance cost.
6. The preferred method of protec-

tion when rock is readily available at a low cost.

DISADVANTAGES

1. Heavy equipment required for construction.
2. Subject to flanking and moderate scour.
3. Limits access to beach.
4. Moderately high first cost.
5. Difficult construction where access is limited.

COST/LIN FT — $28 to $100

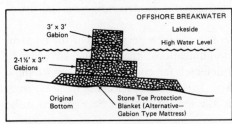

OFFSHORE BREAKWATER

ADVANTAGES

1. Beneficial effect can extend over a considerable length of shoreline.
2. Maintains or enhances recreational value of a beach.
3. Not subject to flanking — can be built in separate reaches.
4. Structure maintenance costs are

lower than those of similar structures designed for other purposes.

DISADVANTAGES

1. May modify beachline and cause erosion in downdrift areas.
2. Structure is subject to foundation and scour failures. Floating plant and heavy equipment maybe required for construction.

COST/LIN FT — $40 to $50

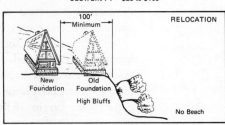

RELOCATION

ADVANTAGES

1. It is permanent. In the long run it may be the best method of protection.
2. Adaptable to short reaches of shoreline.
3. Can be accomplished by the individual without coordination through contract with a house mover.
4. No maintenance required.

DISADVANTAGES

1. Special skills and equipment required.
2. Area must be available for relocation of the house.
3. Does not stop erosion.

COST/LIN FT — $70

FIGURE 8.19 *Example of several methods of shoreline protection from wave erosion. (Source: U.S. Army Corps of Engineers.)*

probably the most common individual adjustment. Attempts to modify the potential loss include *strengthening* the environment by protective structures such as groins, sea walls, and land stabilization; and *adapting* behavior by better land-use zoning, evacuation, and warning (9).

Inhabitants in areas with a serious tropical cyclone hazard, such as coastal Bangladesh and the Gulf Coast of the United States, are very much aware of the hazard (10, 11). In Bangladesh, the appraisal of perception of the cyclone hazard involved the economic and cultural backgrounds of the people. It was found that the people are not particularly concerned about the hazard, and the three socioeconomic classes perceive the cyclones in much the same way. However, the upper class tends to favor earning a living in a less hazardous area.

Although repeatedly devastated by tropical cyclones, coastal areas in Bangladesh do not have an integrated system of private or public adjustment. Although people there are aware of the hazard, the adoption of adjustments is not a function of the frequency of cyclones. Even when shelter is available, people are reluctant to leave their homes. During a 1970 cyclone, approximately 38 percent of the people saved their lives by climbing trees, 5 percent took shelter in a two-story "community center," and 8 percent remained on top of an embankment designed to protect the land against saltwater intrusion. The embankment was incomplete when the storm hit, and it was severely eroded and overlapped by storm-surge waves. It is hard to believe that 22 people in a family died only 1/4 of a mile from the community center that offered safety (10).

People's attitudes toward hurricane damage and the possibility of preventing or modifying damage are related to their educational level and their hurricane experience. Therefore, public awareness programs in areas likely to experience tropical cyclones should be matched to the educational level of the target group. Furthermore, since better-educated people tend to be more positive toward damage prevention, increased attention should be given to the less-educated population to counteract their tendency toward lack of confidence in damage-prevention measures. Awareness programs should also be designed differently depending on the areas' history of hurricane experience (11).

Little information is available concerning the perception of the tsunami hazard. In developed areas with good communications, the established warning system is adequate provided the tsunami is detected far enough out to sea to allow time to warn people in the path of the wave. In undeveloped areas or areas very close to the origin of the seismic sea wave, the loss of life can be expected to be greater.

Perception of coastal erosion as a natural hazard depends primarily on an individual's past experience and proximity to the coastline and the probability that he will suffer property damage. One study of coastal erosion of sea cliffs near Bolinas, California, 15 miles north of the entrance to the San Francisco Bay, established that people living very near the coast in an area likely to experience damage in the near future are generally very well informed and see the erosion as a direct and serious threat (12). People living a few hundred feet from a possible hazard, although aware of the hazard, know little about its frequency of occurrence, severity, and predictability. People located still further inland are aware that coastal erosion exists but have little perception of the hazard.

The Bolinas study is especially interesting because of the controversy that surrounded various alternatives the community considered. Before 1971, coastal-erosion damages and expenditures at Bolinas exceeded $300,000 (Table 8.2). The rate of cliff retreat is approximately 1.5 feet per year under natural conditions and somewhat less when sea walls and other protective measures are used to protect the cliffs. Alternatives the community considered were fourfold. First, *zoning* to prevent development in the hazard zone was considered. No structures to slow the erosion of the cliffs would be provided. This, however, would be effective only for undeveloped land, and in several areas, homes and roads are close enough to the retreating cliffs to be damaged eventually regardless of zoning. Second, the community considered *public land acquisition* in the hazard zone. The land would be purchased for open space and recreation. Natural erosion of the cliffs would be allowed, and damage would be restricted to low-cost recreation facilities. Third, *construction of a sea wall* of broken rock at the base of the cliffs to stabilize erosion was considered. This would cost $3 million to $4 million, and there would still be up to 65 feet of erosion at the top of the cliffs unless other protective measures were taken. Fourth, the community considered a *combination of groins, beachfill, and an energy dissipator* (Figure 8.20). The groins and energy dissipator would be constructed of broken rock at an initial cost of $4 million to $6 million. This would stabilize the base of the cliff, but erosion on the top would continue until a stable angle of repose is reached. As above, this would be up to 65 feet unless protective measures were taken (12).

The people of Bolinas include weekend and summer residents, retired persons who are permanent residents, long-time agricultural residents, and activist

TABLE 8.2 *Coastal erosion damages and expenditures at Bolinas, California.*

	Value ($)[a]
Damages	
Homes (three)	75,000
Real property (the equivalent of 15 lots at $5000 each)	75,000
Public utilities (roads, pipelines)	50,000
Total	200,000
Expenditures	
Seawall construction	75,000
Cliff drainage and planting	25,000
Rip-rap of road area at top of cliffs	5,000
Other road repairs	10,000
Moving of homes (two)	10,000
Problem studies[b]	10,000
Total	135,000

SOURCE: *Natural Hazards: Local, National, Global* edited by Gilbert F. White. Copyright © 1974 by Oxford University Press, Inc. Reprinted by permission.
[a]1971 dollar equivalent.
[b]U.S. Army Corps of Engineers, Bolinas Beach and Erosion Study, $5000 per year for two years.

FIGURE 8.20 *Proposed adjust-ments for cliff stabilization at Bolinas, California. (Source: R. A. Rowntree,* Natural Hazards: Local, National, Global, *edited by G. F. White. New York: Oxford University Press, 1974.)*

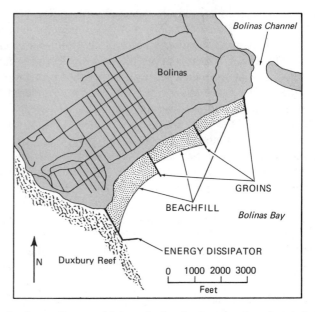

environmentalists. Each, depending on his particular background and proximity to the hazard, had a different perception of the problem. The town has formed a planning group to consider the alternatives, and the likely action appears to be a rejection of the expensive stabilization plans that would be an inducement to growth and tourism, something the community fears more than cliff erosion. The community has, in effect, chosen to accommodate itself to the hazard, taking the position that coastal erosion is a natural process rather than a natural hazard. It is likely that the community as a whole will maintain present cliff stabilization measures, but eventually the cliffs will retreat through the first row of houses regardless of interim measures.*

SUMMARY AND CONCLUSIONS

The coastal environment is one of the most dynamic areas on earth, and rapid change is a regular occurrence.

Migration of people to coastal areas is a continuing trend, and approximately 75 percent of the population in the United States now live in coastal states.

The most serious coastal hazard is tropical cyclones. These violent storms bring high winds, storm surges (wind-driven oceanic waves), and river flooding. They continue to take thousands of lives and cause billions of dollars in property damage.

Tsunamis, or seismic sea waves, are produced by earthquakes or other phenomena that displace oceanic water and generate long waves that travel at speeds as great as 600 miles per hour. The wave, only a foot or

*From *Natural Hazards: Local, National, Global* edited by Gilbert F. White. Copyright © 1974 by Oxford University Press, Inc. Reprinted by permission.

so high in deep water, may grow to a height of 100 feet on reaching the coastline. The loss of the majority of lives in the 1964 Alaskan earthquake was attributed to tsunamis.

Although it causes a relatively small amount of damage compared to other natural hazards such as river flooding, earthquakes, and tropical cyclones, coastal erosion is a serious problem along the East, West, and Great Lakes coasts of the United States.

Man's interference with natural coastal processes by building groins, artificial dunes, and other structures is sometimes successful, but in some cases it has caused considerable coastal erosion. Most problems are in areas where the population density is high, but sparsely populated areas along the Outer Banks in North Carolina are also having trouble with coastal erosion.

Perception of and adjustment to coastal hazards depend on factors such as the magnitude and frequency of the hazard, and the individual's educational level, experience, and proximity to the hazard.

With tropical cyclones, the most common individual adjustment is to do nothing and bear the loss. Community adjustments in developed countries generally attempt to modify the environment by building protective structures designed to lessen potential damage or encourage change in people's behavior by better land-use zoning, evacuation, and warning.

Adjustment to tsunamis is generally to bear the loss. The tsunami warning system is effective in communities that have good communication systems and that are far enough from the location where the wave is generated to allow the necessary time to warn people in its path.

Adjustments to coastal erosion in developed areas are usually the "technological fix": building sea walls, groins, and other structures to stabilize the beach and prevent erosion. These have had mixed success, and they often cause additional problems in adjacent areas. One community in California takes the position that coastal erosion of sea cliffs is a natural process rather than natural hazard, and it has rejected engineering solutions to the problem.

References

1. Coates, D. R., ed. 1973. *Coastal geomorphology*. Binghamton, New York: Publications in Geomorphology, State University of New York.
2. White, A. U. 1974. Global summary of human response to natural hazards: tropical cyclones. In *Natural hazards: local, national, global*, ed. G. F. White, pp. 255–65. New York: Oxford University Press.
3. Office of Emergency Preparedness. 1972. *Disaster preparedness* 1, 2.
4. El-Ashry, M. T. 1971. Causes of recent increased erosion along United States shorelines. *Geological Society of America Bulletin* 82: 2033–38.
5. Krynine, D. P., and Judd, W. R. 1957. *Principles of engineering geology and geotechnics*. New York: McGraw-Hill.
6. Godfrey, P. J., and Godfrey, M. M. 1973. Comparison of ecological and geomorphic interactions between altered and unaltered barrier island systems in North Carolina. In *Coastal geomorphology*, ed. D. R. Coates, pp. 239–58. Bing-

hamton, New York: Publications in Geomorphology, State University of New York.

7. Dolan, R. 1973. Barrier islands: natural and controlled. In *Coastal geomorphology*, ed. D. R. Coates, pp. 263–78. Binghamton, New York: Publications in Geomorphology, State University of New York.

8. Larsen, J. I. 1973. *Geology for planning in Lake County, Illinois.* Illinois State Geological Survey Circular 481.

9. Baumann, D. D., and Sims, J. H. 1974. Human response to the hurricane. In *Natural hazards: local, national, global,* ed. G. F. White, pp. 25–30. New York: Oxford University Press.

10. Islam, M. A. 1974. Tropical cyclones: coastal Bangladesh. In *Natural hazards: local, national, global,* ed. G. F. White, pp. 19–25. New York: Oxford University Press.

11. Baker, E. J. 1974. Attitudes toward hurricane hazards on the Gulf Coast. In *Natural hazards: local, national, global,* ed. G. F. White, pp. 30–36. New York: Oxford University Press.

12. Rowntree, R. A. 1974. Coastal erosion: the meaning of a natural hazard in the cultural and ecological context. In *Natural hazards: local, national, global,* ed. G. F. White, pp. 70–79. New York: Oxford University Press.

Part Three

MAN AND ENVIRONMENT

In this latter part of the twentieth century, Americans are a metropolitan people. Today, nearly 70 percent of the American people live in or near metropolitan areas that have populations greater than 50,000. It is expected that by the year 2000, 85 percent of Americans will be urban residents. Therefore, metropolitan population growth is now a basic feature of the transformation from agrarian to industrial life in the United States.

      The evolution of urban areas has been from farm to small town, to city, to large metropolitan areas, to what is now referred to as the "urban region." An urban region is an area with at least one million people, in which there is a zone of metropolitan areas with intervening counties never far from a city (1). In 1920, there were 10 urban regions that contained approximately 1/3 of the total population of the country. By 1970, there were 16 urban regions with 3/4 of the population, and by the year 2000, there will probably be at least 23 urban regions with 5/6 of the nation's people. If these projections are correct, 54 percent of all Americans will be living in the two largest urban regions—41 percent in the metropolitan belt extending from the Atlantic seaboard and westward past Chicago, and 13 percent in the California region between San Francisco and San Diego.

It is projected that by the year 2000, urban regions will occupy 1/6 of the total land area of the continental United States, and the environmental impact of this will be staggering. Urbanization renders a large portion of the soils within the urbanized area impervious, modifies, if not builds, rivers, and otherwise changes the previous landscape to suit the use of man. However, this process need not totally destroy the land. If growth is properly planned and directed, and if regular review and reevaluation of the goals and objectives are conducted, urban areas should be able to maintain at least a rough balance between acceptable environmental degradation and constructive use of the environment.

The stages of conversion from a rural to an urban area are rural, early urban, middle urban, and late urban (2). During the *rural* stage, the landscape may consist of its virgin state and/or modest to intensive cultivation. Characteristically, there may be occasional farm buildings and dwellings in which the water supply is from wells, springs, streams, or ponds. Sewage is disposed of through outhouses, cesspools, or, perhaps, septic tanks, and garbage may be fed to domestic animals. The *early urban* stage is characterized by city-type homes built on small to large plots of land in a semiorderly to orderly manner. Schools, churches, and shopping facilities are interspersed. The water supply is generally from individual wells. The rubbish is buried or burned, and sewage is disposed of in septic tanks or cesspools. The *middle urban* stage is characterized by large-scale housing developments. There are more schools and shopping centers and some industrial sites. During this stage, systems to provide centralized water systems and sewers to dispose of waste waters may be built. Solid wastes may be collected by trucks. This is the first stage where environmental degradation due to urbanization is easily recognized. The *late urban* stage is characterized by a large number of homes, apartments, commercial and industrial buildings, sidewalks, and parking lots. A large part of the land surface is made impervious or nearly so. Sanitary waste water systems and storm sewers attempt to remove human and industrial wastes and the storm water runoff. It is in this stage that environmental degradation is obvious to many and may, or has, become a serious threat.

Chapters 9 through 11 consider the relations between man and his physical environment. Because these relations are often most stressed in the urban areas, special attention is given to the problems stemming from man's activities and inactivities that create our special urban landscape. Several relationships between hydrology and human use are reviewed in Chapter 9. Especially important here are shallow aquifer contamination, sediment pollution, and stream channelization.

Chapter 10 is concerned with the treatment and disposal of wastes, particularly sanitary landfills; ocean dumping; radioactive wastes; septic systems; and deep-well disposal. Recent advancements involving the relations between health and the fields of geology and geography are discussed in Chapter 11. This subject promises to become more significant as our understanding of the relations among earth processes, earth materials, and health problems increases.

References

1. Commission on Population Growth and the American Future. 1972. *Population and the American future*. Washington, D.C.: U.S. Government Printing Office.
2. Savini, J., and Kammerer, J. C. 1961. *Urban growth and the water regimen*. U.S. Geological Survey Water Supply Paper 1591A.

9

HYDROLOGY AND HUMAN USE

The more obvious and well-known detrimental aspects of human use of the surface water, groundwater, and atmospheric water in the hydrologic cycle include fouled rivers, persistent "suds" on rivers or from wells, and unusual odors from surface water or groundwater. Problems with water supplies and waste waters were among the first environmental problems recognized and attacked on a scientific basis. Traditionally, water and waste waters have been considered and treated as separate entities. While convenient and understandable, such segmentation of complex subjects can compound problems, obscure solutions, and confuse the general public. It has been only within the last fifty years that the interrelationships among and between surface water, groundwater, and atmospheric water have been determined in a reasonable, acceptable manner, but all too frequently, they are still considered as separate entities. Consequently, many local, state, and federal agencies split jurisdictional responsibilities along artificial lines. Fortunately, within the last decade there has been a distinct trend to bridge these jurisdictional splits.

       The fields of engineering, geology, biology, chemistry, and meteorology have each progressed through philosophical ponderings and debates on certain aspects of the "nature of things" in the founding stage of their particular science. Only within this century have the various disciplines joined together to investigate and pursue quantification of elusive segments of the hydrologic cycle. Although a good deal of information has been gathered and studied, much more remains unsolved, and therefore a precise, quantitative definition of each component of the hydrologic cycle is presently impossible. This is especially true of the evapotranspirative, infiltrative, and storage segments of the hydrologic cycle, as shown in Figure 9.1. Methods and techniques to determine approximately what is happening have been developed to avert the need for precise, quantitative definitions of the more elusive components of the hydrologic cycle.

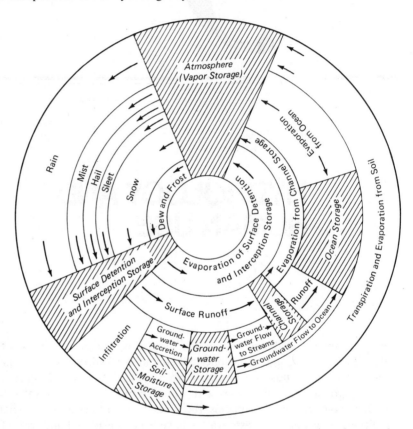

FIGURE 9.1 *Idealized hydrologic cycle. The shaded areas represent storage; arrows indicate flow. Diagram is to be read counterclockwise. (Reprinted, by permission, from C. O. Wisler and E. E. Brater,* Hydrology, *John Wylie & Sons, Inc., 1951.)*

Surface Water Hydrology

Surface water is the portion of precipitation that dislodges soil and rock particles on impact with the land surface (Figure 9.2); is rejected from infiltrating; runs overland

FIGURE 9.2 *Raindrops strike the earth with enough force to tear apart unprotected soil and separate soil particles from one another. They wash the soil and generally move soil particles downslope. (Photos courtesy of U.S. Department of Agriculture.)*

in rivulets or streams to rivers; picks up minerals, physically and chemically; and erodes the earth. Such waters function within the hydrologic and rock cycles simultaneously. They drain the land but not uniformly or consistently, as shown by Figure 9.3 This figure shows the total runoff pattern for the United States and the wide variation across the country. The variations reflect differences in the amounts of precipitation, evaporation, transpiration, and other climatic, soil, and geologic conditions. The runoff values include groundwaters and bank storage waters released to the rivers.

The amount of surface-water runoff from a watershed or a drainage basin and the amount of sediment carried vary significantly. Both of these characteristics of surface waters are composite results of the interactions between the rock and hydrologic cycles. The many and varied effects are produced by the diverse climatic, geologic, and physiographic characteristics of the particular drainage basin, together with variations of these factors with time. For instance, even the most casual observer is aware of the difference in the amount of sediment carried by a stream in flood stage compared to what appears to be carried at low flow conditions. Until recently, however, it was generally believed that the relationship between rainfall and runoff for watersheds of similar size was essentially the same regardless of other factors such as topography (1). If this were true, the science of hydrology would be very much simplified, and one river would be very much like another

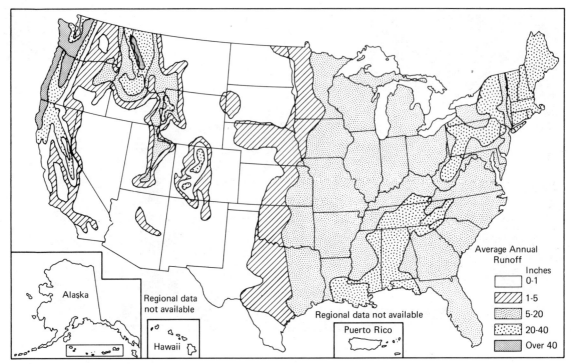

FIGURE 9.3 *Average annual runoff. (Source:* The Nation's Water Resources, *Water Resources Council, 1968.)*

except for velocity (gradient), suspended and dissolved solids, and alterations caused by human use.

The principal geologic factors affecting surface-water runoff and sedimentation include rock and soil type, mineralogy, degree of weathering, and structural characteristics of the soil and rock. Fine-grained, dense, clay soils and exposed correlative rock types with few fractures generally have little infiltration capacity. Hence, the runoff of precipitation falling upon such materials is comparatively rapid. Conversely, sandy and gravelly soils, well-fractured rocks, and soluble rocks such as those in karst areas generally have high infiltration capacities and absorb a larger amount of precipitation. This condition retards overland runoff to streams until the soils become saturated or the rate of precipitation exceeds the infiltrating capacity of the soils.

The structural characteristics of the soil and rock affect the runoff and sedimentation characteristics in several ways. Structure often controls the direction of flow, gradient, and drainage pattern of surface waters and groundwaters. Structural control of surface waters is common in areas where the bedrock is fractured or folded, providing variations in resistance to erosion. Similar control of the direction of movement and the point of discharge of groundwaters is possible, as illustrated in Figure 9.4. There, a portion of the infiltrated waters (recharge to the groundwaters) is diverted from one surface watershed (drainage basin) to an entirely different stream basin. This figure also explains why groundwater basins do not automatically coincide with surface watersheds as is frequently assumed.

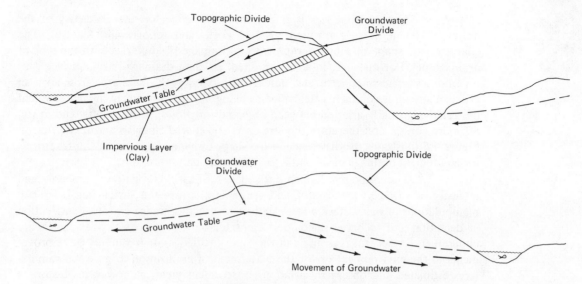

FIGURE 9.4 *Idealized diagrams showing that the topographic divide may not necessarily coincide with the groundwater divide. Hence, a portion of the infiltrated waters is diverted from one surface watershed to an entirely different stream basin.*

Physiographic factors that affect runoff and sediment transport include climate; the shape of the drainage basin; the relief and slope characteristics; the basin's orientation to prevailing storms; and the drainage pattern (spatial arrangement of the natural stream channels).

Climatic factors that influence runoff and stream sedimentation include the type of precipitation that occurs, the intensity with which precipitation occurs, duration of precipitation with respect to the total annual climatic variation, and the types of storms (whether cyclonic or thunderstorm).

The *shape of a drainage basin* is greatly affected by the geologic conditions, and a principal effect of basin shape upon runoff and sedimentation is its role in governing the rate at which water is supplied to the main stream. Basins that are long and narrow and have short tributaries receive the flow (especially stream flows) from the tributaries much more rapidly than basins that have long, sinuous tributaries.

The factors of *relief* and *slope* are interrelated: The greater the relief, the more likely the stream is to have a steep gradient and a high percentage of steep, sloping land. Relief and slope are also important because they affect the rate of overland flow; the velocity of water in a stream; the rate at which water infiltrates into the soil or rock; and the rate at which surplus soil and groundwater enter into a stream.

Orientation of the stream basin to prevailing storms influences the rate of flow, the peak flow, the duration of surface runoff, and the amount of transpiration and evaporation losses. The latter is a factor because of the influence that basin orientation has on the amount of heat received from the sun.

The *drainage net* is important in its effect on the efficiency of the drainage system. It is controlled by soil, rock, and topographic features. The characteristic shape of a hydrograph from a drainage basin varies with the type of drainage net. For instance, in a well-drained, efficient drainage basin, surface runoff from a rainstorm concentrates quickly to a flood peak. Discharge from such an efficient stream is commonly described as being *flashy*. Conversely, in a stream that has relatively small variation between high and low flows, storm waters tend to run off more slowly, and therefore the stream is considered sluggish and less efficient. Figure 9.5 illustrates the difference in the shape of hydrographs from similar storms for an efficient drainage basin and a less efficient one.

The principal human influences affecting the pattern, amount, and intensity of surface-water runoff and sedimentation are the varied land uses and manipulations of our surface waters such as artificial drainage and the construction of detention and recharge ponds and dams for various purposes. Such works are undertaken for several reasons, and they are justified most frequently as improvements to the environment or for the betterment of civilization. Figure 9.6 summarizes estimated and observed variations in sediment yield under various historical changes in land use.

Variations in the natural sediment yield for the different river basins (Figure 9.7) within the United States are summarized in Table 9.1. The amount of sediment carried by rivers as a part of their work within the rock cycle varies with geologic, climatic, topographic, physical, vegetative, and other conditions. Hence, some rivers are consistently and noticeably different in their clarity and appearance, as can be inferred from Table 9.2. Although this table reflects varying degrees of human influence, it demonstrates the sizable variations of the sediment load per unit area in various parts of the world. For instance, on the average, the Lo River of China carries nearly 200 times more suspended load than does the Nile River of Egypt.

FIGURE 9.5 *Idealized diagram illustrating the difference between an efficient drainage basin which removes surface runoff quickly and a less efficient one which takes a longer period of time to remove the water. Many natural stream basins would have the less efficient hydrograph, while man-made channels tend to be more efficient but may damage the environment.*

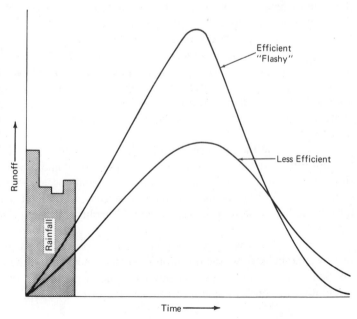

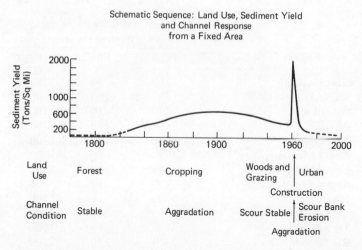

Schematic Sequence: Land Use, Sediment Yield
and Channel Response
from a Fixed Area

FIGURE 9.6 *Graph showing the effect of land-use change on sediment yield in a Piedmont region before the beginning of extensive farming and continuing on through a period of construction and urbanization. (From M. Gordon Wolman,* Geografiska Annaler *49A, 1966).*

Within the United States, the Mississippi is not as "muddy" as the Missouri and the Colorado rivers.

The general relationship between size of drainage basin and sediment load suggests that as the basin size increases, the sediment yield per unit area decreases (Table 9.3). This results from the increase in probability of sediment entrapment and/or deposition with increased basin size and the decreased probability of total basin coverage by a single storm event with increased basin size.

The effects of human use on the natural hydrologic processes have long been recognized. Deforestation, tillage, urbanization, and many other activities of "civilization" have caused changes in surface-water hydrology that sooner or later are evaluated on a cause-and-effect basis. Not infrequently, these evaluations are skewed by "cost effectiveness," "nostalgia," "before and after" visions, and other

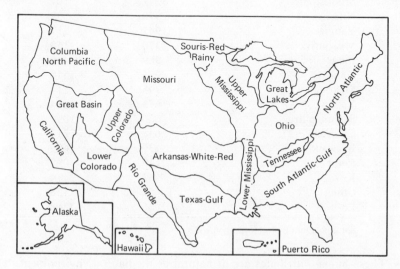

FIGURE 9.7 *Water resource regions of the United States. (Source:* The Nation's Water Resources, *Water Resources Council, 1968.)*

TABLE 9.1 *Estimated ranges in sediment yield from drainage areas of 100 square miles or less.*

Region	Estimated Sediment Yield (tons/sq mi/yr)		
	High	Low	Average
North Atlantic	1210	30	250
South Atlantic-Gulf	1850	100	800
Great Lakes	800	10	100
Ohio	2110	160	850
Tennessee	1560	460	700
Upper Mississippi	3900	10	800
Lower Mississippi	8210	1560	5200
Souris-Red-Rainy	470	10	50
Missouri	6700	10	1500
Arkansas-White-Red	8210	260	2200
Texas-Gulf	3180	90	1800
Rio Grande	3340	150	1300
Upper Colorado	3340	150	1800
Lower Colorado	1620	150	600
Great Basin	1780	100	400
Columbia-North Pacific	1100	30	400
California	5570	80	1300

SOURCE: *The Nation's Water Resources,* Water Resources Council, 1968.

TABLE 9.2 *Some major rivers of the world ranked by sediment yield per unit area.*

River	Drainage Basin (10^3 km²)	Sediment Load per Year (tons/km²)
Amazon	5776	63
Mississippi	3222	97
Nile	2978	37
Yangtze	1942	257
Missouri	1370	159
Indus	969	449
Ganges	956	1518
Mekong	795	214
Yellow	673	2804
Brahmaputra	666	1090
Colorado	637	212
Irrawaddy	430	695
Red	119	1092
Kosi	62	2774
Ching	57	7158
Lo	26	7308

SOURCE: Data from Holman, 1968.

human frailties. These human attributes are as natural as the natural environment. With this in mind, let us consider the serious problems associated with sediment pollution and channelization.

Watershed-Size Range (sq mi)	Number of Measurements	Average Annual Sediment-Production Rate (acre-ft/sq mi)
Under 10	650	3.80
10–100	205	1.60
100–1000	123	1.01
Over 1000	118	0.50

TABLE 9.3 *Arithmetic average of sediment-production rates for various groups of drainage areas in the United States.*

SOURCE: From *Handbook of Applied Hydrology*, by Ven Te Chow. Copyright 1964 by McGraw-Hill. Used with permission of McGraw-Hill Book Company.

Sediment Pollution

Sediment may be our greatest pollutant. In many areas it is choking the streams; filling in lakes, reservoirs, ponds, canals, drainage ditches, and harbors; burying vegetation; and generally creating a nuisance that is difficult to remove. It is truly a resource out of place in that it depletes a land resource (soil) at its site of origin (Figure 9.8), reduces the quality of the water resource it enters, and may deposit sterile materials on productive croplands or other useful land (Figures 9.9 and 9.10).

FIGURE 9.8 *Severe soil erosion and loss of a valuable resource. (Photo by F. M. Roadman, courtesy of U.S. Department of Agriculture.)*

FIGURE 9.9 *Sediment pollution can damage productivity of farmland. (Photo by R. B. Brunstead, courtesy of U.S. Department of Agriculture.)*

Most natural pollutional sediments consist of rock and mineral fragments from sand-size particles less than 2 millimeters in diameter to silt, clay, or very fine colloidal particles. Examples of the variations in the sediment size found in reservoirs in different climatic areas are shown in Table 9.4. Rates of storage depletion caused by sediment filling in ponds and reservoirs are shown in Table 9.5. These data suggest that small reservoirs will fill up with sediment in a few tens of years, while large reservoirs will take hundreds of years to fill.

FIGURE 9.10 *Sediment pollution can also spoil the visual amenities of the urban environment. (Photo by W. B. King, courtesy of U.S. Department of Agriculture.)*

TABLE 9.4 *Average-grain-size distribution of reservoir deposits.*

Climatic Area	Drainage Area (sq mi)	Average-Grain-Size Components		
		Clay (%)	Silt (%)	Sand (%)
Humid				
W. Frankfort Reservoir, Ill.	4	19	78	3
Lancaster Reservoir, S.C.	9	28	56	16
Lake Bracken, Ill.	9	52	42	1
Lake Lee, N.C.	51	36	57	7
High Point Reservoir, N.C.	63	45	35	20
Lake Marinuka, Wis.	139	30	64	6
Lake of the Ozarks, Mo.	14,000	46	50	4
Subhumid				
Ardmore Club Lake, Okla.	4	40	24	36
Moran Reservoir, Kans.	5	52	44	4
Mission Lake, Kans.	8	27	17	56
Lake Merritt, Tex.	12	56	41	3
Lake Claremore, Okla.	56	51	46	3
Lake Brownwood, Tex.	1 544	70	28	2
Great Salt Plains Reservoir, Okla.	3 156	51	41	8
Semiarid				
Wellfleet Reservoir, Nebr.	15	9	14	77
Sheridan Reservoir, Kans.	463	85	15	
Tongue Reservoir, Wyo.	1 740	35	46	18
Canton Reservoir, Okla.	12,483	41	39	20
Arid				
Conchos Reservoir, N.M.	7 350	22	52	26
Lake Mead	168,000	28	25	47

SOURCE: From *Handbook of Applied Hydrology,* by Ven Te Chow. Copyright 1964 by McGraw-Hill. Used with permission of McGraw-Hill Book Company.

Some pollutional sediments are man-made debris resulting from the disposal of industrial, manufacturing, and public wastes. Such sediments include trash (litter is not confined to roadside and parks) directly or indirectly discharged into surface waters. Most of the man-induced sediments are very fine-grained and are difficult to distinguish from the naturally occurring sediments unless they contain unusual minerals or other distinguishing characteristics.

The principal sources of man-induced pollutional sediments result from disruption of the land surface for construction, farming, deforestation, and channelization works. In short, a great deal of sediment pollution is the result of human use of the environment and civilization's continual change of plans and direction. It cannot be eliminated, but only ameliorated.

In this century in the United States, emphasis on soil and water conservation has grown considerably. First to receive emphasis were programs to stabilize areas of excessive wind and water erosion and to provide flood control and

TABLE 9.5 *Summary of reservoir capacity and storage depletion.*

Reservoir Capacity (acre-ft)	Number of Reservoirs	Total Initial Storage Capacity (acre-ft)	Total Storage Depletion (acre-ft)	Total Storage Depletion %	Individual Reservoir Storage Depletion Average %/yr	Individual Reservoir Storage Depletion Median %/yr	Average Period of Record (yr)
0–10	161	685	180	26.3	3.41	2.20	11.0
10–100	228	8 199	1 711	20.9	3.17	1.32	14.7
100–1000	251	97,044	16,224	16.7	1.02	.61	23.6
1000–10,000	155	488,374	51,096	10.5	.78	.50	20.5
10,000–100,000	99	4,213,330	368,786	8.8	.45	.26	21.4
100,000– 1,000,000	56	18,269,832	634,247	3.5	.26	.13	16.9
Over 1,000,000	18	38,161,556	1,338,222	3.5	.16	.10	17.1
Total or average	968	61,239,020	2,410,466	3.9	1.77	.72	18.2 [a]

SOURCE: F. E. Dendy, "Sedimentation in the Nation's Reservoirs." *Journal of Soil and Water Conservation* 23, 1968.
[a]The capacity-weighted period of record for all reservoirs was 16.1 years.

irrigation water for arid and semiarid land. As the programs progressed, adverse environmental effects from some of the work, such as accelerated stream erosion and rapid reservoir siltation, were discovered. Simultaneous with implementation of new conservation programs, additional needs produced demands for broader and expanded usage of completed improvements for recreation, water supply, and navigation, and this necessitated research and development to find solutions to new or rediscovered problems. To correct the erosion and sedimentation problems, design changes were developed and applied. These changes included contour plowing, changes in farming practices, and construction of small dams and ponds to hold runoff or trap sediments.

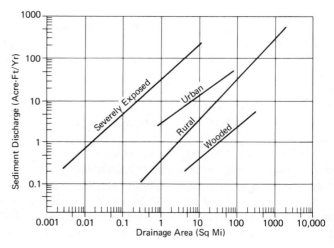

FIGURE 9.11 *Relationship between sediment yield and drainage area for various land uses. (Source: Putnam, U.S. Geological Survey Open File Report, 1972.)*

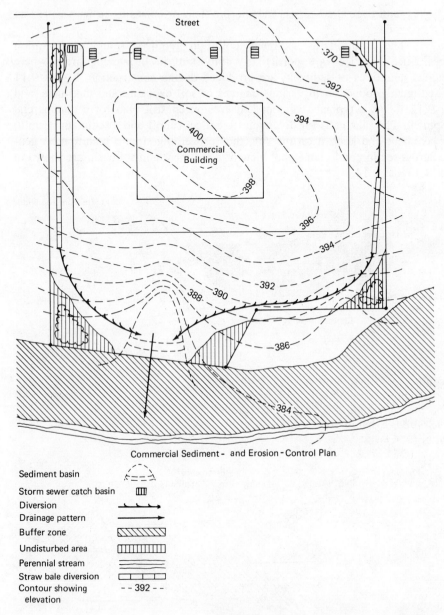

Commercial Sediment - and Erosion - Control Plan

Sediment basin

Storm sewer catch basin

Diversion

Drainage pattern

Buffer zone

Undisturbed area

Perennial stream

Straw bale diversion

Contour showing
  elevation   - - 392 - -

FIGURE 9.12 *Example of a sediment- and erosion-control plan for a commercial development. (Courtesy of Braxton Williams, Soil Conservation Service.)*

Currently, the same basic problems—erosion control and sediment pollution—occupy a significant portion of the public's attention. The sources of the sediment have expanded, however, to include those from land development, highways, mining, and production of new, unusual chemical compounds and products, all in the interest of a better life. Sediment pollution affects rivers, streams, the

Great Lakes, and even the oceans, and the problem promises to be with us for an indefinite period.

The solution to sediment pollution requires sound conservation practices. Such practices are particularly important in urbanizing areas where tremendous quantities of sediment are produced during construction. Figure 9.11 shows sediment production in various degrees of soil exposure and land uses, and Figure 9.12 shows a typical sediment and erosion control plan for a commercial development. The plan calls for diversions to collect runoff and a sediment basin to collect sediment and keep it on the site, thus preventing stream pollution. A generalized cross section and plan (map) view of a sediment control basin are shown in Figures 9.13 and 9.14.

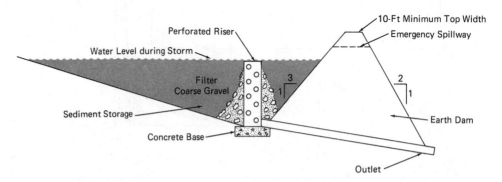

FIGURE 9.13 *Cross section of a sediment basin. Storm water carrying sediment runs into the sediment basin where the sediment settles out and the water filters through loose gravel and into a pipe outlet. (Source:* Erosion and Sediment Control, *Soil Conservation Service, 1974.)*

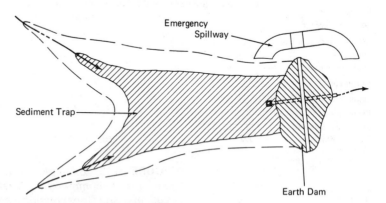

FIGURE 9.14 *Plan view of a sediment basin. Sediment enters the basins from the several different streams and overland flow to be trapped behind the dam. At times of high flow, water may bypass the dam by way of the emergency spillway. Sediment basins must be periodically cleaned out to remain efficient. (Source:* Erosion and Sediment Control, *Soil Conservation Service, 1974.)*

Channelization

For over 200 years, Americans have lived and worked on floodplains. Advantages and enticement to do so have stemmed from the abundance of rich, alluvial soil found on the floodplain; conveniences such as abundant water supply and ease of waste disposal; and proximity to communications, transportation, and commerce that developed along the rivers. Of course, building of houses, industry, public buildings, and farms on the floodplain invite disaster, but floodplain residents have refused to recognize the natural floodway of the river for what it is: part of the natural river system. As a result, flood control and drainage of wetlands became prime concerns. It is not too great an oversimplification to infer that as the pioneers moved West, they had a rather set procedure for modifying the land: first, clear the land by cutting and burning the trees; and, second, modify the natural drainage. From that history came two parallel trends: an accelerating program to control floods matched by an even greater growth of flood damages (3).

Historically, the response to flooding has been to attempt to control the water by constructing dams or modifying the stream by constructing levees or even rebuilding the entire stream to more efficiently remove the water from the land. Every new project has the effect of luring more people to the floodplain in the false hope that the flood hazard is no longer significant. However, man has yet to build a dam or channel capable of controlling the heaviest rainwaters, and when the water finally exceeds the capacity of the structure, flooding may be extensive (3).

Channelization of streams consists of straightening, deepening, widening, clearing, or lining of existing stream channels. Basically, it is an engineering technique, the objectives of which are to control floods, drain wetlands, control erosion, and improve navigation (3). Of the four objectives, flood control and drainage improvement are the two most often cited in channel improvement projects.

Thousands of miles of streams in the United States have been modified, and thousands of more miles of channelization are being planned or constructed. Federal agencies alone have carried out several thousands of miles of channel modification in the last twenty years (4). Considering that there are approximately 3.5 million miles of stream in the United States and the average yearly rate of channel modification is only a very small fraction of one percent of the total stream miles, the impact may appear small. Nevertheless, nearly one thousand miles of channel per year is a significant amount of environmental disruption and, therefore, merits evaluation.

Channelization is not necessarily a bad practice. However, inadequate consideration is being given to its adverse environmental effects. In fact, the picture emerging is that little is known about these effects and that furthermore little is being done to evaluate them (3).

Adverse Effects of Channelization   Opponents of channelizing natural streams emphasize that the practice is completely antithetical to the production of fish and wetland wildlife and that furthermore the stream suffers from extensive aesthetic degradation. The argument is as follows: Drainage of wetlands adversely affects plants and animals by eliminating habitats necessary for survival of certain species. Cutting of trees along the stream eliminates shading and cover for fish while exposing the stream to the sun, resulting in damage to plant life and heat-sensitive aquatic organisms. Cutting of bottomland (floodplain) hardwood trees eliminates the habitats of many animals and birds and also facilitates erosion and siltation of the stream. Straightening and maintaining of the stream bed destroys the diversity of flow patterns, changes peak flow, and destroys feeding and breeding areas for aquatic life. Finally, the conversion of wetlands with a meandering stream to a straight, open ditch seriously degrades the aesthetic value of a natural area (3).

It is commonly believed that channelization increases the flood hazard downstream from the modified channel. Although in many cases this is true, it is far from the entire story. One study concluded that on the contrary, increases of normal and peak flows due to channelization are not particularly significant, and in some cases the flood peaks are actually reduced (4). This results partly because the contribution of runoff from the modified channels tends to be a small fraction of the total basin runoff; peak runoff from channelized streams may not coincide with basinwide runoff, and thus the quicker flow from a modified stream may pass prior to the natural flood flow from the entire basin, thus reducing the normal aggregated peak flow; and many streams that are modified have gradients so low that no amount of straightening could significantly increase the downstream flow. However, it is emphasized that channel modification, especially straightening, can increase flooding directly downstream from the project, and the problem may be compounded if there are a number of projects in the same basin.

Examples of channel work projects that have adversely affected the environment are well known. The Willow River in Iowa and Blackwater River in Missouri emphasize some of the possible adverse impacts to streams that are channelized.

The channelization of the Willow River in Iowa has been extensively studied (5, 6). As early as 1851, the river was known for its frequent flooding that caused extensive damage to crops. As a result, the valley was generally considered to be too risky for agriculture. At that time, the river was confined at low flow to a meandering channel a few feet below the floodplain. In order to alleviate flooding, channelization in three stages from 1906 until 1920 converted about 28 miles of the Willow River into a straight drainage ditch. This had the effect of producing a much shorter channel and steeper gradient. Changes in the size and shape of the drainage ditch in the Willow River Valley since the construction ended in 1920 show that there has been progressive widening and deepening of the channel (Figure 9.15). In 1958 the channel was characteristically U-shaped with nearly vertical walls (Figure 9.16). The ditch in this vicinity in 1920 was only about 15 feet deep and less than 50 feet wide. In 1958, it was over 35 feet deep and 100 feet wide, as shown by the cross sections in Figure 9.15, and the excessive bank erosion is not unique along the ditch. The history is one of continuous enlargement since the project was completed.

The channelization designed for the Willow River was inadequate and resulted in excessive channel erosion. The drainage ditch response is a classic example of adverse channel change due to man's reshaping of a natural channel (6).

Channelization of the Blackwater River in Missouri has also resulted in environmental degradation, including channel enlargement, reduced biological productivity, and increase in downstream flooding (7). The natural meandering channel was dredged and shortened in 1910, nearly doubling the natural gradient.

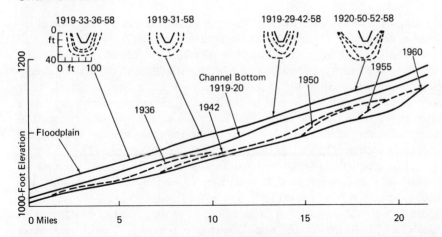

FIGURE 9.15 *Longitudinal (long) and cross-channel profiles of Willow Drainage ditch during specific years from 1919 to 1960. (Reprinted, by permission, from R. U. Ruhe, "Stream Regimen and Man's Manipulation." In* Environmental Geomorphology, *ed. D. R. Coates, p. 1, State University of New York, 1971.)*

This evidently initiated a cycle of channel erosion which is still in progress, and the channel has shown no tendency to resume meandering (7). Maximum increase in channel cross-sectional area exceeds one thousand percent, and as a result, many bridges have collapsed and have had to be replaced. One particular bridge over the Blackwater collapsed due to bank erosion in 1930 and was replaced by a new bridge 27 meters wide. This bridge had to be replaced in 1942 and again in 1947. The 1947 bridge was 70 meters wide, but it too collapsed from bank erosion. Since channelization, erosion has progressed at an average rate of 1 meter in width and 0.16 meter in depth per year (7).

Biologic productivity in the Blackwater River was reduced due to channelization. This reduction resulted because the channel was excavated in easily erodible shale, sandstone, and limestone, producing a smooth stream bed that does

FIGURE 9.16 *Willow Drainage ditch near the mouth of Thompson Creek. The ditch here is about 35 feet deep and 100 feet wide. (Photo courtesy of R. B. Daniels.)*

not easily support bottom fauna after it has been disturbed. As a result, the channelized reaches contain far fewer fish than the natural reaches (7).

The downstream end of the channelization project of the Blackwater terminates where the rock type changes to a more resistant limestone, making dredging financially unfeasible. Since the channelized portion has a larger channel, it can carry more water without flooding than the natural, downstream channel. As a result, when heavy rains fall, the runoff that is carried by the upstream channelization section is discharged into the unchannelized downstream section; and because the channel cross section is much less there, frequent flooding results (7).

The channelization of the Blackwater River is a good example of possible trade-offs and unforeseen degradations caused by channelization. The upper reaches benefited from the utilization of more floodplain land. However, this benefit must be weighed against loss of farmland to erosion, cost of bridge repair, loss of biological life in the stream, and downstream flooding. A more reasonable trade-off might have been to straighten the channel less, still providing a measure of flood protection while not causing such rapid environmental degradation. The difficult question is, How much straightening can be done before unacceptable damage to the riverine system is induced?

Benefits of Channelization   Not all channelization causes serious environmental degradation, and in some cases drainage projects can even be beneficial. Such benefits are probably best observed in urban areas subject to flooding and in rural areas where previous land use has caused drainage problems. In addition, other examples can be cited to show where channel modification has improved navigation or reduced flooding and has not caused environmental disruption.

Many streams in urban areas scarcely resemble natural channels. The process of constructing roads, utilities, and buildings with associated sediment production is sufficient to disrupt most small streams. A channel restoration project, the objective of which is to clean out urban waste from the channel, allowing the stream to flow freely, is often needed. The channel should be made as attractive as possible by allowing the stream to meander and provide for variable, low-water flow conditions, fast and shallow (riffle) alternating with slow and deep (pools). Where lateral bank erosion must be absolutely controlled, the outside of bends may be defended with large stone (riprap).

An example of an attempt to improve the aesthetic appeal of an open cut ditch known as Whiskey Creek near Grand Rapids, Michigan, emphasizes that engineers are beginning to consider channel restoration seriously. The idea was to make the open drain appear as a country stream meandering through substantial residential and commercial properties. Several of the meander bends of the channel during construction are shown in Figure 9.17. Although it may be argued that open cut drains are unlikely to be considered as pleasing sights under any circumstances, certainly the Whiskey Creek project is something out of the ordinary and is an improvement over a straight ditch designed only for flood control.

Another example of channel restoration is found in the Coastal Plain of North Carolina where many streams have been channelized. As part of a mosquito-abatement program in Onslow County, the state is attempting to drain some areas and thus destroy the insect's habitat. Examination of some of the streams before improvement reveals that they are nearly blocked by debris and logs. Evidence such

FIGURE 9.17 *Channelization of Whiskey Creek near Grand Rapids, Michigan. This is an attempt to improve the aesthetic appeal of a stream project by constructing a series of meandering bends. (Photo courtesy of Williams and Works.)*

as buried, hand-cut cyprus logs and handmade cyprus shingles along the stream bank suggests that streams that were once free-flowing have been choked with debris left from logging practices years ago. The channel restoration involves excavating the old prelogging channel (Figure 9.18). Engineers on the project state that the old channel can be followed by the high organic soils that form along the channel bank. The resulting restoration is a meandering stream that is allowed to flow freely again (Figure 9.19). The material dredged in restoring the channel is left on the bank and quickly supports new vegetation. Small cyprus seedlings are present only a few months after the work is completed. Although this particular practice of channel work is in need of further study and evaluation, it seems to be a step in the right direction. Furthermore, it appears to be popular with landowners who request that the work be done rather than have to be convinced that it should be, as is often the case with drainage work.

Channel modification on the Missouri River is an interesting example of river training in which the power of running water is used to maintain the desired channel. The project involves engineering improvements to produce a series of relatively narrow, gentle meanders rather than a succession of wide, straight reaches with a steep gradient and abrupt changes in flow direction that previously existed along some parts of the river. That is, before modification, the channel was in places semibraided and contained numerous islands and bars that formed as a result of deposition of large amounts of sediment. This was especially true downstream from the Platte River where it enters the Missouri River (6).

The engineering project that was started in the 1930s to induce the Missouri River to narrow and meander rather than braid involves construction of permeable pile dikes (Figures 9.20) that reduce the stream's velocity, causing depo-

FIGURE 9.18 *Channel restoration on the coastal plain of North Carolina. The dragline operator is instructed to excavate the old prelogging channel as much as possible.*

sition of sediment in desired locations. The water passing around the dikes, on the other hand, is constricted, causing an increase in velocity and scour in desired areas. As erosion and deposition progress, the dikes silt in, and natural growth of willows is established. A section of the river showing the improvements and results from 1930 to 1940 is shown in Figure 9.21. This project is one case where man's interference has improved the area by providing additional land for agriculture and facilitating navigation (6). However, the dikes have caused a reduction in aquatic habitat by filling in sloughs and marshland downstream from the dikes (Figure 9.22). As result, the Corps of Engineers have designed a new type of dike, with a notch at the top, that will allow more water to pass through and hopefully minimize the loss of valuable wildlife and fish habitat. Furthermore, there is concern that the river training has reduced the total channel capacity and thus is facilitating more frequent flooding.

The Future of Channelization   Although there are considerable controversy and warranted anxieties surrounding channelization, it is expected that channel modification will remain a necessity as long as land-use changes such as urbanization or conversion of wetlands to farmland continue. Therefore, we must strive to design

FIGURE 9.19 *Channel restoration on the coastal plain of North Carolina. The channel produced is a meandering stream in which vegetation soon becomes reestablished. This photograph was taken only several days after the channel had been re-formed.*

channels that reduce adverse effects. This can be accomplished if project objectives of flood control or drainage improvement are carefully tailored to specific needs.

FIGURE 9.20 *Upstream view of the Treville Bend of the Missouri River. The permeable pile dikes on the inside of the bend effectively converge the flow of water toward the outside, where the faster water scours the channel. Thus the power of running water is used to maintain the channel. (Photo courtesy of the U.S. Army Corps of Engineers, Omaha District.)*

FIGURE 9.21 *Otoe Bend. The diagram on the left shows the channel as it was in 1930 with the 1940 design channel superposed on it. The diagram on the right shows the systems of permeable dikes used to train the river. Notice that Schemmel Island has developed precisely where deposition was planned. (Reprinted, by permission, from R. U. Ruhe, "Stream Regimen and Man's Manipulation." In* Environmental Geomorphology, *ed. D. R. Coates, p. 16, State University of New York, 1971.)*

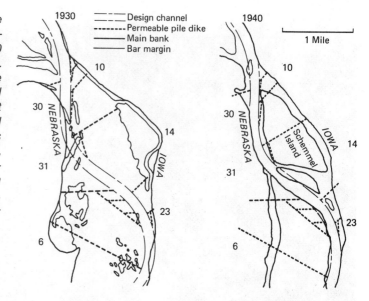

If the primary objective is drainage improvement in areas where natural flooding is not a hazard, then there is no need to convert a meandering stream into a straight ditch. Rather, design might involve cleaning the channel and maintaining a sinuous stream. In addition, a series of relatively deep areas (pools) and shallow areas (riffles) might be constructed by changing the channel cross-sectional shape, asymmetrically to form pools and symmetrically to form riffles, as found in natural streams (Figure 9.23). Where this is possible, the resulting channel would tend to duplicate nature rather than be alien to it. In addition, cutting of trees along the channel bank would be minimized and new growth would be encouraged. This plan would tend to minimize adverse effects by producing a channel that is more biologically productive and aesthetically pleasing.

FIGURE 9.22 *Upstream view of Browers Bend on the Missouri River. Notice the Iowa Public Service power plant on the right bank. A new design of permeable pile dikes is now under experimentation. It will allow water to flow through notches in the dikes, thus causing less filling-in of aquatic habitat (marsh and slough), shown on the left bank behind the dikes. The dikes shown here are of the conventional type, and filling of the marsh is clearly reducing wildlife and aquatic habitat. (Photo courtesy of the U.S. Army Corps of Engineers, Omaha District.)*

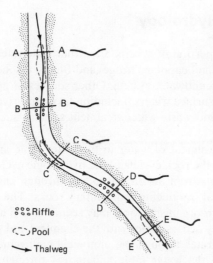

FIGURE 9.23 *Idealized diagram showing pools and riffles and alternating asymmetric and symmetric cross-channel profiles. (Reprinted, by permission, from E. A. Keller, "Channelization: A Search for a Better Way," Geology 5, The Geological Society of America, 1975.)*

Channel design for flood control is more involved because natural channels are maintained at flows with a recurrence interval of one to two years, whereas flood-control projects may need to carry the 25- or even 100-year flood. A 100-year channel cannot be expected to be maintained by a 2-year flood. Such a channel would probably be braided and choked with migrating sandbars. Therefore, it is suggested that a pilot channel be constructed where possible (Figure 9.24). A meandering channel designed to be maintained by the 2-year flood is superposed on the larger flood-control channel. Addition of pools and riffles in the pilot channel when possible would help provide fish habitat and better low flow conditions. The large channel may be vegetated, and the untrained observer might not easily recognize it. It is obvious that the above plan will not work in urbanizing areas where sediment production is high and property not available to be purchased for the larger channel. However, if sediment is reduced by good conservation practice and the property is available to be purchased, the urban area will benefit from a more aesthetically pleasing and more useful stream.

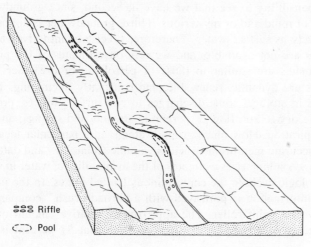

FIGURE 9.24 *Pilot channel (channel within a channel) suggested for some flood-control projects. Pools and riffles are constructed in the meandering channel within the larger flood-control channels. (Reprinted, by permission, from E. A. Keller, "Channelization: A Search for a Better Way," Geology 5, The Geological Society of America, 1975.)*

Groundwater Hydrology

Groundwater is the portion of precipitation that penetrates the soil; moves through the zone of aeration and capillary fringe; and finally reaches the water table and becomes part of the groundwater system. Other sources of groundwater include water that infiltrates from surface waters, including lakes and rivers, storm water retention or recharge ponds, and waste-water treatment systems such as cesspools and septic tank drain fields.

The movement of water to the water table and through aquifers is an integral part of both the rock cycle and the hydrologic cycle. For example, the water may dissolve minerals from material it moves through and deposit them elsewhere as cementing material, producing sedimentary rocks. The water may also transport sediment and/or microorganisms, or it may seemingly do none, some, or all of these chores. What actually occurs varies with the chemical and physical characteristics of the water as it infiltrates through the biological environments and the soil-horizon environments above the water table and moves through the groundwater system below the water table.

Movement of groundwater in the hydrologic cycle may render the water more or less suitable for human use; it may even be rendered toxic by natural processes. For instance, it may become toxic by dissolving sufficient amounts of one or more elements or minerals, such as arsenic. Most often, the groundwaters dissolve a mixture of minerals and some gases which, at worst, are nuisances to some human uses. Iron as ferrous hydroxide, calcium carbonate (which creates so-called hardness), and hydrogen sulphide (which creates a "rotten egg" odor) are examples.

Vast amounts of potable (drinkable) groundwaters are available, although not uniformly distributed, as shown in Figure 9.25. However, like surface water, groundwater is not always as readily available for human use as the population concentrated within an area might desire. Three problems concerning how people perceive the availability of groundwater and our understanding of this resource are apparent. First, people tend to assume that water is available when, where, and in the amounts they want. We turn on a faucet and expect water—it is somebody else's responsibility to see that we have it. Second, since groundwater is out of sight, it is out of mind and/or mysterious. Third, groundwater is not as easily measured quantitatively as surface water. Therefore, precise quantitative values of groundwater reserves are not available, and we rely on estimates of the probable reserves. These estimates are similar to those of petroleum or ore reserves, but groundwater reserves are dynamic, renewable, and constantly fluctuating. That is, some groundwater is in storage; some discharges to our surface waters, providing water to maintain the dry-season flow of perennial streams and springs; and some is lost in evapotranspiration. How much groundwater is in a particular location at any one time defies accurate measurement with available technology and data. With this limitation, Table 9.6 summarizes estimates of the availability of water in various portions of the hydrologic cycle in the conterminous United States. In the parlance of the planner, this information is portrayed with a "broad brush," comparable to painting a picture with a paint roller. The approximate relative amount of water circulated yearly, excluding atmospheric waters, is such that 57 percent is in the

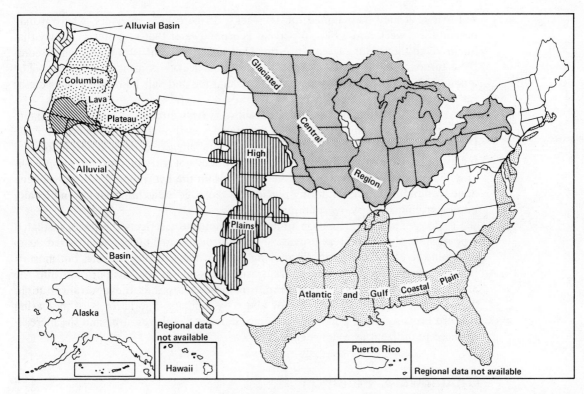

FIGURE 9.25 *Areas where potential groundwater development may be possible. (Source:* The Nation's Water Resources, *Water Resources Council, 1968.)*

	Annual Circulation (million acre-ft per year)	Detention Period (yr)	Relative Abundance (circulated/yr, percent)
Frozen water			
Glaciers	1.3	40	0.03
Liquid water			
Freshwater lakes (U.S. part of Great Lakes only)	150	100	3.10
Salt lakes	4.6	10	0.10
Average in stream channels	1500	.03	34.60
Groundwater			
Shallow	250	200	5.66
Deep	5	10,000	0.12
Soil moisture (3-ft root zone)	2500	.2	56.69

TABLE 9.6 *Annual circulation and detention of water resources in the United States, excluding atmospheric water.*

SOURCE: Data from U.S. Geological Survey.

soil, where it may be detained for 10 weeks; 35 percent is in rivers, where it may remain for a week and a half; 6 percent is groundwater that migrates so slowly that hundreds to thousands of years are involved; and 3 percent is in lakes, where complete turnover for large lakes such as the Great Lakes is around 100 years. The amounts of water circulated annually in glacial ice and salt lakes are minuscule as far as the United States is concerned.

The amount of all of the above waters circulated annually probably does not equal the amount of water circulated yearly in the atmosphere over the United States. Thus rainmaking becomes a very attractive method for increasing the available supply, but on the other hand, it does not truly increase the supply—it only tampers with the location of the water within the hydrologic cycle as has been done by building dams, rendering portions of recharge areas impervious, and locally mining groundwater by overpumping.

The long detention time for groundwaters reflects the deep, insulated type of storage our aquifers provide. Not all groundwater takes one hundred years to rejoin the other, more rapidly moving parts of the hydrologic cycle, but most of it is well below the influence of transpiration by plants and evaporation into the atmosphere. Where it is not that deep, it is most susceptible to evapotranspiration, discharge to streams, and use and/or abuse by humans. The latter is of increasing concern because of the potential long-term damage to this resource and the increasing need to utilize it as water increases.

Groundwater Pollution

Chemical and bacterial pollution of aquifers have been documented and are receiving more attention in the hope of averting widespread degradation. The problem is compounded by the fact that these waters are hidden and by long-held opinions that filtration through soils magically purifies waters or dilutes the wastes, as can occur in rapidly moving and mixing surface waters.

A pollutant is any substance, biological or chemical, in which an identifiable excess is known to be deleterious to other desirable living organisms. Within this framework, excessive amounts of the heavy metals such as lead and mercury; certain radioactive isotopes; nitrogen, phosphorous, and sodium; and other useful, even necessary, elements as well as certain pathogenic bacteria and viruses are pollutants. In some instances, a material may be considered as a pollutant to a particular segment of the world's population although it is not harmful to other segments. For instance, excessive nitrogen in the proper compound is deleterious to infants, but much less so to adults, if at all. In like manner, excessive sodium as in salt is not generally harmful, but it can be to certain people on diets restricting intake of salt for medical reasons.

The differences in the physical, geologic, and biologic environments associated with groundwater pollution compared with those of surface-water pollution are significant. In the latter, hydraulics of flow and availability of oxygen and sunlight (which may tend to degrade pollutants), along with the rapidity with which processes of dilution and dispersion of pollutants occur, are markedly different than for groundwater, where the opportunity for bacterial degradation of pollutants is generally confined to the soil or a few feet below. Furthermore, the channels through which groundwaters move are very small and variable, as shown schematically in Figure 9.26. From this, it is obvious that the rate of movement is much reduced, except, perhaps, in large solution channels within limestones, and the opportunity for dispersion and dilution is very limited. In addition, groundwaters are usually devoid of oxygen, which is helpful in killing aerobic types of micro-organisms but which may provide a "happy home" for anaerobic varieties.

Like surface water, groundwater flows in response to the force of gravity. The preference for horizontal groundwater motion is related to layering characteristics of consolidated and unconsolidated sediments. However, for fractured, folded, and faulted rocks, structural characteristics can be the dominant factors that control the direction of flow.

The ability of most soils and rocks to physically filter out solids, including pollutional solids, is well recognized. However, this ability varies with different sizes, shapes, and arrangements of filtering particles, as evidenced in the use of selected sands and other materials in water filtration plants. Also known, but perhaps not so generally, is the ability of clays and other selected minerals to capture and exchange some elements and compounds when they are dissociated in solutions as positively or negatively charged elements or compounds. Such exchanges, along

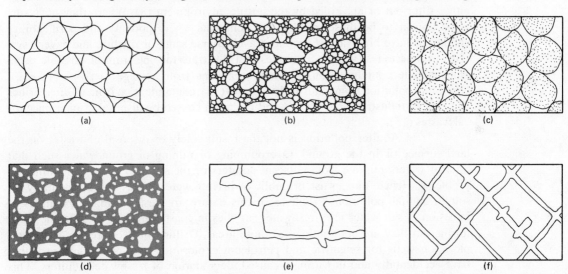

FIGURE 9.26 *Idealized diagram showing several types of soil and rock openings and the relation to porosity. (a) Well-sorted sediment with high porosity. (b) Poorly sorted sediment with low porosity. (c) Well-sorted sediment in which the particles themselves are also porous, resulting in a higher total porosity. (d) Well-sorted sediment in which the openings between grains are partially filled with cementing material, reducing the porosity. (e) Rock in which porosity is increased due to solutional openings. (f) Rock that is porous due to open fractures. (Source: Meinzer, U.S. Geological Survey Water Supply Paper 489, 1923.)*

with sorption and precipitation processes, are important in the capture of pollutants. These processes have definable units of capacity and are reversible. They also can easily be overlooked in designing facilities to correct pollution problems by relying on soils and rocks of the geologic environment for treatment; this oversight can possibly result in groundwater pollution. This is especially significant in land application of waste waters.

Much conjecture, debate, and research have occurred regarding groundwater pollution. The principal concerns of groundwater pollution are the introduction of chemical elements, compounds, and microorganisms that do not naturally occur in aquifers that can be or are used for drinking water. Not only because of the degradation of the aquifer but also because of the difficulty in detection, the long-term residency, and the difficulty and expense of aquifer recovery, strong arguments can be made that no wastes or possible pollutants should be allowed to enter any part of the groundwater system. This is an impossible dream. Rather, the answer lies in knowing more about how natural processes treat wastes to insure that when soil and rocks are not capable of treating, storing, or recycling wastes, we can develop processes to make the pollutants treatable, storable, or recyclable.

Studies and case histories of groundwater pollution generally show that microorganisms seldom survive for distances greater than 100 feet, because they either filter out or are killed by the hostile environment (8). Where they have survived appear to be those geologic environments where porosity and permeability conditions have high values, such as gravel, coarse sand, or fracture and cavern systems in limestones (9, 10). Examples that illustrate the spectrum of possible problems include contamination by microorganisms from sewage disposal (11, 12); chromium pollution from industrial processes (13); contamination from oil-field brines and related industrial wastes (14); and pollution caused by disposal of pharmaceutical wastes (15).

Aquifer pollution is not the result solely of disposal of wastes on the land surface or in the ground. Overpumping or mining of groundwater such that inferior waters migrate from adjacent aquifers or the sea also cause contamination problems. Hence, human use of public or private water supply can accidentally result in aquifer pollution. Problems of this nature are best shown by examples of intrusion of salt water into freshwater supplies in coastal areas of Florida and California, but they are not confined to those areas. Drilling of deep wells, such as exploration wells for scientific and petroleum exploration, can cause, and probably have accidentally and unwittingly caused, degradation of freshwater aquifers. This resulted because of the unrealized hazards during the early phases of petroleum and other mineral water (especially for salt) well drilling and the lack of information about the total depths at which potable waters are known to occur in many parts of the United States and throughout the world.

Correction of aquifer contamination is not impossible, but it is often seemingly so because of the complex nature of some of the wastes and the time, expense, and difficulties in locating, monitoring, and confining the pollutants. Efforts are being made to improve the methods, but in general, the techniques of locating and developing groundwater supplies are the basic tools of correcting groundwater pollution.

SUMMARY AND CONCLUSIONS

Obvious and well-known detrimental aspects of human use of surface water, groundwater, and atmospheric water include pollution of rivers and groundwaters. We now understand that solutions to many hydrologic problems require integrating all aspects of the water cycle.

Interactions between the rock and hydrologic cycles greatly affect the amount of surface-water runoff and sediment yield of a particular drainage basin. Geologic and physiographic factors that control runoff and sediment yield include structure and type of earth material; climate; relief and slope; drainage pattern; and orientation and exposure of the drainage basin.

Principal human influences affecting runoff and sediment production include the varied land uses (especially urban) and construction projects to control floods, such as channel improvement and reservoirs.

Sediment pollution, natural or man-induced, is certainly one of the most significant pollution problems. A great deal of sediment pollution is a result of human use, particularly construction, agriculture, and urbanization. Although the problem cannot be completely eliminated, it can be minimized.

Channelization is the straightening, deepening, widening, cleaning, or lining of existing streams. The most commonly cited objectives of channel modification are flood control and drainage improvement. Although many cases of environmental degradation resulting from channelization can be cited, it need not necessarily be a bad practice. In fact, there are several cases where channel modification was clearly a beneficial practice. Nevertheless, the present state of channel modification technology is not sufficient for us to be reasonably sure that environmental degradation will not result. Therefore, until new design criteria compatible with natural stream processes are developed and tested, channelization should be considered only as a solution to a particular flood or drainage problem where the impact and alternatives are carefully studied to minimize environmental disruption.

The movement of water down to the water table and through aquifers is an integral part of the rock and hydrologic cycles. In moving through an aquifer, groundwater may improve in quality. However, it may also be rendered unsuitable for human use by natural or man-made contaminants.

A pollutant is any substance in which an identifiable excess is known to be harmful to desirable living organisms. In the case of groundwater pollution, the physical, biologic, and geologic environments are considerably different from those of surface water. The ability of many soils and rocks to physically or otherwise degrade pollutants is well known, but not so generally known is the ability of clays and other earth materials to capture and exchange certain elements and compounds.

Man-induced aquifer pollution is not the result solely of disposal of wastes on the land surface or in the ground. It can also result from over-pumping of groundwater in coastal areas, resulting in intrusion of salt water into freshwater aquifers.

References

1. Wisler, C. D., and Brater, E. F. 1949. *Hydrology*. New York: John Wiley & Sons.
2. Robinson, A. R. 1973. Sediment, our greatest pollutant? In *Focus on environmental geology,* ed. R. W. Tank, pp. 186–92. New York: Oxford University Press.
3. House Report No. 93-530. 1973. *Stream channelization: what federally financed draglines and bulldozers do to our nation's streams.* Washington, D.C.: U.S. Government Printing Office.
4. Arthur D. Little, Inc. 1972. *Channel modifications: an environmental, economic and financial assessment.* Report to the Council on Environmental Quality.
5. Daniels, R. B. 1960. Entrenchment of the Willow drainage ditch, Harrison County, Iowa. *American Journal of Science* 225: 161–76.
6. Ruhe, R. V. 1971. Stream regimen and man's manipulation. In *Environmental geomorphology,* ed. D. R. Coates, pp. 9–23. Binghamton, New York: Publications in Geomorphology, State University of New York.
7. Emerson, J. W. 1971. Channelization: a case study. *Science* 173: 325–26.
8. Mallman, W. L., and Mack, W. N. 1961. A review of biological contamination of ground water. Proceedings of Symposium on Ground Water Contamination, Taft Sanitary Engineering Center, Cincinnati, Ohio.
9. LeGrand, H. E. 1968. Environmental framework of ground water contamination. *Journal of Ground Water* 6: 14–18.
10. Deutsch, M. 1965. Natural controls involved in shallow aquifer contamination. *Journal of Ground Water* 3: 37–40.
11. Bogan, R. H. 1961. *Problems arising from ground water contamination by sewage lagoons at Tieton, Washington.* Public Health Service Technical Report W61-5.
12. Vogt, J. E. 1961. *Infectious hepatitis outbreak in Posen, Michigan.* Public Health Service Technical Report W61-5.
13. Deutsch, M. 1961. *Incidents of chromium contamination of ground water in Michigan.* Public Health Service Technical Report W61-5.
14. Pettyjohn, W. A. 1971. Water pollution by oil-field brines and related industrial wastes in Ohio. *Ohio Journal of Science* 71: 257–69.
15. Burt, E. M. 1972. The use, abuse and recovery of a glacial aquifer. *Journal of Ground Water* 10: 65–72.

10

WASTE DISPOSAL

One view of waste disposal might consider the ultimate technology to be a system capable of accepting an unlimited amount of waste and safely containing it forever outside man's sphere of life. In the first century of the Industrial Revolution, the volume of waste produced was relatively small, and the concept of "dilute and disperse" was adequate. Factories were located near rivers because the water provided easy transport of materials by boat, ease of communication, sufficient water for processing and cooling, and easy disposal of waste into the river. With few factories and sparse population, "dilute and disperse" seemed to remove the waste from the environment (1).

As industrial and urban areas expanded, the concept of "dilute and disperse" became inadequate, and new concepts such as "concentrate and contain" became popular. However, it became apparent that containment was and is not always achieved. Containers, natural or artificial, may leak or break and allow waste to escape. Thus, another new concept developed—wastes are converted to useful material, in which case they are no longer wastes but resources. However, even with our technology, there are large volumes of waste that cannot be economically converted or that are essentially indestructible; hence, we still have waste-disposal problems (1).

Disposal or treatment of liquid and solid wastes by state and municipal agencies costs billions of dollars every year, and it is the most expensive environmental expenditure that governments must pay. Figure 10.1 shows that total environmental expenditures more than doubled in 12 years from less than $3 billion in 1958 to approximately $7 billion in 1970. Of this, solid-waste management and waste-water treatment accounted for about 55 percent of the expenditures (2).

All societies produce waste, but industrialization and urbanization have caused an ever-increasing affluence and have greatly compounded the problem of waste management. Although tremendous quantities of liquid and solid waste from municipal, industrial, and agricultural sources are being collected and recycled, treated, or disposed of, new and innovative programs remain a necessity. The popular notion is to consider all so-called wastes as essentially "resources out of place." Although we will probably never be able to recycle all wastes, it seems apparent that increasing costs of raw materials, energy, transportation, and land will make it financially feasible to recycle more resources as well as reuse the land where wastes not recycled are buried, creating new land resources for development.

For discussion purposes, it is advantageous to break the management, treatment, and disposal of waste into several categories: solid-waste disposal, deep-well disposal, radioactive waste management, ocean dumping, septic-tank sewage disposal, and waste-water treatment. Although there is some obvious overlap in some of the categories, for example, ocean dumping of solid waste, each has sufficiently different aspects to warrant separate discussion.

FIGURE 10.1 *State and munici-pal expenditures for environmental purposes from 1958 to 1971. [Source: Government Finances, U.S. Department of Commerce, Bureau of the Census (Washington, D.C.: U.S. Government Printing Office, 1958–1970/71).]*

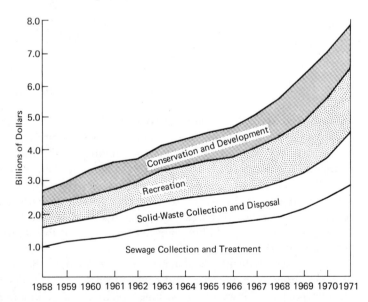

Solid-Waste Disposal

Disposal of solid waste is primarily an urban problem. In the United States alone, urban areas produce about 1400 million pounds of solid waste each day. That amount of waste is sufficient to cover more than 400 acres of land every day to a depth of 10 feet (3). Figure 10.2 summarizes major sources and types of solid waste (refuse), and Table 10.1 lists their compositions. It is no surprise that paper is by far the most abundant of these solid wastes. However, it is emphasized that this is only an average content, and considerable variation can be expected as such factors as land use, economic base, industrial activity, climate, and season of the year vary.

The common methods of solid-waste disposal summarized from a U.S. Geological Survey report are on-site disposal; feeding of garbage to swine; composting; incineration; ocean dumps; and sanitary landfills (3). Of these, the geologic factor is most significant in planning and siting of the sanitary landfill. Without careful consideration of the soils, rocks, and hydrogeology, a landfill program may not function properly.

On-Site Disposal   By far the most common on-site disposal method in urban areas is the mechanical grinding of kitchen food waste. Garbage disposal devices are installed in the waste-water pipe system from a kitchen sink, and the garbage is ground and flushed into the sewer system. This effectively reduces the amount of handling and quickly removes food waste. However, final disposal is transferred to the sewage treatment plant where solids such as sewage sludge must still be disposed of (3).

Another method of on-site disposal is incineration. This method is common in institutions and apartment houses (3). It requires constant attention and

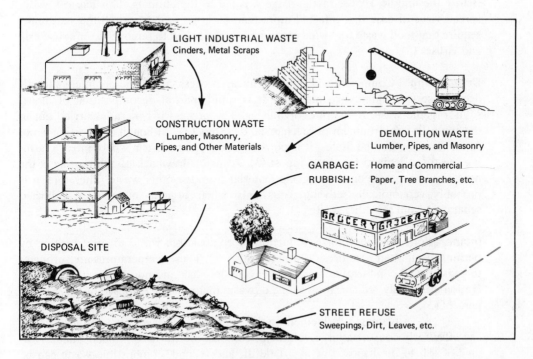

FIGURE 10.2 *Types of materials or refuse commonly transported to a disposal site.*

TABLE 10.1 *Generalized compo-*
sition of solid waste likely to end up
at a disposal site.

Type of Waste	Average (%)
Paper products	43.8
Food wastes	18.2
Metals	9.1
Glass and ceramics	9.0
Garden wastes	7.9
Rock, dirt, and ash	3.7
Plastics, rubber, and leather	3.1
Textiles	2.7
Wood	2.5
Total	100.0

SOURCE: *Ocean Dumping: A National Policy,*
Council on Environmental Quality, 1970.

periodic maintenance to insure proper operation. In addition, the ash and other residue must be removed eventually and transported to a final disposal site.

Feeding of Garbage to Swine   The use of garbage from urban areas as food for swine remains a common practice, and as recently as 1966, about 1/4 of the total quantity of food waste produced was still used in the United States to feed swine. Before the middle 1950s, raw garbage was fed to the animals. This led to a widespread virus disease that infected more than 400,000 swine, and all the states now require that food waste for swine consumption be cooked to remove harmful bacteria and viruses (3).

Composting   Composting is a biochemical process in which organic materials decompose to a humuslike material. It is rapid, partial decomposition of moist, solid, organic waste by aerobic organisms. The process is generally carried out in the controlled environment of mechanical digesters (3). Although composting is not common in the United States, it is popular in Europe and Asia where intense farming creates a demand for the compost (3). A major drawback of composting is the necessity to separate out the organic material from the other waste. Therefore, it is probably economically advantageous only when organic material is collected separately from the other waste (4).

Incineration   Incineration is reduction of combustible waste to inert residue by burning at high temperatures (1700° to 1800° F). These temperatures are sufficient to consume all combustible material, leaving only ash and noncombustibles. Incineration effectively reduces the volume of waste that must be disposed of by 75 to 95 percent (3).

Advantages of incineration at the municipal level are twofold. First, it can effectively convert a large volume of combustible waste to a much smaller volume of ash to be disposed of at a landfill; and second, combustible waste can be used to supplement other fuels in generating electrical power. Disadvantages are that it requires high capital outlay, high maintenance cost, and additional handling

Floodplain of the Mississippi River. The numerous meanders of the river and its tributaries attest to the ever-changing character of rivers. (ERTS photo)

Missouri River. This upstream view shows a series of permeable pile dikes used to train the river. (Photo courtesy of the U.S. Army Corps of Engineers, Omaha District)

to remove materials that do not burn, and it can cause air pollution. Because of the disadvantages, it seems unlikely that incineration will become a common way for urban areas to dispose of waste (4). Nevertheless, there are about 600 central incinerator plants operating in the United States burning about 150,000 tons of waste per day (3).

Open Dumps   Open dumps (Figure 10.3) are the oldest and most common way of disposing of solid waste, and although in recent years thousands have been closed, many still are being used. In many cases, they are located wherever land is available, without regard to safety, health hazard, and aesthetic degradation. The waste is often piled as high as equipment allows. In some instances, the refuse is ignited and allowed to burn. In others, the refuse is periodically leveled and compacted (3).

Although well known, the potential for water pollution caused by indiscriminate dumping generally has not been studied sufficiently to determine the long-range effects of dumping and total area of extent of the pollution. One interesting example was reported from Surrey County, England, where household wastes were dumped into gravel pits, polluting the groundwater (3). Refuse was dumped directly into 20-foot-deep pits with about 12 feet of water in the bottom. The maximum rate of dumping over a six-year period (1954 to 1960) was about 100,000 tons per year. Observation of water quality after the dumping was terminated revealed that chloride concentrations at the dump were 800 mg/liter (milligrams per liter), which is over three times the maximum concentration of 250 mg/liter considered safe for a public water supply. Adjacent gravel pits in the down-gradient direction revealed chloride concentrations of 290 mg/liter, and pits as far as 1/2 mile away had concentrations of 70 mg/liter (3). In addition, bacterial pollution was also detected within 1/2 mile of the disposal site.

As a general rule, open dumps tend to create a nuisance by being unsightly, breeding pests, creating a health hazard, polluting the air, and sometimes polluting groundwater and surface water. Fortunately, the day of the open dumps is

FIGURE 10.3   *Open dump. (Photo by W. J. Schneider, courtesy of U.S. Geological Survey.)*

apparently giving way to the better planned and managed sanitary landfills that will eventually take their place.

Sanitary Landfill The sanitary landfill as defined by the American Society of Civil Engineering is a method of solid-waste disposal that functions without creating a nuisance or hazard to public health or safety. Engineering principles are utilized to confine the waste to the smallest practical area, reduce it to the smallest practical volume, and cover it with a layer of compacted soil at the end of each day of operation, or more frequently if necessary. It is the covering of the waste with compacted soils that makes the sanitary landfill "sanitary." The compacted layer effectively denies continued access to the waste by insects, rodents, and other animals. It also isolates the refuse from the air, thus minimizing the amount of surface water entering into and gas escaping from the wastes (4). (See Figure 10.4.)

   The sanitary landfill as we know it today emerged in the late 1930s, and by 1960 over 1400 cities disposed of their solid waste by this method. Two types are used: *area landfill* on relatively flat sites; and *depression landfill* in natural or man-made gullies or pits. The depths of landfills vary from about 7 feet to 40 feet. Normally, refuse is deposited, compacted, and covered by a minimum of 6 inches of compacted soil at the end of each day. A 1 to 4 cover ratio is common, that is, 1 foot of soil for every 4 feet of compacted waste. The finishing cover is at least 2 feet of compacted soil designed to mimimize infiltration of surface water (3).

   The most significant possible hazard from a sanitary landfill is groundwater or surface-water pollution. If waste buried in a landfill comes into contact with water percolating down from the surface water or groundwater moving laterally through the refuse, obnoxious, mineralized liquid capable of transporting bacterial pollutants called *leachate* is produced (5). The nature and strength of the leachate depend on the composition of the waste, the length of time that the infiltrated water

FIGURE 10.4 *Sanitary landfill near Charlotte, North Carolina.*
(Photo courtesy of R. R. Phelps.)

is in contact with the refuse, and the amount of water that infiltrates or moves through the waste (3). Table 10.2 shows a generalized analysis of landfill leachates at three stages of stabilization compared to raw sewage and slaughter-house waste. Note that landfill leachate has a much higher concentration of pollutants. Fortunately, however, the amount of leachate produced from urban waste disposal is much less than the amount of raw sewage produced.

   Factors controlling the feasibility of sanitary landfills include topographic relief, location of the groundwater table, amount of precipitation, type of soil and rock, and the location of the disposal zone in the surface-water and groundwater flow system. The best sites are those in which natural conditions insure reasonable safety in disposal of solid waste; conditions may be safe because of climatic, hydrologic, or geologic conditions, or combinations of these (6).

   The best sites are in arid regions (Figure 10.5). Disposal conditions are relatively safe there because in a dry environment, regardless of whether the burial material is permeable or impermeable, there is essentially no leachate produced. On the other hand, in the humid environment, some leachate will always be produced, and therefore an acceptable level of leachate production must be established to determine the most favorable sites. What is acceptable varies with local water use, local regulations, and the ability of the natural hydrologic system to disperse, dilute, and otherwise degrade the leachate to a harmless state.

   The most desirable site in a humid climate is shown in Figure 10.6. There the waste is buried above the water table in relatively impermeable clay and silt soils. Any leachate produced will remain in the vicinity of the site and be degraded by natural filtering action and base exchange of some ions between the clay and the leachate. This would also hold for high water table conditions, as is often the case in humid areas, provided the impermeable material is present. Ideally, a mini-

Constituents	Landfill Leachate			Raw Sewage [a]	Slaughter-House [b] Wastes
	Less than 2 yr Old	6 yr Old	17 yr Old		
BOD [c]	54,610	14,080	225	104	3 700
COD [d]	39,680	8 000	40	246	8 620
Total solids	19,144	6 794	1 198		2 690
Chloride	1 697	1 330	135		320
Sodium	900	810	74		
Iron	5 500	6.3	0.6	2.6	
Sulfate	680	2	2		370
Hardness	7 830	2 200	540		66
Misc. heavy metals	15.8	1.6	5.4	1.3	

TABLE 10.2 *Comparison of liquid wastes: landfill leachate, raw sewage, and slaughter-house waste in ppm (parts per million) or mg/l (milligrams per liter).*

SOURCE: G. M. Hughes and K. Cartwright, "Scientific and Administrative Criteria for Shallow Water Disposal," *Civil Engineering* 42 (New York: American Society of Civil Engineers, 1972).
[a] Data provided by the Metropolitan Sanitary District of Greater Chicago.
[b] Data from the files of the Illinois Department of Public Health.
[c] Biological oxygen demand.
[d] Chemical oxygen demand.

FIGURE 10.5 *Sanitary landfill site in a semiarid environment. (Modified from Bergstrom, Environmental Geology Notes No. 20, Ilinois State Geological Survey, 1968.)*

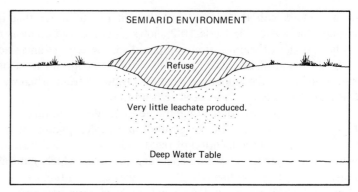

mum of 30 feet of relatively impermeable material is desired between the base of the landfill and the shallowest aquifer. If less than 30 feet of relatively impermeable material is present, then the nature of the bed rock must be considered (7).

If the refuse is buried above the water table over a fractured-rock aquifer, as shown in Figure 10.7, the potential for serious pollution is low because the leachate is partly degraded by natural filtering as it moves down to the water table. Furthermore, the dispersion of contaminates is confined to the fracture zones (3). However, if the water table were higher or the cover material thinner and permeable, then widespread groundwater pollution of the fractured-rock aquifer might result.

If the geologic environment of a landfill site is characterized by an inclined limestone-rock aquifer, overlain by permeable sand and gravel soils as shown in Figure 10.8, considerable contamination of the groundwater might result. This happens because the leachate quickly moves through the permeable soil and enters the limestone, where open fractures or cavities may transport the pollutants with little degradation other than dispersion and dilution. Of course, if the inclined rock is all shale, which is impermeable, little pollution will result (3).

In summary, the following guidelines should be followed in site selection of sanitary landfills (7). First, limestone or highly fractured rock quarries and most sand and gravel pits make poor landfill sites because these earth materials are good aquifers. Second, swampy areas, unless properly drained to prevent disposal into standing water, make poor sites. Third, clay pits, if kept dry, may provide satisfactory sites. Fourth, flat upland areas are favorable sites provided an adequate layer of impermeable material such as clay is present above any aquifer. Fifth, flood-

FIGURE 10.6 *Most desirable landfill site in a humid environment. Waste is buried above the water table in a relatively impermeable environment. (Source: W. J. Schneider, U.S. Geological Survey Circular 601F, 1970.)*

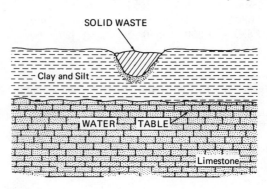

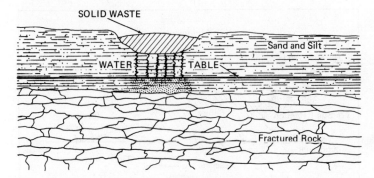

FIGURE 10.7 *Waste-disposal site where the refuse is buried above the water table over a fractured rock aquifer. Potential for serious pollution is low because leachate is partially degraded by natural filtering as it moves down to the water table. (Source: W. J. Schneider, U.S. Geological Survey Circular 601F, 1970.)*

plains likely to be periodically inundated by surface water should not be considered as acceptable sites for refuse disposal. Sixth, any permeable material with a high water table is probably an unfavorable site. Last, in rough topography, the best sites are near the heads of gullies where surface water is at a minimum. It is emphasized that while the above general criteria are useful, they are not intended to preclude a hydrogeological investigation including drilling to obtain samples, permeability testing, and other tests to predict the movement of leachate from the buried refuse (4).

Once a site is chosen for a sanitary landfill, monitoring of the movement of groundwater should begin before filling commences. After the operation starts, continued monitoring of the movement of leachate and gases should be continued as long as there is any possibility of pollution. This is particularly important after the site is completely filled and the permanent cover material is in place because there is always a certain amount of settlement after a landfill is completed, and if small depressions form, then surface water may collect, infiltrate, and produce leachate. Therefore, monitoring and proper maintenance of an abandoned landfill will reduce its pollution potential (4).

Hazardous waste pollutants from a solid-waste-disposal site may enter the environment by as many as six paths (Figure 10.9). First, compounds volatilized in the soil and fill after placement such as methane, ammonia, hydrogen sulfide, and nitrogen gases may enter the atmosphere. Second, heavy metals such as lead, chromium, and iron are retained in the soil. Third, soluble material such as chloride, nitrate, and sulfate readily pass through the fill and soil to the groundwater system.

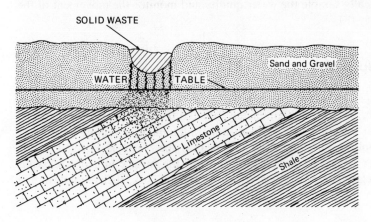

FIGURE 10.8 *Solid-waste disposal site where the waste is buried above the water table in permeable material in which leachate can migrate down to fractured bed rock (limestone). The potential for groundwater pollution may be high due to many open and connected fractures in the rock. (Source: W. J. Schneider, U.S. Geological Survey Circular 601F, 1970.)*

FIGURE 10.9 *Idealized diagram showing several ways that hazardous waste pollutants from a solid-waste disposal site may enter the environment.*

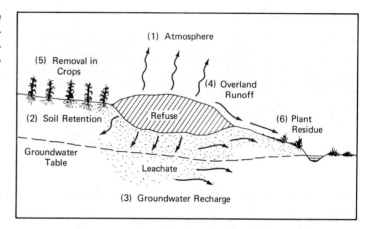

Fourth, overland runoff may also pick up leachate and transport it into the surface-water network. Fifth, some crops and cover plants growing in the disposal area may selectively take up heavy metals and other toxic materials to be passed up the food chain as people and animals eat them. Sixth, if the plant residue left in the field contains toxic substances, it will return these materials to the environment through soil-forming and runoff processes (8).

       A thorough monitoring program will consider all six possible paths by which pollutants enter the environment. Generally, atmospheric pollution by gas from landfills is not a problem, and this trend will probably continue as more and more food wastes are ground and flushed into the sewage system. Many landfills have no surface runoff, and therefore monitoring of surface water is not necessary. However, thorough monitoring is required if overland runoff does occur. Monitoring of soil and plants should include periodic chemical analysis at prescribed sampling locations. A minimum of nine sample sites is recommended. It is advisable to collect plant samples at harvest time or, if plants are left in the field, after they become dormant. Soil sample sites should be analyzed twice a year: in early summer and late fall. Monitoring of groundwater quality through observation wells is minimal if the landfill is in relatively impermeable soil overlying dense, permeable rock. In this case, leachate and groundwater movement may be less than one foot per year. However, if permeable, water-bearing zones exist in the soil or bed rock, then observation wells to periodically sample the water quality and monitor the movement of the leachate are necessary (8).

Deep-Well Disposal

In the practice of waste disposal by injection into deep wells, the term *deep* refers to rock, not soil, which is below and completely isolated from all freshwater aquifers, thereby assuring that injection of waste will not contaminate or pollute existing or potential water supplies. This generally means that the waste is injected into a permeable rock layer several thousands of feet below the surface in geologic basins confined above by relatively impervious, fracture-resistant rock, such as shale or salt deposits (1).

Deep-well injection of oil-field brine (salt water) has been important in the control of water pollution in oil fields for many years, and huge quantities of liquid waste (brine) pumped up with oil has been injected back into the rock. Today, more than one billion gallons per day are pumped into subsurface rocks (9). In recent years, the technique has been used more commonly to permanently store industrial waste deep underground, and by mid-1973, there were nearly 300 deep-well disposal units pumping industrial waste underground (9). A typical well is about 2000 feet deep, and wastes are pumped into a 200-foot-thick zone at a rate of about 100 gallons per minute (10).

Deep-well disposal of industrial wastes should not be viewed as a quick and easy solution to industrial waste problems (11). Even where geologic conditions are favorable for deep-well disposal, there are natural restrictions because there are a limited number of suitable sites, and within these sites there is limited space for disposal of waste. Possible injection zones of porous rock are usually already filled with natural fluids, usually brackish or brine water. Therefore, to pump in waste, some of the natural fluid must be displaced by compression (even very slight amounts of compression of the natural fluids in a large volume of permeable rock can provide considerable storage space) and by slight expansion of the reservoir rock as the waste is being injected (10).

The most favorable sites from a geologic consideration are synclinal basins and coastal plains because they generally contain thick sequences of sedimentary rocks bearing salt water. Figure 10.10 shows geologic features significant in evaluation of sites for deep waste-injection wells and location of some existing injection systems. The most favorable basins are tectonically stable and range from tens to hundreds of miles in width. Since natural flow of fluids is slow, less than about 3 feet per year, deep-well disposal should be restricted vertically and unrestricted horizontally (1).

Problems Associated with Deep-Well Disposal   Several problems associated with disposal of liquid waste in deep wells have been reported (10, 11). Perhaps the best known are the earthquakes of 1962 to 1967 caused by injection of waste from the Rocky Mountain Arsenal near Denver, Colorado. The injection zone was fractured gneiss at a depth of 12,000 feet, and the increased fluid pressure evidently initiated movement along the fractures. This is not a unique case. Similar initiations of earthquakes have been reported in oil fields in western Colorado, Texas, and Utah (10). A similar activation of faults in Southern California caused by injection of fluids into the Inglewood oil field for secondary recovery is thought to have contributed to the failure of the Baldwin Hills Reservoir. In a different sort of incident, in 1968 a disposal well on the shore of Lake Erie in Pennsylvania "blew out," releasing several millions of gallons of harmful chemicals from a paper plant into the lake (11). Regardless of these incidents, most oil-field brine and industrial disposal wells are functioning as designed and are not degrading the near-surface environment.

Feasibility and General Site Considerations for Disposal Wells   The feasibility of deep-well injection as the best solution to a disposal problem depends on four factors: the geologic and engineering suitability of the proposed site; the volume and

FIGURE 10.10 *Geologic features significant in deep-well disposal. Notice that most of the industrial-waste injection systems (wells) are located in geologic basins or on coastal plains. Both of these environments are characterized by thick sequences of sedimentary rocks. (Source: D. L. Warner, American Association of Petroleum Geologists Memoir 10, 1968.)*

the physical and chemical properties of the waste; economics; and legal considerations (12).

Geologic consideration for disposal wells are twofold (12). First, the injection zone must have sufficient porosity, thickness, permeability, and areal extent to insure safe injection. Sandstone and fractured limestone are the commonly used reservoir rock (11). Second, the injection zone must be below the level of fresh-water circulation and confined by relatively impermeable rock such as shale or salt, as shown in Figure 10.11.

Optimal use of limited underground storage space is realized when deep-well injection is used only when more satisfactory methods of waste disposal are not available and when the volume of injected wastes are minimized by good waste management (12). Such use includes a thorough evaluation of the physical and chemical properties of the waste to insure that it does not adversely affect the ability of the rock reservoir to accept waste. This may be minimized in some instances by preinjection treatment of the waste to enhance compatibility with the reservoir rock and natural fluids present there. It is also advisable to take advantage of natural

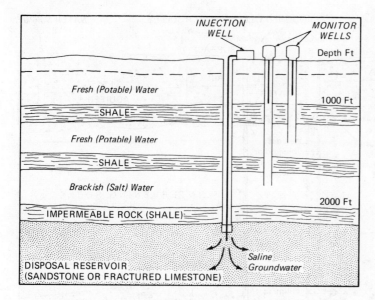

INJECTION WELL

MONITOR WELLS

Depth Ft

Fresh (Potable) Water

1000 Ft

=SHALE=

Fresh (Potable) Water

=SHALE=

Brackish (Salt) Water

2000 Ft

=IMPERMEABLE ROCK (SHALE)=

DISPOSAL RESERVOIR
(SANDSTONE OR FRACTURED LIMESTONE)

Saline Groundwater

FIGURE 10.11 *Idealized diagram showing a deep-well injection system. The disposal reservoir is a sandstone or fractured limestone capped by impermeable rock and isolated from all fresh water. Monitor wells are a safety precaution to insure that there is no undesirable migration of the liquid waste into either the above formation or fresh water.*

buffers. For example, if the waste is acidic, the use of limestone as a reservoir rock may be advantageous because the acid waste tends to increase the permeability of the reservoir by chemically attacking and enlarging natural fractures, thus allowing more waste to be injected.

Construction and operating costs, particularly those related to geologic factors, often determine the feasibility of deep-well injection as the preferred method of waste treatment. Important geologic factors pertaining to construction costs are the depth of the well and the ease with which drilling proceeds. Operating costs depend on the permeability, porosity, and thickness of the injection zone, and the fluid pressure in the reservoir. All of these are important in determining the rate at which the reservoir will accept liquid waste (12).

Consideration of legal aspects of deep-well disposal suggests that existing laws, regulations, and policies are probably adequate. However, regulatory agencies do need more funds and trained personnel to successfully regulate the growing number of disposal wells (9).

Monitoring of Disposal Wells An essential part of any disposal system is monitoring. It is very important to know exactly where the wastes are going, how stable they are, and how fast they are migrating. It is especially important in deep-well disposal where toxic or otherwise hazardous materials are involved.

Effective monitoring requires that the geology be precisely defined and mapped before initiation of the disposal program. Especially important is the locating of all freshwater-bearing zones and old or abandoned oil or gas wells that might allow the waste to migrate up to freshwater aquifers or the surface (Figure 10.12). A system of deep observation-wells drilled into the disposal reservoir in the vicinity of the well can monitor the movement of waste, and shallow observation-wells drilled into freshwater zones can monitor the water quality to quickly identify any upward migration of the waste.

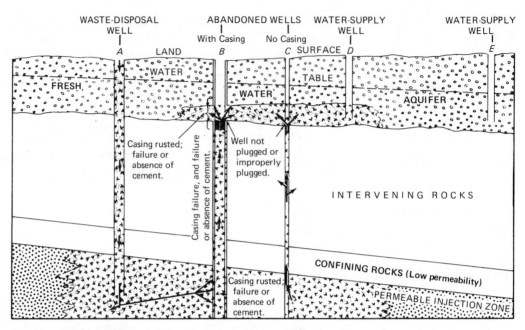

FIGURE 10.12 *Idealized diagram showing how liquid waste might enter a freshwater aquifer through abandoned wells. This diagram illustrates why all abandoned wells should be located, and it emphasizes the necessity of monitoring wells. (Source: Irwin and Morton, U.S. Geological Survey Circular 630, 1969.)*

Radioactive Waste Management

It is predicted that by 1985, uranium will fuel nearly 30 percent of all domestic electrical power generation, and, as shown in Figure 10.13, the nuclear fuel cycle necessary to maintain the industry will generate waste with varying degrees of contamination in several areas of the cycle. Considering waste-disposal procedures, radioactive wastes may be grouped into two general categories: *high-level* wastes and *low-level* wastes.

High-level wastes include gaseous, liquid, or solid wastes from reprocessing highly irradiated nuclear reactor fuels. These wastes are highly radioactive, generate considerable heat for a thousand years or so, and need to be isolated from the biosphere for a quarter of a million years or longer before they no longer pose a threat to life (13).

Low-level wastes contain sufficiently low concentrations or quantities of radioactivity that they do not present a significant environmental hazard if properly handled. Included here are a wide variety of items such as residuals or solutions from chemical processing; solid or liquid plant wastes, sludges, and acids; and slightly contaminated equipment, tools, plastic, glass, wood, fabric, and other materials (13).

Hazardous radioactive materials produced from nuclear reactors include fission products such as krypton-85 (half-life of 10 years), strontium-90 (half-life of 28 years), and cesium-137 (half-life of 30 years). The half-life is the time required for the radioactivity to be reduced to one-half its original level. Generally, at least ten half-lives (and preferably more) are minimal before a material is no longer considered a health hazard. Therefore, a mixture of the above fission products will require hundreds of years of confinement from the biosphere.

Reactors also produce a small amount of plutonium-239 (half-life of 24,000 years), which is a man-made element that may be reprocessed into a fuel for reactors. Unfortunately, it can also be used to produce nuclear weapons. Although most plutonium-239 is extracted along with uranium at reprocessing plants (Figure 10.13), some (less than one ppm) plutonium remains with the fission products to become the high-level wastes. Because plutonium and the fission products must be isolated from the biological environment for very long periods of time, their permanent disposal is basically a geologic problem.

Storage and Disposal of High-Level Wastes   The philosophy for management of high-level waste is "concentrate and contain." The majority of high-level wastes, which totaled approximately 85 million gallons in the United States in 1973, were contained at three federal repositories. Of this, 64 million gallons, or about 75 percent, were stored at the Hanford Reservation near Richland, Washington. At Hanford, the wastes were stored in 151 underground tanks that varied in capacity from 55,000 to 1,000,000 gallons (14).

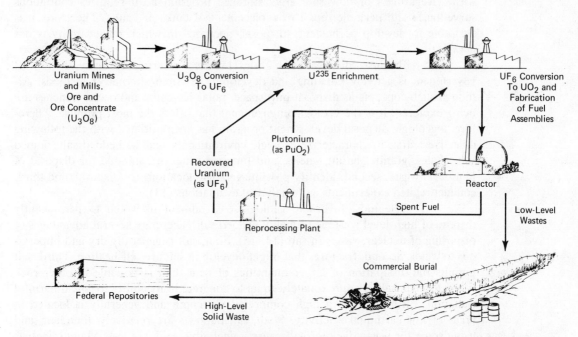

FIGURE 10.13   *Idealized diagram showing the nuclear fuel cycle. (Source:* The Nuclear Industry: 1974, *Office of Industry Relations, 1974.)*

Before high-level liquid wastes are stored or disposed of, it is desirable to convert them to a dry, granular or solid salt cake which is easier to handle and less likely to cause environmental problems. At Hanford, high-level wastes that are self-boiling (that is, the wastes produce heat that boils the liquid) are first treated to remove the strontium-90 and cesium-137, which produces a low heat, nonboiling residue that is aged from 3 to 5 years. At the end of that time, it is evaporated to form a salt cake. The separated cesium and strontium, as of 1973, are being stored as liquid concentrate until facilities to solidify, encapsulate, and store them become fully operational (14).

Research to discover and develop other ways of solidifying liquid waste, such as fixing it in a monolithic, highly insoluble matrix of silicate glass is progressing (13), but regardless of how it is done, commercial processors are required to convert high-level liquid waste to stable solid waste before transporting it to a federal repository.

Most serious problems with radioactive wastes have occurred while the wastes were in a liquid form. Sixteen leaks involving 350,000 gallons were located at Hanford from 1958 to 1973. An incident in 1973 involved a leak of 115,000 gallons of low-temperature waste. Since then, improvements such as stronger, double-shelled storage tanks, reduction of the volume of liquid waste stored through the solidification program, and increased reserve capacity will hopefully reduce the chance of future incidents (14).

Storage of high-level waste is at best a temporary solution that allows the federal government to meet its commitments for accepting solidified commercial waste. Regardless of how safe any "storage" program is, it requires continuous surveillance and periodic repair or replacement of tanks or vaults. Therefore, it is desirable to develop permanent "disposal" methods in which retrievability is not necessary.

Because "disposal" of high-level waste is a certain necessity, the federal government is actively pursuing and developing possible alternative methods. Although such concepts as disposal into space, polar ice caps, and ocean trenches are being explored, it is the geologic environment that offers the most promise. Therefore, a geologic disposal development program has been initiated with the following objectives: first, to characterize geologic environments such as bedded salt, domed salt, shales, granite, basalt, gneiss, and limestone that are suitable for disposal of high-level waste; second, identify possible geologic locations for disposal; and third, conduct related experiments and analytical evaluations (13).

The most favorable geologic environment in which to permanently dispose of high-level waste is probably bedded salt. There are several advantages of disposing of nuclear wastes in salt (14, 15). First, salt is generally dry and impervious to water. Second, fractures that might develop in salt are self-healing. Third, salt permits the dissipation of larger quantities of heat than is possible in other rock types. Fourth, salt is approximately equal to concrete in its ability to shield harmful radiation. Fifth, salt has a high compressive strength and generally is located in areas of low earthquake activity. Sixth, salt deposits are relatively abundant, and using some for waste disposal will cause a negligible resource loss. Many scientists have studied the feasibility of salt as a disposal site for high-level wastes, and the general conclusion is that the salt should effectively isolate the waste from the biosphere for hundreds of thousands of years (14).

Significant salt deposits underlie about 400,000 square miles of the United States (Figure 10.14). However, the salt best suited for waste disposal is that located within 2000 feet from the surface, with a minimum thickness of 200 feet. Bedded salt meeting these criteria are located in Michigan, New York, and Kansas (Figure 11.14) (14).

The government has considered disposal of radioactive waste in salt since 1955, and during 1966 and 1967 it initiated the Project Salt Vault in which high-level solid-waste disposal procedures were simulated in abandoned salt mines in Kansas. The project was considered a success, and in 1970 it was announced that the abandoned Carey Salt Mine near Lyons, Kansas, was to be used as a demonstration site for waste disposal and, if successful, the site would eventually become the National Radioactive Waste Repository. Figures 10.15, 10.16, and 10.17 show the proposed site and some of the equipment used in Project Salt Vault. However, the Lyons site has been abandoned because of scientific, public, and state concern over two factors: The first is *geologic uncertainty*—several thousand abandoned mine shafts and drill holes in the area might allow the waste to migrate up the shafts or drill holes. The second is *land-use conflicts*—present land use of partly urbanized to farming and grazing land is in conflict with the proposed disposal plan (14). Investigation into other sites in Kansas and New Mexico is under way, and it is expected

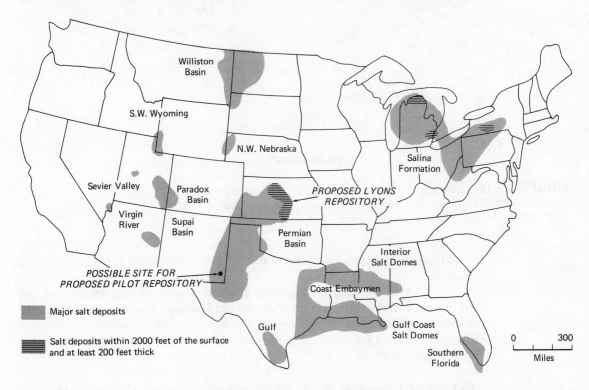

FIGURE 10.14 *Major salt deposits in the United States. In some of these areas we may eventually permanently dispose of high-level radioactive waste. (Source: P. P. Micklin,* Science and Public Affairs *30, 1974.)*

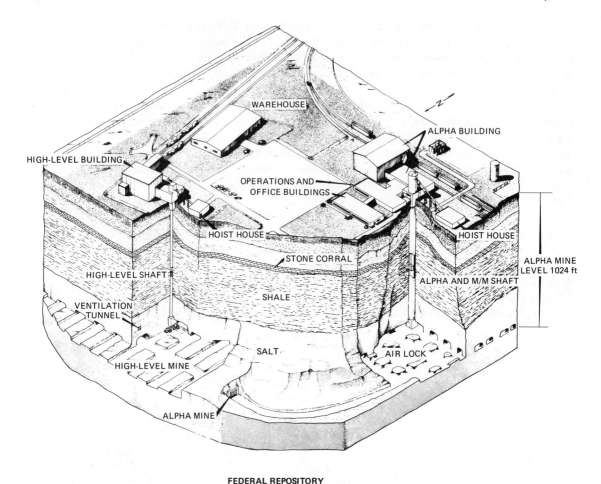

FEDERAL REPOSITORY

FIGURE 10.15 *Schematic diagram showing how under-ground bedded salt may be used as a disposal site for solidi-fied, high-level radioactive wastes. (Courtesy of ERDA.)*

that a pilot waste-disposal program in bedded salt or salt domes is likely in the near future.

Consideration is also being given to disposal of high-level waste into bed rock at several other federal repositories. At Hanford, Washington, disposal into caverns mined in basalt deep under the site is being studied; and at the Savannah River site in South Carolina, the possibility of waste disposal in metamorphic rock is being evaluated (16).

Disposal of Low-Level Waste   The philosophy for management of low-level wastes is "dilute and disperse." Experience suggests that low-level radioactive waste can be buried safely in carefully controlled and monitored near-surface burial areas in which the hydrologic and geologic conditions severely limit the migration of

FIGURE 10.16 *Underground salt mine near Lyons, Kansas, where experiments to determine the possibility of disposing of high-level radioactive wastes were completed. (Photo courtesy of ERDA.)*

radioactivity. Monitoring at existing acceptable sites has confirmed that essentially no migration of radioactivity of any significant health or safety hazard has occurred (13).

Three commercial radioactive-waste-disposal companies are now operating six burial sites on federal and state land (Figure 10.18). Table 10.3 lists

Year	Cubic Feet
1962	36,000
1963	215,000
1964	447,000
1965	490,000
1966	503,000
1967	774,000
1968	667,000
1969	751,000
1970	995,000
1971	1,207,000
1972	1,335,000
1973	1,674,000
1974 (estimated)	1,800,000

TABLE 10.3 *Private disposal of low-level commercial waste from 1962 to 1974.*

SOURCE: *The Nuclear Industry: 1974,* Office of Industry Relations, 1974.

FIGURE 10.17 *How canisters of solidified, high-level wastes will be transported underground to a disposal site by specially designed equipment (a) and experimental chambers for emplacement of highly radioactive wastes 100 feet underground in salt (b). (Photos courtesy of ERDA.)*

(a)

(b)

FIGURE 10.18 *Map showing commercial burial sites for low-level solid waste in the United States. (Source: The Nuclear Industry: 1974, Office of Industry Relations, 1974.)*

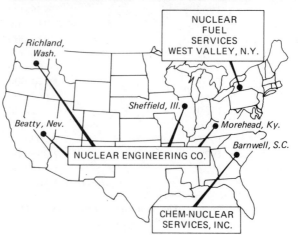

the quantities of private low-level waste buried from 1962 to 1974. The annual volume of low-level waste delivered to the commercial burial sites is expected to reach two million cubic feet in 1975, four million cubic feet in 1980, and six million cubic feet in 1985 (13).

Ocean Dumping

The oceans of the world, 140 million square miles of water, cover over 70 percent of the earth and play a part in maintaining the world's environment by providing the water necessary to maintain the hydrologic cycle, contributing to the maintenance of the oxygen-carbon-dioxide balance in the atmosphere, and affecting global climate. In addition, the oceans are very valuable to man, providing such necessities as foods and minerals.

It seems reasonable that such an important resource as the ocean could receive preferential treatment, and yet in 1968 alone, over 48 million tons of waste at 246 sites were dumped off the coasts of the United States. Federal legislation and regulation by the Environmental Protection Agency in 1972 reduced the number of sites to 118, but ocean dumping continues to degrade the oceanic environment (2). Furthermore, if population growth in coastal regions continues as expected, increased amounts of waste will inevitably result that might end up in ocean dump sites.

The types of wastes that have been dumped in the oceans off the United States include the following: *dredge spoils*—solid materials such as sand, silt, clay, rock, and pollutants deposited from industrial and municipal discharges that are removed from the bottom of water bodies generally to improve navigation; *industrial waste*—acids, refinery wastes, paper mill wastes, pesticide wastes, and assorted liquid wastes; *sewage sludge*—solid material that remains after municipal waste water treatment; *construction and demolition debris*—cinder block, plaster, excavation dirt, stone, tile, etc.; *solid waste*—refuse, garbage or trash; *explosives;* and *radioactive waste* (17).

Figure 10.19 shows the known sites that have been used for ocean dumping off the United States coast, and Table 10.4 summarizes major aspects of ocean dumpings by type and amount for 1968. In that year, approximately 80 percent by weight (over 38 million tons) of all ocean dumping was dredge spoils, 34 percent of which was polluted by agricultural, municipal, and industrial wastes (Table 10.5). It is emphasized that this list of sites and amount of dumping is not complete, and little information is recorded for sites off Alaska, Hawaii, and more than 12 miles off the shore of the contiguous United States. Nevertheless, the sites and volume of dumping are significant for they indicate a rough magnitude of the intensity of recent ocean dumping (17).

The 1972 Marine Protection Research and Sanctuaries Act prohibits ocean dumping of radiological, chemical, and biological warfare agents and any high-level radioactivity waste. Furthermore, it provides for regulation of all other waste disposal in the oceans off the United States by the Environmental Protection

---

The section on ocean dumping is summarized from *Ocean Dumping: A National Policy* (Washington, D.C.: Council on Environmental Quality, 1970), pp. 1–25.

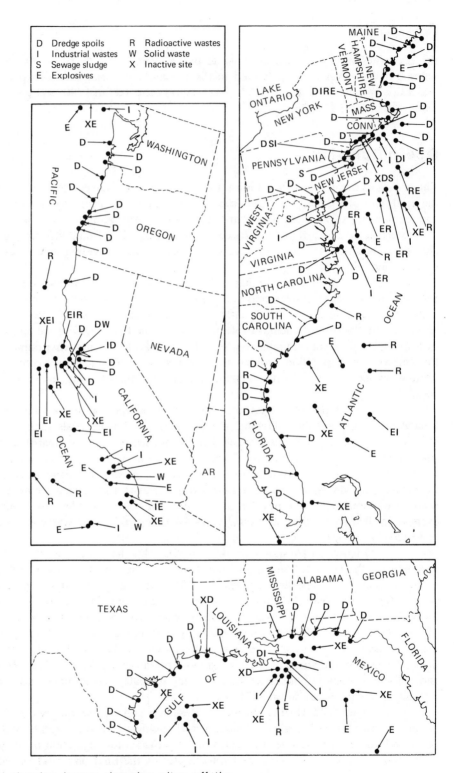

FIGURE 10.19 *Map showing known dumping sites off the United States coasts. (Source:* Ocean Dumping: A National Policy, *Council on Environmental Quality, 1974.)*

TABLE 10.4 *Generalized classification of waste dumped in the ocean in 1968 (amounts given in tons).*

Waste Type	Atlantic	Gulf	Pacific	Total	Percent of Total
Dredge spoils	15,808,000	15,300,000	7,320,00	38,428,000	80
Industrial wastes	3,013,200	696,000	981,300	4,690,500	10
Sewage sludge	4,477,000	0	0	4,477,000	9
Construction and demolition debris	574,000	0	0	574,000	<1
Solid waste	0	0	26,000	26,000	<1
Explosives	15,200	0	0	15,200	<1
Total	23,887,400	15,966,000	8,327,300	48,210,700	100

SOURCE: *Ocean Dumping: A National Policy,* Council on Environmental Quality, 1970.

Agency or in the case of dredge spoil, the U.S. Army Corps of Engineers. In addition to materials prohibited by law from being dumped, the Environmental Protection Agency has prohibited material whose effect on marine ecosystems cannot be determined; persistent inert materials that float or remain suspended, unless they are processed to insure that they sink and remain on the bottom; and material containing more than trace concentrations of mercury and mercury compounds, cadmium and cadmium compounds, organohalogen compounds (organic compounds of chlorine, fluorine, and iodine, etc.) and compounds that may form from such substances in the oceanic environment, crude oil, fuel oil, heavy diesel oil, lubricating oils, and hydraulic fluids. The Environmental Protection Agency also requires special permits and strictly regulated dumping of arsenic, beryllium, chromium, and lead; low-level nuclear waste; organo-silicon compounds; organic and inorganic processing wastes, including cyanide, fluoride, and chlorine; oxygen-demanding waste; organic chemicals; and petrochemicals. A major objective of the act is to prevent ocean dumping of wastes that before recent federal laws were enacted were discharged into rivers and the atmosphere (2).

TABLE 10.5 *Estimation of the extent of dredge spoils polluted.*

Coastal Area	Total Spoils (Tons)	Estimated Percent of Total Polluted Spoils [a]	Total Polluted Spoils (Tons)
Atlantic Coast	15,808,000	45	7,120,000
Gulf Coast	15,300,000	31	4,740,000
Pacific Coast	7,320,000	19	1,390,000
Total	38,428,000	34	13,250,000

SOURCE: *Ocean Dumping: A National Policy,* Council on Environmental Quality, 1970.
[a] Estimates of polluted dredge spoils consider chlorine demand; BOD; COD; volatile solids; oil and grease; concentrations of phosphorus, nitrogen, and iron; silica content; and color and odor of the spoils.

Ocean Dumping and Pollution Problems   Ocean dumping contributes to the larger problem of ocean pollution, which has seriously damaged the marine environment and caused a health hazard to people in some areas. Shellfish have been found to contain pathogens such as polio virus and hepatitis, and at least 20 percent of the nation's commercial shellfish beds have been closed due to pollution. Beaches and bays have been closed to recreational uses. Lifeless zones in the marine environment have been created. There have been heavy kills of fish and other organisms as well as profound changes in marine ecosystems (17).

Major impacts of marine pollution on oceanic life include the following: killing or retarding of growth, vitality, and reproductivity of marine organisms due to toxic pollutants; reduction of dissolved oxygen necessary for marine life due to increased oxygen demand from organic decomposition of wastes; biostimulation by nutrient-rich waste, causing excessive blooms of algae in shallow waters of estuaries, bays, and parts of the continental shelf, resulting in depletion of oxygen and subsequent killing of algae that may wash up and pollute coastal areas; and habitat change caused by waste-disposal practices that subtly or drastically change entire marine ecosystems (17).

Major impacts on people and society caused by marine pollution include the following: production of a public health hazard caused by marine organisms transmitting disease to people; loss of visual and other amenities as beaches and harbors become polluted by solid waste, oil, and other materials; and economic loss. In 1969 alone, loss of shellfish due to pollution in the United States was $63 million. In addition, a great deal of money is being spent cleaning solid waste, liquid waste, and other pollutants in the coastal areas (17).

Alternatives to Ocean Dumping   The ultimate solution to the problem of ocean dumping is development of economically feasible and environmentally safe alternatives. To emphasize some of the possibilities, we will discuss some of the alternatives to ocean dumping of dredge spoils and sewage sludge.

Disposal of dredge spoils represents the vast majority of all ocean dumping. They are removed primarily to improve navigation and are usually redeposited only a few miles away. Approximately 1/3 is seriously polluted with heavy metals such as cadmium, chromium, lead, and nickel (Table 10.6), as well as other industrial, municipal, and agricultural wastes. As a result, disposal in the marine environment can be a significant source of pollution (17).

TABLE 10.6 *Concentrations of heavy metals in dredge spoils.*

Metal	Concentrations in Dredge Spoils (ppm)	Natural Concentrations in Seawater (ppm)	Concentrations Toxic to Marine Life (ppm)
Cadmium	130	0.08	0.01–10.0
Chromium	150	0.00005	1.0
Lead	310	0.00003	0.1
Nickel	610	0.0054	0.1

Source: *Ocean Dumping: A National Policy,* Council on Environmental Quality, 1970.

The long-range alternative to disposal of dredge spoils is phasing out ocean disposal of polluted dredge spoils. However, at present, this is not possible because of the great volume involved. Until land base disposal can handle the necessary volume, interim techniques should be developed, such as determining which spoils are polluted before disposal and then hauling the polluted spoils further from the dredging site to a safe disposal site. The main disadvantage here is the increased cost of longer hauls (Table 10.7) (17).

Environmental problems from disposal of sewage sludge at sea are important in terms of their volume, toxic and possible pathogenetic materials involved, and possible effects on marine life. Table 10.8 shows the minimum, average, and maximum concentrations of three heavy metals found in sewage sludge compared to normal concentrations in seawater and concentrations toxic to marine life. These data suggest that sewage sludge, like polluted dredge spoils, may be very harmful to marine life. Based on a very conservative estimation of 0.12 pound of sludge generated per person per day, by the year 2000 over 2 million tons of sludge will be generated in the coastal regions, representing an increase of 50 percent in 30 years. Most sewage sludge is currently disposed of on land or is incinerated, but in some areas, tremendous quantities are disposed of by ocean dumping. For example, in 1968, approximately 200,000 tons of sludge were disposed of at sea compared with about 3 million tons which were disposed of on land (17).

Alternatives to ocean dumping of sewage sludge in coastal regions include various methods of land disposal. Table 10.9 summarizes the cost of land-based sewage disposal compared to that of near-shore ocean dumping. These figures show that land disposal is considerably more expensive than near-shore marine disposal. However, if the sludge is digested (treated to control odors and pathogens)

TABLE 10.7 *Cost of hauling sewage sludge by barge.*

City	Distance (Miles)	Cost per Ton-Mile[a]	Cost per Ton[a]
New York City	25	$0.30	$7.50
Elizabeth, Md.	30	0.23	6.90
Baltimore, Md.	230	0.08	18.40
Philadelphia, Pa.	300	0.04	12.00

SOURCE: *Ocean Dumping: A National Policy,* Council on Environmental Quality, 1970.
[a] 1970 dollars.

TABLE 10.8 *Concentration in parts per million (ppm) of heavy metals in sewage sludge.*

Metal	Concentrations in Sewage Sludge (ppm)			Natural Concentrations in Sea-water (ppm)	Concentrations Toxic to Marine Life (ppm)
	Min.	Avg.	Max.		
Copper	315	643	1980	0.003	0.1
Zinc	1350	2459	3700	0.01	10.0
Manganese	30	262	790	0.002	—

SOURCE: *Ocean Dumping: A National Policy,* Council on Environmental Quality, 1970.

TABLE 10.9 *Estimated cost (1970 dollars) of disposal of sewage sludge on land compared to cost of ocean dumping.*

Location	Method	Cost per Ton
Land	Digestion and lagoon storage (Chicago)	$45
	Digestion and land disposal [a]	22
	Composting	35–45
	Processing into granular fertilizer (net cost)	35–50
	High-temperature incineration	35–60
Ocean	Barging undigested sludge	3–18
	Barging digested sludge	8–36
	Piping disposal	12–30

SOURCE: *Ocean Dumping: A National Policy,* Council on Environmental Quality, 1970.
[a] At Chicago, with a 7-mile pipeline to the land-disposal site.

or barged long distances from shore, the cost becomes comparable, and land disposal may even be cheaper (17).

Perhaps the best long-range alternative to ocean dumping is the use of digested sludge for land reclamation, especially strip-mined land reclamation. Many strip mines in other areas are in great need of reclamation, and the nutrient value and organic material in this sludge can improve lands with undesirable low-productivity soils by improving soil texture, composition, and nutrient levels.

Septic-Tank Sewage Disposal

Population movement in the United States continues to be from rural to urban, or urbanizing, areas. In many cases, construction of an adequate sewage system has not been able to keep pace with growth. As a result, the individual septic tank (soils absorption system) which has been used for many years in both rural and urban areas continues to be an important method of sewage disposal for millions of Americans. However, all land is not suitable for installation of a septic-tank disposal system, and evaluation of individual sites is necessary and often required by law before a permit to construct a subsurface sewage disposal system can be issued. Furthermore, it is emphasized that the most satisfactory method of sewage disposal is the use of municipal sewers and sewage treatment facilities.

The basic parts of a septic-tank disposal system are shown in Figure 10.20. The sewer line from the house leads to an underground septic tank in the yard. The tank is designed to separate solids from the liquid, digest and store organic matter through a period of detention, and allow the clarified liquids to discharge into the seepage bed, or absorption field, which is a system of piping through which treated sewage may seep into the surrounding soil.

Geologic factors that affect the suitability of a septic tank disposal system include type of soil, depth to the water table, depth to bed rock, and topography. These variables are generally listed with soil descriptions associated with a soil survey of a county or other area. Soil surveys are generally published by the Soil

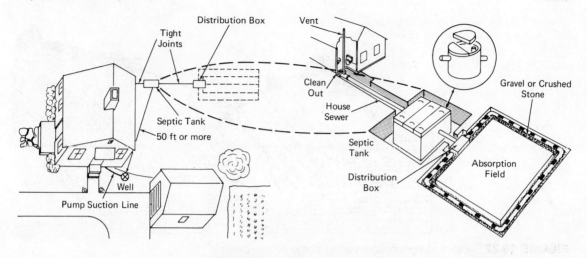

FIGURE 10.20 *Generalized diagram showing septic-tank sewage disposal system (above) and location of the absorption field with respect to the house and well. (Source: Indiana State Board of Health.)*

Conservation Service and, when available, are extremely valuable in interpreting possible land use, such as suitability for a septic system. However, since the reliability of a soils map for predicting limitations of soils is limited to an area larger than an acre or so, and because soil types can change in a few feet, it is often desirable to

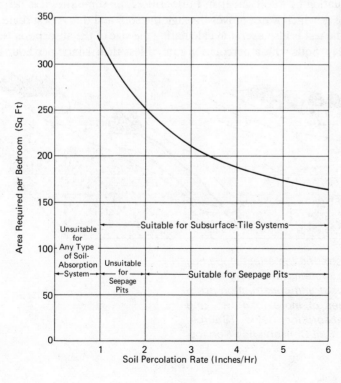

FIGURE 10.21 *Graph used to determine the size of absorption field needed for a septic system for a private residence (Source: Agriculture Information Bulletin 349, Soil Conservation Service, 1971.)*

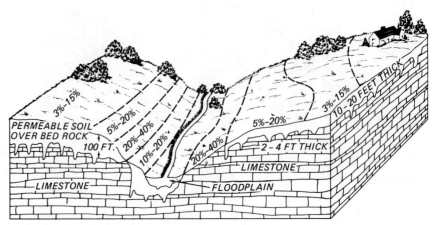

FIGURE 10.22 *Block diagram illustrating some of the limita-tions of a septic system due to bed rock and soil cover. The deep soil over the limestone at the left has moderate limitations, and construction may be difficult on the steeper slopes. The soil that is 10 to 20 feet thick on the extreme right may not be suitable for a septic system if the water supply is to come from a well. The shallow soil 2 to 4 feet thick has severe limitations, and if a septic system is placed there, the stream at the bottom of the hill may become polluted. (Source: Agriculture Informa-tion Bulletin 349, Soil Conservation Service, 1971.)*

have an on-site evaluation by a soil scientist. Furthermore, on some sites it is helpful to determine the rate at which water moves through the soil, and this is best done by a percolation test. The test is also useful in calculating the size of the absorption field needed (Figure 10.21). Soils with a percolation rate of less than 1 inch per hour are

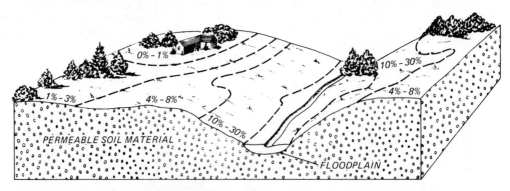

FIGURE 10.23 *Block diagram illustrating the effect slope has on usage of a septic system. Septic systems work very well in deep, permeable soils. However, layout and construction prob-lems may be encountered on slopes of more than 15°, and floodplains are not suitable for absorption fields. (Source: Agriculture Information Bulletin 349, Soil Conservation Service, 1971.)*

FIGURE 10.24 *Effluent from a septic-tank sewage disposal system rising to the surface in a backyard. The soil is poorly drained, and septic systems in such soils will not function well during wet weather. (Photo by W. B. Parker, courtesy of U.S. Department of Agriculture.)*

always undesirable for a septic-tank disposal system. Other criteria are as follows: The maximum seasonal elevation of the groundwater table should be at least 4 feet below the bottom of the absorption field. Bed rock should be at a depth greater than 4 feet below the absorption field. The surface slope should generally be less than 15 percent, or a 15-foot drop per 100-foot horizontal distance (18, 19). Figures 10.22 and 10.23 demonstrate some of these limitations.

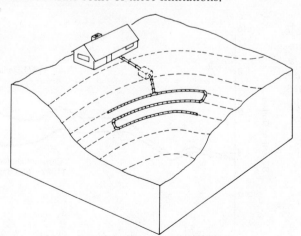

FIGURE 10.25 *Block diagram illustrating how a septic field may be constructed on a steeper slope. The tile lines are laid out on the contour. This type of configuration is necessary on most sloping fields or in fields where there may be a change in soil type. (Source: Agriculture Information Bulletin 349, Soil Conservation Service, 1971.)*

Sewage absorption fields may fail for several reasons. The most common cause is poor soil drainage (Figure 10.24), which allows effluent to rise to the surface in wet weather. Common reasons for poor drainage are the existence of the following: clay or compacted soils with low permeability, high water table, impermeable rock near the surface, and frequent flooding. Another factor causing failure is steep topography. However, if the absorption field is laid on the slope contours, a steeper slope might be accommodated (Figure 10.25) provided there are no horizontal, impervious clay layers in the soil near the surface. Such layers allow the effluent to flow above them until it is discharged on the surface and runs unfiltered down the slope (18).

Waste-Water Treatment

Waste-water treatment is one of the greatest expenses that state and municipal governments pay for environmental purposes. In 1972, the bill was over $3 billion. Figure 10.26 shows the per capita state and local investments for waste-water collection and treatment facilities from 1958 to 1971.

The main purpose of waste-water treatment is to reduce the amount of suspended solids, bacteria, and oxygen-demanding materials in waste water. In addition, new techniques are being developed to remove harmful dissolved inorganic materials that may be present. However, these methods will most likely reinforce rather than replace present technology (20).

Existing waste-water treatment generally falls into two broad classes (Figure 10.27): *primary treatment,* involving removal of grit, screening, grinding, flocculation and sedimentation; and *secondary treatment,* involving controlled biological assimilation and degradation processes that occur in nature by microorganisms. The name *tertiary treatment* is sometimes used for chemical purification or other additional treatment measure designed to further purify the water (20).

The most troublesome aspect of waste-water treatment is the handling and disposal of sludge produced in the treatment process. The amount of sludge produced is conservatively estimated at about 0.12 to 0.25 pound per person per

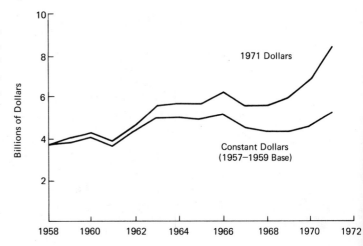

FIGURE 10.26 *Per capita state and local investment for waste-water collection and treatment facilities from 1958 to 1971. (Source: Environmental Quality — 1973, Council on Environmental Quality, 1973.)*

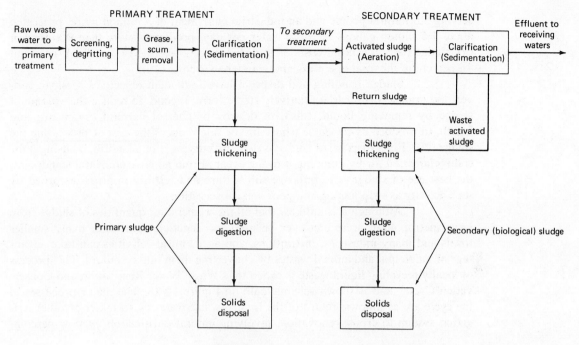

FIGURE 10.27 *Flow chart showing various procedures in primary and secondary treatment of waste water. (Reprinted from* Cleaning Our Environment—The Chemical Basis for Action, *a report by the Subcommittee on Environment Improvement, Committee on Chemistry and Public Affairs, American Chemical Society, 1969, p. 107. Reprinted by permission of the copyright owner.)*

day, and disposal of it accounts for 25 to 50 percent of the capital and operating cost of a treatment plant. An example from the Metropolitan Sanitary District of Greater Chicago emphasizes the problem. Waste-water treatment there in the late 1960s,

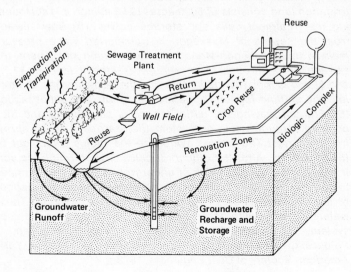

FIGURE 10.28 *Block diagram showing the waste-water renovation and conservation cycle. (Source: Parizek and Myers, American Water Resources Association, 1968.)*

serving 5.5 million people and an industrial equivalent of 3 million more, processed about 1.5 billion gallons of waste water per day, producing more than 900 tons of dry sludge daily. Disposal costs of this sludge were about $14.5 million, or about 46 percent of the total operating and maintenance costs (20).

Sludge handling and disposal have four main objectives: first, to convert the organic matter to a relatively stable form; second, to reduce the volume of sludge by removing liquid; third, to destroy or control harmful organisms; and fourth, to produce by-products whose use or sale reduces the cost of processing the sludge (20). Final disposal of sludge is done by burying it in a landfill, by using it for soil reclamation, or by dumping it in the ocean. From an environmental standpoint, the best use of sludge is to improve soil texture and fertility in areas disturbed by such activity as strip mining and poor soil conservation.

Although it is unlikely that all the tremendous quantities of sludge from large metropolitan areas can ever be used for beneficial purposes, many smaller towns and many industries, institutions, and agricultural activities can take advantage of municipal and animal wastes by converting them into resources. This process of locally recycling liquid waste is called the "Waste Water Renovation and Conservation Cycle" and is shown schematically in Figure 10.28. The major processes in the cycle are as follows: return of the treated waste water by sprinkler or other irrigation system to crops; renovation, involving natural purification by slow percola-

FIGURE 10.29 *Opposite. (a) Two effluent-application areas, one near Muskegon, the other near Whitehall, will eliminate the need to discharge inadequately treated industrial and municipal wastes to surface waters. These two subsystems will replace four municipal treatment facilities [total flow, 11 mgd (41,500 m³/d)] and eliminate direct discharges of five industrial plants [combined flow of 19 mgd (71,800 m³/d)]. The effluent-on-land system will handle the 1992 requirements of the county, with a population of 170,000 generating a flow of 43.4 mgd (165,000 m³/d), including an industrial flow of 24 mgd (90,600 m³/d). Irrigation land consists of 6000 acres (24.3 million m²) surrounding the treatment cells and storage basins. (b) Raw sewage from the sewers of Muskegon is transported via a pressure main to this Muskegon-Mona Lake reclamation site. There, the sewage flows through three aerated treatment lagoons (1, 2, and 3) and is then diverted either into the large storage basin or into the settling lagoon (4) before going into the outlet lagoon (5). Effluent leaving the outlet lagoon is chlorinated and then pumped through a piping system that carries it to 55 spray-irrigation rigs. The circles symbolize the radius swept out by the spray rigs. After the effluent trickles down through the soil, it is collected by a network of underground pipes and pumped to surface waters. A series of berms around the lowland edges of the irrigation circles will contain storm-water runoff. These waters will pond within the circles until handled by the drainage system. [Reprinted, by permission, from E. I. Chaiken, S. Poloncsik, and C. D. Wilson,* Civil Engineering 42 *(New York: American Society of Civil Engineers, 1973.)]*

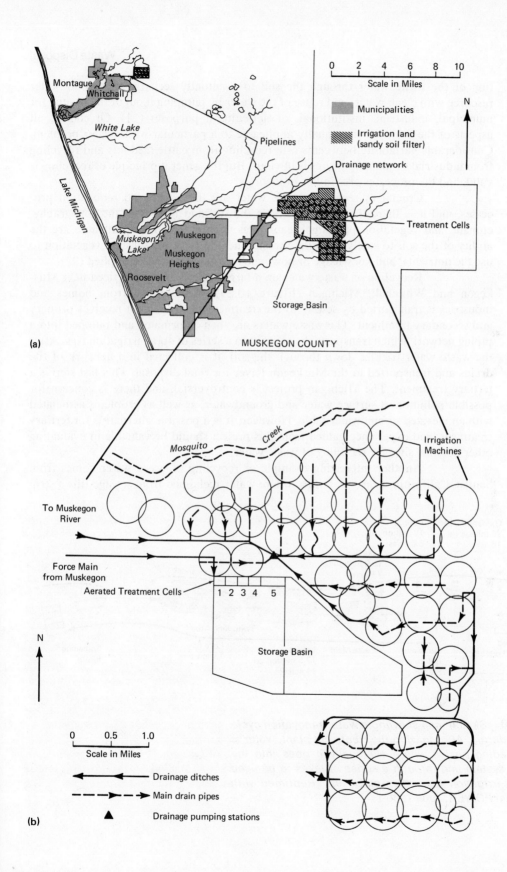

Montague
Whitchall

White Lake

Pipelines

Drainage network

Lake Michigan

Muskegon
Lake

Muskegon

Muskegon
Heights

Roosevelt

Storage Basin

Treatment Cells

MUSKEGON COUNTY

Scale in Miles
0 2 4 6 8 10

N

Municipalities

Irrigation land
(sandy soil filter)

(a)

Mosquito   Creek

Irrigation
Machines

To Muskegon
River

Force Main
from Muskegon

Aerated Treatment Cells

1 2 3 4 5

Storage Basin

N

0   0.5   1.0
Scale in Miles

Drainage ditches

Main drain pipes

Drainage pumping stations

(b)

261

tion of the waste water through the soil to eventually recharge the groundwater resource with clean water; and reuse of the water by pumping it out of the ground for municipal, industrial, institutional, or agricultural purposes (21). Of course, all aspects of the cycle are not equally applicable to a particular waste-water problem. Considerable variability between waste recycling from cattle feed lots and recycling from industrial or municipal sites is obvious. But the general principle of recycling is valid, and the processes are similar in theory.

Crucial to recycling waste water are the return and renovation processes, and it is there that such factors as the soil and rock type and topography, climate, and vegetation play significant roles. Particularly important here are the ability of the soil to safely assimilate waste; the ability of the selected vegetation to use the nutrients; and knowledge of how much waste water can be applied (22).

Recycling of waste water on a large scale is being practiced near Muskegon and Whitehall, Michigan (Figure 10.29). Raw sewage from homes and industries is transported by sewers to the treatment plant where it receives primary and secondary treatment. The waste waters are then chlorinated and pumped into a piping network which transports the effluent to a series of spray irrigation rigs. After the waste water trickles down through the soil, it is collected in a network of tile drains and transported to the Muskegon River for final disposal. This last step is a tertiary treatment. The Michigan project is controversial, and there is concern for possible pollution of surface water and groundwater, as well as problems associated with an elevated groundwater table. However, it is a possible alternative for tertiary treatment, and experience gained from this project should be valuable in evaluating other possible sites for recycling of waste water.

Another interesting example of recycling of waste water comes from Santee, California, near San Diego, where water reclaimed from sewage fills a strip

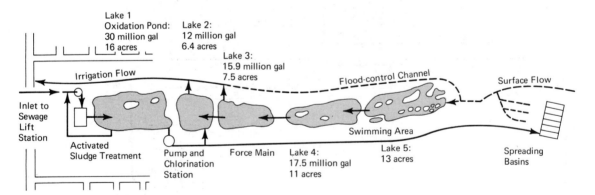

FIGURE 10.30 *Sketch map showing a water renovation cycle at Santee, California. After leaving the treatment plant, water is taken to spreading basins where eventually it goes into the surface-water system and through a series of lakes to be used eventually for irrigation and other purposes. (Reprinted with permission from POWER, June 1966.)*

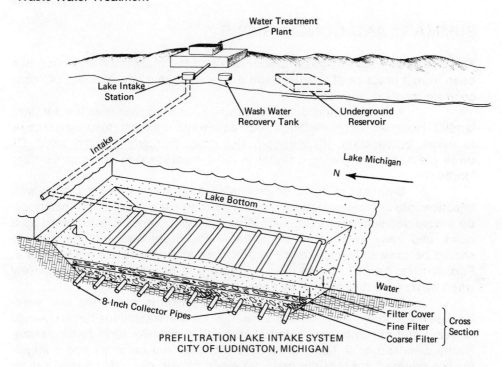

PREFILTRATION LAKE INTAKE SYSTEM
CITY OF LUDINGTON, MICHIGAN

FIGURE 10.31 *Idealized diagram showing how the water supply for Ludington, Michigan, is derived from the waters of Lake Michigan, utilizing the natural, sandy bottom of the lake as a filter. (Courtesy of William and Works.)*

of man-made lakes (Figure 10.30). After regular treatment of the sewage, the waste water is stored for a month in a holding pond (lake 1 in Figure 10.30) to allow further oxidation. Then, after chlorination, the water is pumped to a spreading area, where it trickles into the ground and flows through rocky, sandy soil, constituting a tertiary treatment. The water surfaces again and enters the lakes and overflow waters from a stream which is used to irrigate a golf course. Microbiological investigation has shown the lake waters to be safe, and in addition to providing wildlife habitat, the area is enthusiastically accepted by 12,500 residents and other people who know where the water comes from and who use the lake area for a variety of recreational activities (23).

A final example of an innovative system to advantageously use naturally occurring earth materials to purify water for public consumption is reported from a Michigan community on the shore of Lake Michigan. Here, the city of Ludington, with a summer population of 11,000, uses sands and gravels below the lake bottom to prefilter and thus treat the lake water for municipal use. The procedure is shown schematically in Figure 10.31. A system of lateral intakes is buried in sand and gravel 12 to 15 feet below the lake bottom, where the water depth is at least 15 feet. The water is pumped out for municipal use, and in some cases the only additional treatment is chlorination.

SUMMARY AND CONCLUSIONS

The history of waste-disposal practices since the Industrial Revolution has been from a practice of "dilution and dispersion" to a new concept of "concentrate, contain, and recycle."

A fast-growing disposal practice for solid waste is the sanitary landfill. However, other methods such as on-site disposal, feeding garbage to swine, composting, incineration, and open dumping are still used. Of these methods, the geologic factor is most significant in siting of sanitary landfills.

Disposal of toxic or otherwise hazardous waste by deep-well injection into an appropriate geologic environment can, in some instances, be a wise decision. However, such disposal should not be considered as a quick and easy solution to industrial-waste problems, and the method should be used only when more appropriate alternatives are not available. Furthermore, careful, continuous monitoring is necessary to insure safety when wastes are disposed of by deep-well injection.

Radioactive-waste management presents a serious and ever-increasing problem. High-level wastes that remain hazardous for thousands of years and currently are being stored will eventually have to be permanently disposed of. A likely method is disposal in a carefully and continuously monitored appropriate geologic environment, possibly bedded salt or salt dome. Low-level waste apparently can be safely buried and carefully controlled and monitored at near-surface sites.

Ocean dumping can be a significant source of marine pollution, and efforts to control indiscriminate dumping are in effect. Alternatives to ocean dumping of materials such as polluted dredge spoils, sewage sludge, and other potentially hazardous materials are being developed, but in many cases they are not yet practical or economically feasible.

Continued urban expansion will insure that the septic-tank sewage disposal system remains an important and common method of waste disposal. However, suitability of soil and topography is necessary to reduce the chance of the systems failing, which would result in pollution of the environment and the possible production of a public health problem.

The costs of waste-water treatment are the highest environmental costs facing state and local governments, and this trend is expected to continue. One of the largest expenses of operating and maintaining a waste-water treatment facility is that of the handling and disposal of sewage sludge.

An exciting possibility for some municipal, industrial, and agricultural waste water is the concept of a "Waste Water Renovation and Conservation Cycle." The basic idea is to convert treated waste water into a resource and recycle it locally.

References

1. Galley, J. E. Economic and industrial potential of geologic basins and reservoir strata. In *Subsurface disposal in geologic basins: a study of reservoir strata,* ed.

Lavendar Pit copper mine, Bisbee, Arizona. (Photo courtesy of Peter L. Kresan)

Salt Lake City, Utah. This area contains the Bingham copper mine as well as evaporite resources associated with the Great Salt Lake, which is only a remnant of the once much larger Lake Bonneville. (ERTS photo)

J. E. Galley, pp. 1–19. American Association of Petroleum Geologists, Memoir 10.

2. Council on Environmental Quality. 1973. *Environmental quality—1973*. Washington, D.C.: U.S. Government Printing Office.

3. Schneider, W. J. 1970. *Hydraulic implications of solid-waste disposal*. U.S. Geological Survey Circular 601F.

4. Turk, L. J. 1970. Disposal of solid wastes—acceptable practice or geological nightmare? In *Environmental geology*, pp. 1–42. Washington, D.C.: American Geological Institute Short Course, American Geological Institute.

5. Hughes, G. M. 1972. *Hydrologic considerations in the siting and design of landfills*. Environmental Geology Notes, no. 51. Illinois State Geological Survey.

6. Bergstrom, R. E. 1968. *Disposal of wastes: scientific and administrative considerations*. Environmental Geology Notes, no. 20. Illinois State Geological Survey.

7. Cartwright, K., and Sherman, F. B. 1969. *Evaluating sanitary landfill sites in Illinois*. Environmental Geology Notes, no. 27. Illinois State Geological Survey.

8. Walker, W. H. 1974. Monitoring toxic chemical pollution from land disposal sites in humid regions. *Ground Water* 12: 213–18.

9. McKenzie, G. D., and Pettyjohn, W. A. 1975. Subsurface waste management. In *Man and his physical environment*, eds. G. D. McKenzie and R. O. Utgard, pp. 150–56. Minneapolis: Burgess Publishing.

10. Piper, A. M. 1970. *Disposal of liquid wastes by injection underground: neither myth nor millennium*. U.S. Geological Survey Circular 631.

11. Committee of Geological Sciences. 1972. *The earth and human affairs*. San Francisco: Canfield Press.

12. Warner, D. L. 1968. Subsurface disposal of liquid industrial wastes by deep-well injection. In *Subsurface disposal in geologic basins: a study of reservoir strata*, ed. J. E. Galley, pp. 11–20. American Association of Petroleum Geologists, Memoir 10.

13. Office of Industry Relations. 1974. *The nuclear industry, 1974*. Washington, D.C.: U.S. Government Printing Office.

14. Micklin, P. P. 1974. Environmental hazards of nuclear wastes. *Science and Public Affairs* 30: 36–42.

15. National Academy of Sciences—National Research Council. 1970. *Disposal of solid radioactive wastes in bedded salt deposits*. Washington, D.C.: U.S. Government Printing Office.

16. Atomic Energy Commission. 1972. *Plan for the management of AEC-generated radioactive wastes*. Washington, D.C.: U.S. Government Printing Office.

17. Council on Environmental Quality. 1970. *Ocean dumping: a national policy*. Washington, D.C.: U.S. Government Printing Office.

18. Soil Conservation Service. 1971. *Soils and septic tanks*. Agriculture Information Bulletin 349. Washington, D.C.: U.S. Government Printing Office.

19. U.S. Department of Health, Education, and Welfare. 1967. *Manual of septic tank practice*. Rockville, Maryland: Bureau of Community Management.

20. American Chemical Society. 1969. *Clean our environment: the chemical basis for action*. Washington, D.C.: U.S. Government Printing Office.

21. Parizek, R. R., and Myers, E. A. 1968. Recharge of ground water from renovated sewage effluent by spray irrigation. Proceedings of the Fourth American Water Resources Conference, pp. 425–443.

22. Fogg, C. E. 1972. Waste management—nationally. *Soil Conservation*, May 1972.

23. Power. 1966. *Water: a special report*. New York.

11

THE GEOLOGIC ASPECTS OF ENVIRONMENTAL HEALTH

The geologic cycles—tectonic, rock, geochemical, and hydrologic—are responsible on a global scale for the geographic pattern of continents, ocean basins, and climates. On a smaller scale, the cycles produce the spatial arrangement of all landforms, rocks, minerals, soils, groundwater, and surface water. As a member of the biological community, man has carved a niche with other members of the community, a niche highly dependent on complex interrelations among the biosphere, atmosphere, hydrosphere, and lithosphere. We are only beginning to inquire into and gain a basic understanding of the total range of factors that affect our health and well-being. As we continue our exploration of the geologic cycle from the scale of minute quantities of elements in soil, rocks, and water to regional patterns of climate, geology, and topography, we are making startling discoveries of how these factors might influence the incidence of certain diseases and death rates. In the United States alone, the chances of dying in some areas varies considerably. Generally, the highest death rates are near the East Coast, the rates being about twice those in the Great Plains and some mountain areas (1).

      Disease is defined as an imbalance resulting from a poor adjustment between an individual and his environment (2). It seldom has a one-cause, one-effect relationship. The contribution of the geologist is to help isolate aspects of the geologic environment that might influence the incidence of disease. This is a tremendously complex task that requires sound scientific inquiry coupled with interdisciplinary research involving physicians and other scientists. Although the picture is now rather vague, the possible rewards of the emerging field of medical geology are exciting and may eventually play a significant role in environmental health.

Determining the Factors

To study the geologic aspects of environmental health, one must consider cultural and climatic factors associated with patterns of disease and death rates. This is important if the geologic influence is to be isolated. Cultural aspects of a society reflect the sum total of concepts and techniques that a group of people have developed to survive in their environment; these aspects influence the occurrence of disease by linking or separating disease and people. The nature and extent of these links depend on such factors as local customs and degree of industrialization. Primitive societies living directly off the land and water are obviously plagued by a different variety of health problems than urban society. Industrial societies have nearly eliminated diseases such as cholera, typhoid, hookworm, and dysentery, but they are more likely to suffer from lung cancer and other diseases related to air, soil, and water pollution.

The high incidence of stomach cancer in Japanese people is an example of a relationship between culture and disease (3). The Japanese prefer rice that is polished and powdered. Unfortunately, the powder contains asbestos as an impurity. Asbestos, a fibrous mineral, either is a true carcinogen (cancer-causing material) or else takes a passive role as a carrier of trace-metal carcinogens.

Incidence of lead poisoning also suggests cultural, political, and economic relations with patterns of disease. Effects of lead poisoning may include anemia, mental retardation, and palsy. It is found in some moonshine whiskey and has resulted in lead poisoning among adults and even unborn or nursing infants whose mothers indulged in the drink (4). It has even been suggested that one of the reasons for the fall of the Roman empire was widespread lead poisoning. Some estimates are that the Romans produced about 60,000 tons of lead per year for 400 years. Lead was used in pots used for processing grape juice into a syrup for a preservative and sweetener of wine, in cups from which the Romans drank wine, and in cosmetics and medicines. The ruling class also had water piped into their homes through lead pipes. The argument is that a gradual lead poisoning of the upper class resulted in their eventual demise through widespread stillbirths, deformities, and brain damage. This hypothesis is reportedly supported by a high lead content found in the bones of ancient Romans (4).

Climatic factors such as temperature, humidity, and amount of precipitation are sometimes intimately related to disease patterns. Serious health hazards exist in the tropics where two of the worst climatically controlled diseases, schistosomiasis and malaria, are found. Schistosomiasis, called *snail fever,* is a significant cause of death among children and saps the energy of millions of people in the world. Its effects have tremendous socioeconomic consequences, and therefore some researchers consider it to be the world's most important disease (2).

Malaria and schistosomiasis, among other diseases, are clearly related to climatic factors because the vectors necessary to carry the disease, that is, mosquitoes and snails, are climatically controlled. Beyond this, there are other diseases such as Burkett's tumor, a debilitating, but seldom fatal, cancerous disease of the lymph system which in Africa is associated with three climatic conditions: annual rainfall greater than 20 inches; elevations lower than 5000 feet; and mean annual temperature of at least 60 degrees Fahrenheit (3). Figure 11.1 shows the climatic conditions and the occurrence of Burkett's tumor in Africa. It is emphasized that the relationship between climate and the disease may not be a cause-effect relationship. In fact, some workers believe that Burkett's tumor is caused by a virus. Nevertheless, the association of the disease with particular climatic conditions is evident.

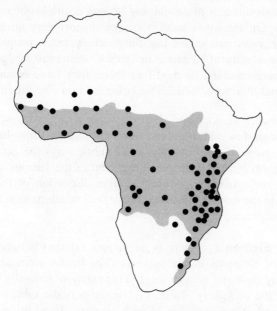

FIGURE 11.1 *Map of Africa showing areas below 5000 feet in elevation with mean annual temperature greater than 60°F and more than 20 inches of rainfall superposed on sites where Burkett's tumor has been identified. (Source: Armed Forces Institute of Pathology.)*

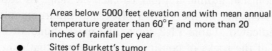

Areas below 5000 feet elevation and with mean annual temperature greater than 60°F and more than 20 inches of rainfall per year

● Sites of Burkett's tumor

Assumed relations between culture and climate and incidence of disease must be viewed with some skepticism because there is seldom a simple answer to environmental health problems. For example, if schistosomiasis were only climatically controlled, then all areas with an appropriate climate, such as much of the Amazon River Basin, would have the disease. Fortunately, this is not the case, and in some instances the reason is geologic. The conditions for the disease in the Amazon River Basin are nearly optimal, and yet it occurs only in two very limited areas. The main reason for this is that there is not a sufficient amount of calcium in the water to support the snail, the intermediate host. In other areas, the acidity of the water in presence of copper and other heavy metals may be responsible for the absence of the necessary snails in an otherwise suitable environment for schistosomiasis (1).

The above discussion establishes some of the complex relations between disease patterns and environment. With this in mind, let us now focus on the geologic aspects of health problems. To accomplish this we will discuss geologic factors of environmental health, occurrence and effects of trace elements on health, and the significance of the geologic environment to the incidence of heart disease and cancer, the leading causes of deaths in the United States.

Some Geologic Factors of Environmental Health

The very soil that we cultivate to provide nutrients for subsistence, the rock that we build our homes and industries on, the water that we drink and the air that we breathe all can influence our chances of developing serious health problems such as cancer and heart disease. On the other hand, the same factors may influence our chances of living a longer, more productive life. Surprisingly, many people still believe that soil and water in a "natural," "pure," or "virgin" state must be "good" and that if man's activities have changed or modified them, they have become "contaminated," "polluted," and therefore "bad." This belief is in no way the entire story (5).

Relations between geology and health are very significant research and discussion topics. Although few valid cause-and-effect relationships have been isolated, we are learning more all the time about the subtle ways the geologic environment affects the general health of people. Treatment of the various aspects of medical geology at even the introductory level requires discussion of the natural distributions of elements in the earth's crust and the ways in which natural and man-induced processes can concentrate or disperse elements.

Natural Abundance of Elements   There is an inverse relation between atomic number and abundance of elements in the universe. The lighter elements are encountered more frequently than the heavier ones. This relation generally holds for both the lithosphere and the biosphere. Table 11.1 is the periodic table of the elements showing position and atomic number of each element. In addition, some of the more abundant elements in the earth's crust and environmentally important trace elements are indicated. Table 11.2 shows the most abundant elements in the average composition of the rocks of continental crust. Note that over 99 percent of these rocks by weight are concentrated in the first 26 elements of the periodic table. Table 11.3 shows the distribution of the more abundant elements in the total body of the standard man. Note that over 99 percent of the body by weight is composed of elements in the first 20 elements of the periodic table.

Living tissue, animal or vegetable, is composed primarily of 11 elements, so-called bulk elements. Five of these are metals: hydrogen, sodium, magnesium, potassium, and calcium. Six are nonmetals: carbon, nitrogen, oxygen, phosphorus, sulfur, and chlorine. For those species that have hemoglobin, iron is added to the list. In addition to the bulk elements, there are several other elements which living tissue requires in order to function properly. These are trace-element metals (present in minute quantities) which assist in regulating the dynamic processes of life. Trace elements that have been studied and shown essential for nutrition include fluorine, chromium, manganese, iron, cobalt, copper, zinc, selenium,

TABLE 11.1 Periodic table of the elements with examples of environmentally important trace elements and elements that are relatively abundant in the earth's crust highlighted.

Legend:

- Atomic Number → 20
- * Element Relatively Abundant In Earth's Crust
- ** Environmentally Important Trace Elements
- **Ca** — Calcium ← Name

1	2											13	14	15	16	17	18
H 1 Hydrogen																	**He** 2 Helium
Li 3 ** Lithium	**Be** 4 Beryllium											**B** 5 * Boron	**C** 6 Carbon	**N** 7 Nitrogen	**O** 8 * Oxygen	**F** 9 ** Fluorine	**Ne** 10 Neon
Na 11 Sodium	**Mg** 12 * Magnesium											**Al** 13 * Aluminum	**Si** 14 * Silicon	**P** 15 Phosphorus	**S** 16 Sulfur	**Cl** 17 Chlorine	**Ar** 18 Argon
K 19 * Potassium	**Ca** 20 * Calcium	**Sc** 21 Scandium	**Ti** 22 Titanium	**V** 23 Vanadium	**Cr** 24 ** Chromium	**Mn** 25 ** Manganese	**Fe** 26 * Iron	**Co** 27 ** Cobalt	**Ni** 28 ** Nickel	**Cu** 29 ** Copper	**Zn** 30 ** Zinc	**Ga** 31 Gallium	**Ge** 32 Germanium	**As** 33 Arsenic	**Se** 34 ** Selenium	**Br** 35 Bromine	**Kr** 36 Krypton
Rb 37 Rubidium	**Sr** 38 Strontium	**Y** 39 Yttrium	**Zr** 40 Zirconium	**Nb** 41 Niobium	**Mo** 42 ** Molybdenum	**Tc** 43 Technetium	**Ru** 44 Ruthenium	**Rh** 45 Rhodium	**Pd** 46 Palladium	**Ag** 47 Silver	**Cd** 48 ** Cadmium	**In** 49 Indium	**Sn** 50 Tin	**Sb** 51 Antimony	**Te** 52 Tellurium	**I** 53 ** Iodine	**Xe** 54 Xenon
Cs 55 Cesium	**Ba** 56 Barium	**La** 57 • Lanthanum	**Hf** 72 Hafnium	**Ta** 73 Tantalum	**W** 74 Wolfram	**Re** 75 Rhenium	**Os** 76 Osmium	**Ir** 77 Iridium	**Pt** 78 Platinum	**Au** 79 Gold	**Hg** 80 ** Mercury	**Tl** 81 Thallium	**Pb** 82 ** Lead	**Bi** 83 Bismuth	**Po** 84 Polonium	**At** 85 Astatine	**Rn** 86 Radon
Fr 87 Francium	**Ra** 88 Radium	**Ac** 89 •• Actinium															

58	59	60	61	62	63	64	65	66	67	68	69	70	71
Ce Cerium	**Pr** Praseodymium	**Nd** Neodymium	**Pm** Promethium	**Sm** Samarium	**Eu** Europium	**Gd** Gadolinium	**Tb** Terbium	**Dy** Dysprosium	**Ho** Holmium	**Er** Erbium	**Tm** Thulium	**Yb** Ytterbium	**Lu** Lutetium
90	91	92	93	94	95	96	97	98	99	100	101	102	103
Th Thorium	**Pa** Protactinium	**U** Uranium	**Np** Neptunium	**Pu** Plutonium	**Am** Americium	**Cm** Curium	**Bk** Berkelium	**Cf** Californium	**Es** Einsteinium	**Fm** Fermium	**Md** Mendelevium	**No** Nobelium	**Lw** Lawrencium

TABLE 11.2 *The relative abundance of the most common elements in the rocks of the earth's crust.*

Atomic No.		Element	Weight (%)
8	O	Oxygen	46.4
14	Si	Silicon	28.15
13	Al	Aluminum	8.23
26	Fe	Iron	5.63
20	Ca	Calcium	4.15
11	Na	Sodium	2.36
12	Mg	Magnesium	2.33
19	K	Potassium	2.09
		Total	99.34

TABLE 11.3 *Distribution of the more abundant elements in the adult human body.*

Atomic No.	Element		Weight (%)
8	Oxygen	(O)	65.0
6	Carbon	(C)	18.0
1	Hydrogen	(H)	10.0
7	Nitrogen	(N)	3.0
20	Calcium	(Ca)	1.5
15	Phosphorus	(P)	1.0
16	Sulfur	(S)	0.25
19	Potassium	(K)	0.2
11	Sodium	(Na)	0.15
17	Chlorine	(Cl)	0.15
12	Magnesium	(Mg)	0.05
	Total		99.30

molybdenum, and iodine. This list is not complete, and it would not be surprising to learn that many more are essential or at least active in life processes (6, 7). Other elements, such as nickel, arsenic, aluminum, and barium, accumulate as tissues age and are known as "age elements." The physiological consequences of the accumulation of some elements in living tissue are known in some cases but completely unknown or poorly understood in many others (7).

Concentration and Dispersion of Chemical Components   The movement of chemical compounds among various paths through the lithosphere, hydrosphere, atmosphere, and biosphere constitutes the geochemical cycle. Natural processes, such as the releasing of gas by volcanic activity or the weathering of rock and rock debris, release chemical material into the environment. In addition, human use may result in the release of material and substances leading to pollution or contamination of the environment. Figure 11.2 shows some of the paths that trace elements may take to become concentrated in man, possibly causing health problems.

   Once released by natural or man-induced processes, elements and other substances are cycled and recycled by geochemical and rock-forming processes. Thus, a certain concentration of a particular trace element in igneous rocks may have quite a different concentration in sedimentary rocks formed from the weathered products of the igneous rock. Whether the concentration has increased or decreased

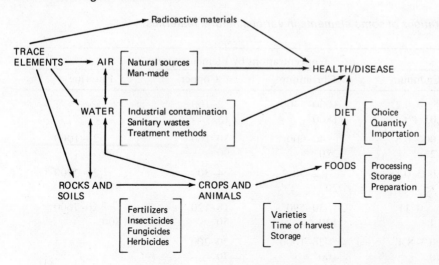

FIGURE 11.2 *A schematic drawing showing mechanisms by which trace elements may find their way to man and other animals, thus influencing the quality of health or producing disease. (Source: K. E. Beeson,* Geochemistry and the Environment, *vol. 1, 1974. Reproduced with permission of the National Academy of Sciences.)*

depends upon the nature of the geochemical and rock-forming processes. Table 11.4 lists concentrations of selected elements in igneous and sedimentary rocks. Although this information is not detailed, it is useful because it indicates a change in the relative abundance of elements due to rock-forming and biological processes, as, for example, the approximately tenfold increase in selenium from the original weathering of igneous rocks to the formation of shale. With the exception of coal and phosphorites (rock rich in calcium phosphate), other types of sedimentary rocks do not show the similar increase of selenium. This example and others suggest that geochemical and rock-forming processes such as weathering, leaching, accretion, deposition, and biological activity effectively sort, concentrate, and disperse elements and other substances throughout the environment.

Weathering is the physical and chemical breakdown of rock material. It is a major process in the formation of soil. If weathering were the only process that removed and modified the concentration of common and trace elements, then soils would have essentially the same natural concentration of trace elements as the parent material from which the soil formed. Regardless of whether the parent material is bed rock or is transported and deposited by running water, wind, or ice, weathering is a natural process that frees trace elements to be utilized by the biosphere in life processes.

The man-induced counterpart of weathering is pollution or contamination. These processes release trace elements into the environments, for example, lead released into the environment when lead additives in gasoline are emitted through exhaust systems. Other examples include release of mercury, cadmium, nickel, zinc, and other metals into the atmosphere and water through industrial and mining operations.

TABLE 11.4   *Concentrations of some elements in various natural materials.*

Type of Material	Concentrations by Elements (ppm)			
	Cadmium	Chromium	Copper	Fluorine
Ultramafic igneous[a,b]	0–0.2 0.05 (?)	1000–3400 1800	2–100 15	—
Basaltic igneous[a,b]	0.006–0.6 0.2	40–600 220	30–160 90	20–1060 360
Granitic igneous[a,b]	0.003–0.18 0.15	2–90 20	4–30 15	20–2700 870
Shales and clays[a,b,c]	0–11 1.4	30–590 120	18–120 50	10–7600 800
Black shales (high C)[d]	0.3–8.4 1.0	26–1000 100	20–200 70	—
Deep-sea clays[a,b]	0.1–1 0.5	— 90	— 250	— 1300
Limestones[a,b,c,e]	— 0.05	— 10	— 4	0–1200 220
Sandstones[a,b,c]	— 0.05	— 35	— 2	10–880 180
Phosphorites[f]	0–170 30	30–3000 300	1.0–100 30	24,000–41,500 31,000
Coals (ash)[a]	— 2	10–1000 20	2–40 15	40–480 80

SOURCE: M. Fleischer, U.S. Geological Survey. Reproduced with permission of the National Academy of Science.
NOTE: The upper figure is the range usually reported; the lower figure, the average.
[a] Turekian and Wedepohl (1961).
[b] Parker (1967).
[c] Becker et al. (1972).
[d] Vine, J. D., and Tourtelot, E. B., *Econ. Geol.* 65, 223 (1970).
[e] Wedepohl (1970).
[f] Gulbrandsen, R. A., *Geochim. Cosmochim. Acta* 30, 769 (1966).
[g] U.S. Geological Survey (1972).

Because some studies have shown a dramatic increase of lead in the air, water, and soil, and even the Antarctic ice, it is evident that a closer examination of the possible adverse effects of chronic lead exposure is needed. However, with the exceptions of cases of lead poisoning (such as the poisoning of people by lead-base paint and plant damage caused by exposure to lead close to highways or by heavy mineral concentrations released in mining areas), documented adverse health effects caused by lead are somewhat speculative. In fact, one author states that there is little difference between the amounts of lead in blood and urine samples from urban and rural residents (8).

Leaching, accretion, deposition, biologic activity, and other processes may concentrate or disperse elements after they are released by natural and man-

		Concentrations by Elements (ppm)			
Iodine	Lead	Lithium	Molybdenum	Selenium	Zinc
0.06–0.3	—	—	—	—	—
0.1	1	0.5	0.3	0.05	40
—	2–18	3–50	0.9–7	—	48–240
0.5	6	20	1.5	0.05	110
—	6–30	10–120	1–6	—	5–140
0.5	18	35	1.4	0.05	40
212–380	16–50	4–400	—	—	18–180
5 (?)	20	80	2.6	0.6	90
—	7–150	—	1–300	—	34–1500
	20		10		100 (?)
11–50	—	—	—	—	—
35 (?)	80	57	27	0.17	165
0.4–29	—	5–10	—	—	—
5	9	7	0.4	0.08	20
—	1–31	7–90	—	—	2–41
1.7	7	30	0.2	0.05	16
—	10–30	—	3–300	1–100	20–300
	10		30	18	50
1–11	2–50	2–300	0.2–16	0.4–3.9	7–108
4	15	50	5	2 [g]	50

induced processes. *Leaching* is a natural washing or draining that moves material from the upper to lower soil horizons. Material that leaches out of soil may enter into the groundwater system and be dispersed or diluted. If the material is sufficiently abundant or toxic (or otherwise harmful), it may also pollute the groundwater. Leaching is most prevalent in warm, humid climates where the soil may be nutrient-poor because the nutrients are washed out. Furthermore, trace elements that remain may be concentrated at undesirable levels.

Accumulation refers to processes that cause or increase retention of material in soil. Examples include salts that due to evaporation processes may accumulate on the surface and the upper zones of soils, and materials that have been removed by leaching from the A horizon and accumulate in the B horizon. An example of the latter is found in semiarid regions where accumulation of calcium carbonate (caliche) is found in the B horizon of some soils (see Chapter 3 for a discussion of soil horizons).

Deposition of earth materials has two environmentally important aspects: first, materials deposited in streams, lakes, and oceans cause biologic disruptions; and second, problems develop in areas with a deficiency of needed trace elements that exists because the elements were not originally deposited along with other sediments.

Absorption of mercury to suspended sediments and bottom sediments may lead to high concentrations of mercury in the aqueous environments. This may lead to biological disruption because inorganic substances in the water are basic to the life processes of low life-forms in aquatic environments, and these organisms assimilate the mercury which is passed on at a higher and higher concentration through the food chain.

Deficiencies that result when a certain material is not deposited with sediments moved by water, ice, and wind are less well understood because of the possibility of interactions of other processes such as leaching. Erosion and deposition by wind are particularly susceptible to selective removal, and one author attributes the lack of lead, iron, copper, cobalt, and other materials in the sand hills of Nebraska to the fact that these metals occur in grains smaller and heavier than quartz and therefore were not moved with the quartz grains that formed the sand hills (3).

Trace Elements and Health

Every element has a whole spectrum of possible effects on a particular plant or animal. For example, selenium is toxic in seleniferous areas, has no observable effect in most conditions, and is beneficial to animal production (raising of cattle, sheep, etc.) in some areas. The apparent contradiction is resolved by recognizing that the first case is one of oversupply of selenium; the second represents a balanced state; and the third case is one of deficiency, which in some cases is rectified by supplementation of the element into the food supply of animals (9).

It was recognized many years ago that the effects of a certain trace element on a particular organism depend on the dose or concentration of the element. This *dose dependency* can be represented by a dose-response curve, as shown in Figure 11.3 (9, 10). When various concentrations of an element present in a biological system are plotted against effects on the organism, three things are apparent. First, while relatively large concentrations are toxic, injurious, and even lethal (*D-E-F* in Figure 11.3), trace concentrations may be very beneficial or necessary for life (*A-B*). Second, the dose-response curve has two maxima (*B-C*) forming a plateau of optimal concentration and maximum benefit to life. Third, the threshold concentration where harmful effects to life start is not at the origin (zero concentration) but varies with concentrations less than at point *A* and greater than at point *D* in Figure 11.3 (9, 10).

Points *A, B, C, D, E,* and *F* in Figure 11.3 are significant threshold concentrations. Unfortunately, points *E* and *F* are known only for a few substances for a few organisms, including man, and the really important point *D* is all but unknown (10). The width of the maximum-benefit plateau (points *B-C*) for a particular life form depends on the organism's particular physiological equilibrium (10). In other words, the different phases of activity, whether beneficial, harmful, or lethal, may differ widely quantitatively and qualitatively for different substances and, therefore, are completely observable only under special conditions (9).

A complete discussion of the geology and environmental effects of all trace substances could be a textbook in itself. Our objective here is to discuss repre-

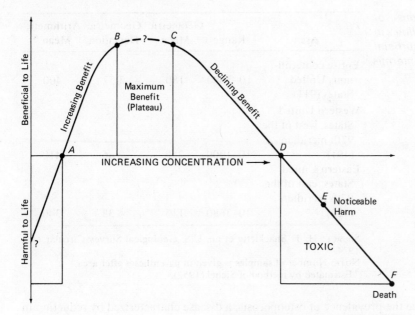

FIGURE 11.3 *Generalized dose-response curve.*

sentative examples to indicate possible effects of imbalances of trace elements. For this purpose, fluorine, iodine, zinc, and selenium are selected. In addition, it is valuable to relate trace-element problems to human use of the land. Examples of mining activity serve this purpose.

Fluorine   Fluorine is fairly abundant in rocks (Table 11.4) and soils (Table 11.5). Most of the fluorine in soils is derived from the parent rock, but it can also be added by volcanic activity which deposits fluorine-rich volcanic ash on the land. In addition, industrial activity and application of fertilizers have also, on a limited basis, contributed locally to an increase in the concentration of fluorine in soils.

   Fluorine is an important trace element that forms fluoride compounds such as calcium fluoride which increases the crystallinity of the apatite (calcium phosphate) crystals in teeth. It helps prevent tooth decay by facilitating the growth of larger, more perfect crystals. The same processes occur in bones where fluoride assists in the development of more perfect bone structure that is less likely to fail with old age.

   Relations between the concentration of fluoride (a compound of fluorine such as sodium fluoride, NaF) and health indicate a specific dose-response curve, as shown in Figure 11.4. The optimum fluoride concentration (point *B*) for the reduction of dental caries (DMF index: decayed, missing, and filled) is about 1 ppm (part per million; 100 ppm = 0.01 wt. percent) (Figures 11.5, 11.6, and 11.7). Figure 11.8 shows areas in the United States known to have fluoride content in the groundwater greater than 1.0 ppm and areas with a fluoride content less than 0.4 ppm. Fluoride levels greater than 1.5 ppm do not significantly decrease the DMF index, but they do increase the occurrence and severity of mottling (discoloration of teeth) (11). Regardless of this, fluoride concentrations of 4 to 6 ppm

TABLE 11.5 *Concentrations of fluorine in parts per million in samples of soils and other surficial earth materials from the contermi-nous United States.*

Area	Range	Geometric Mean	Geometric Deviation	Arithmetic Mean [a]
Entire contermi-nous United States (911)	10–3680	180	3.57	400
Western United States, west of the 97th meridian (491)	10–1900	250	2.66	410
Eastern United States, east of the 97th meridian (420)	10–3680	115	4.38	340

SOURCE: H. T. Shacklette et al., U.S. Geological Survey Circular 692, 1974.
NOTE: Number of samples is given in parentheses after area.
[a] Estimated by method of Sichel (1952).

greatly reduce the prevalence of osteoporosis, a disease characterized by reduction in bone mass (Figures 11.9 and 11.10) and collapsed vertibrae (Figure 11.11) (12). The letters *N.S.* in the figures indicate the relations that are not statistically significant. Based on the study of fluoride in osteoporosis, point *C* on the dose-response curve for fluoride was located (Figure 11.4). Point *E* on the curve was located by the fact that fluoride concentrations of 8 to 20 ppm are associated with excessive bone formation in the periosteum (dense, fibrous, outer layer of bone) and calcification of ligaments that usually do not calcify (7). The position of points *A, D,* and *F*

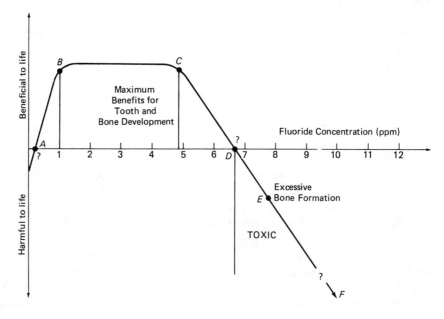

FIGURE 11.4   *Dose-response curve for fluoride.*

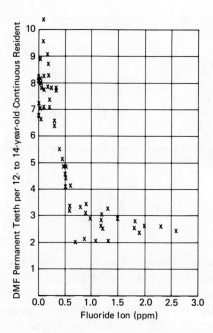

FIGURE 11.5 *Relationship between fluoride concentration and prevalence of dental decay. (Reprinted from F. J. Maier,* Water Quality and Treatment, *3rd ed., by permission of the Association. Copyrighted 1971 by the American Water Works Association, Inc., 6666 West Quincy Avenue, Denver, Colorado 80235. Used with permission of McGraw-Hill Book Company.)*

are not precisely known, but fluoride in massive doses is the main ingredient of some rodent poisons.

Iodine   Thyroid disease, caused by a deficiency of iodine, is probably the best-known example of the relationship between geology and disease. The thyroid gland, located at the base of the neck, requires iodine for normal function. Lack of iodine causes goiter, a tumorous condition involving an enlargement of the thyroid gland (13). Furthermore, a child born to a mother who suffered iodine deficiency during

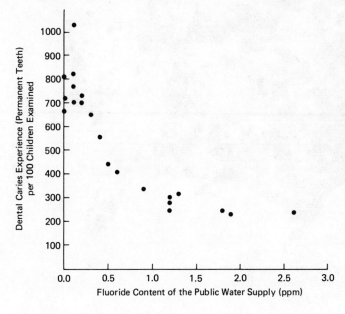

FIGURE 11.6 *Relation between the number of dental caries (in permanent teeth) observed in 7257 selected 12- to 14-year-old school children in 21 cities of four states, and the fluoride content of public water supply. (Reprinted from F. J. Maier,* Water Quality and Treatment, *3rd ed., by permission of the Association. Copyrighted 1971 by the American Water Works Association, Inc., 6666 West Quincy Avenue, Denver, Colorado 80235. Used with permission of McGraw-Hill Book Company.)*

FIGURE 11.7 *Reduction in tooth decay at pioneer cities due to fluoridation of the water. (Reprinted from F. J. Maier,* Water Quality and Treatment, *3rd ed., by permission of the Association. Copyrighted 1971 by the American Water Works Association, Inc., 6666 West Quincy Avenue, Denver, Colorado 80235. Used with permission of McGraw-Hill Book Company.)*

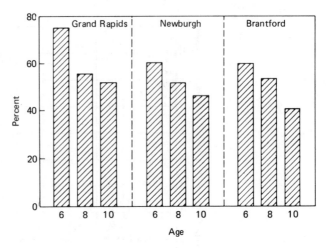

pregnancy may be born a *cretin,* a mentally retarded dwarf. At one time cretinism was fairly common in areas of Mexico and Switzerland with high goiter-rate (14). The incidence of goiter is clearly related to deficiencies of iodine, as shown by the relation between iodine-deficient areas and the occurrence of goiter in the United States shown in Figure 11.12. The use of iodized salt is now common in the goiter

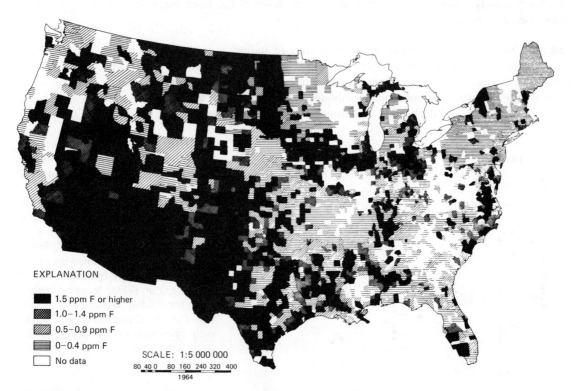

EXPLANATION

- 1.5 ppm F or higher
- 1.0–1.4 ppm F
- 0.5–0.9 ppm F
- 0–0.4 ppm F
- No data

SCALE: 1:5 000 000

80 40 0   80 160 240 320 400
1964

FIGURE 11.8 *The fluoride content of groundwater in the conterminous United States. (Source: M. Fleischer, Geological Society of America Special Paper 90, 1967.)*

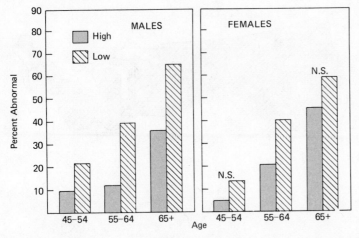

FIGURE 11.9 *Percentage of radiologically detectable calcification of abdominal aorta in males and females residing in high- and low-fluoride areas. (Reprinted from the* Journal of the American Medical Association, *October 31, 1966, Volume 198. Copyright 1966, American Medical Association.)*

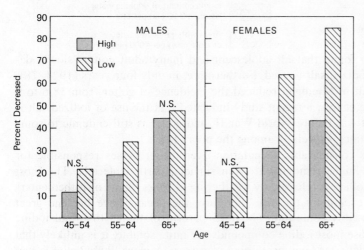

FIGURE 11.10 *Percentage of decreased bone density in subjects from a high-fluoride area (4 to 5.8 ppm) compared to those from a low-fluoride area (0.15 to 0.3 ppm). (Reprinted from the* Journal of the American Medical Association, *October 31, 1966, Volume 198. Copyright 1966, American Medical Association.)*

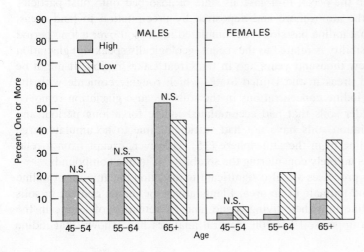

FIGURE 11.11 *Percentage of subjects with one or more collapsed vertibrae in subjects from high- and low-fluoride areas. (Reprinted from the* Journal of the American Medical Association, *October 31, 1966, Volume 198. Copyright 1966, American Medical Association.)*

FIGURE 11.12 *Map of the United States showing relationship between goiter occurrence and iodine deficiencies. (Source: Armed Forces Institute of Pathology.)*

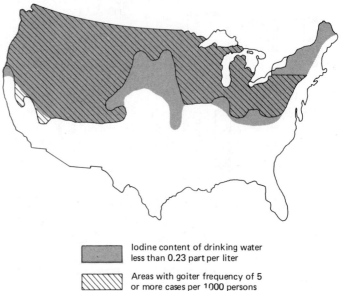

Iodine content of drinking water less than 0.23 part per liter

Areas with goiter frequency of 5 or more cases per 1000 persons

belt, and it has been found that all adolescent and many adult goiters slowly decrease in size when iodized salt is used. Furthermore, in only four years (1924–28), the use of iodized salt in Michigan reduced the incidence of goiter from 38.6 to 9 percent (13). In spite of this, a recent study indicated that the use of iodized salt is not as prevalent as it was before World War II, and goiter is still endemic to some areas in the United States, especially among the poor (14).

There is considerable speculation over the processes responsible for concentrating iodine in or removing it from surficial earth materials. The most popular idea is that iodine is released by weathering of rocks. Some of it then enters into the rivers and eventually the sea; therefore, the oceans have become great iodine reservoirs, possibly containing about 25 percent of the earth's total iodine (15). However, because most iodine compounds are quite soluble, it is unlikely that much iodine is residual after long weathering processes. A more likely belief is that iodine is picked up in the ocean in a gaseous state or adsorbed onto dust particles and transported by the atmosphere and deposited by precipitation to land areas. This idea proposes that iodine has slowly accumulated in the soil over a long period of time. Another possibility is related to the recent (geologically speaking) glaciation which ended just a few thousand years ago in the Great Lakes region. It might be argued that glaciated areas in the United States which roughly coincide with the goiter belt have low iodine concentrations in the soils because glaciation removed and destroyed the older soils that had accumulated iodine for a long period, and the young, post-glaciation soils have not had sufficient time to accumulate large concentrations of iodine from the atmosphere (15). However, except from a relative standpoint, this is unlikely considering the solubility of iodine compounds.

Biologic processes are also significant in affecting the amount of iodine available to plants and animals in an area. Plants affect the iodine content in soils by absorption of iodine into their living tissue, and by retention of iodine in the organic-rich (humus) upper soil horizon. Several analyses have shown that iodine tends to be concentrated in the upper soil horizons. Therefore, biological processes which cycle iodine in soils and plants might be more significant in determining availability of iodine than the amount of iodine in the local bed rock (15).

Zinc   Zinc is a trace element necessary to plants, animals, and man. Although zinc is a heavy metal that in excessive amounts has been associated with disease, it is known primarily from zinc-deficiency studies. The zinc content of surficial materials in the United States is shown in Figure 11.13. Zinc deficiencies are known in 32 states and have resulted in a variety of plant diseases that cause low yields, poor seed development, and even total crop loss (16).

Zinc deficiency in plants is related primarily to three soil conditions: low content of zinc; unavailability of zinc present in the soil; and poor soil management. The amount of zinc is low in areas that are highly leached, as, for example, many coastal areas in the Coastal Plain of North Carolina, Georgia, and Florida (Figure 11.13). Zinc content is also low in areas such as the sand hills of Nebraska (northern Nebraska in Figure 11.13) where wind transport of the sand that formed the hills failed to move the heavy metals with the lighter quartz grains (16).

Zinc is recognized to be essential to all animals and man, especially during early stages of development and growth. Although required concentrations are small, even slight deficiencies can cause disorders of bones, joints, and skin; loss of fertility; and delayed healing. Zinc deficiencies are not unknown in man and may be associated with some chronic arterial disease, lung cancer, and other chronic diseases (13). A problem is that interrelations between zinc deficiencies common in hospital patients and cause of disease are poorly understood. Do these people have the disease partially because they suffer from a zinc deficiency, or do they have a zinc deficiency because they have the disease? If the former is true, then zinc therapy with its known beneficial effects on tissue repair may be useful in treatment of some chronic diseases (16).

Zinc supplements added to the soil may be helpful in preventing retardation of plant and animal growth. However, care must be taken because if the supplementary zinc is not of a very high quality, relatively large amounts of cadmium, which is often associated with zinc, may be inadvertently released into the environment. The result might be an elimination of the growth problem, but a new hazard from the cadmium. Cadmium has been associated with bone disease, heart disease, and cancer.

Selenium   In concentrated amounts, selenium may be the most toxic element in the environment, and it is a good example of why there is increasing concern over the necessity to control health-related trace elements. Selenium is required in the diet of animals at a concentration of 0.04 ppm. It is beneficial to 0.1 ppm and toxic above 4 ppm. Until a few years ago, selenium was of concern to biologists because of its toxicity. However, we now know that greater losses have resulted from selenium deficiency than from selenium toxicity (17, 18).

The primary source of selenium is volcanic activity. It has been estimated that throughout the history of the earth, volcanoes have released about 0.1 gram of selenium for every square centimeter of the earth's surface (17). Selenium ejected from volcanoes is in a particulate form. Therefore, it is easily removed from

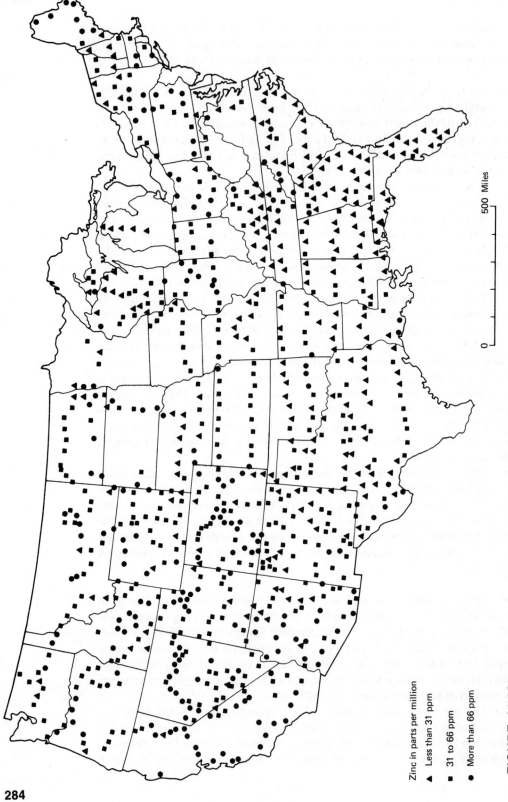

Zinc in parts per million

▲   Less than 31 ppm

■   31 to 66 ppm

●   More than 66 ppm

500 Miles

0

FIGURE 11.13 *Zinc content of surficial earth materials in the conterminous United States. (Source: H. T. Shacklette, U.S. Geological Survey Professional Paper 574 D, 1971.)*

volcanic gas by rain and is usually concentrated near the volcano. This explains why the average concentration of selenium in the crust of the earth is about 0.05 ppm, whereas the soils of Hawaii, which are derived from volcanic material, contain 6 to 15 ppm (17).

Selenium in soils varies from about 0.1 ppm in deficient areas to as much as 1200 ppm in organic, rich soils in toxic areas. However, it is interesting to note that Hawaii, with a selenium content of 6 to 15 ppm in the soil, does not produce toxic seleniferous plants, whereas South Dakota and Kansas with soils containing less than 1 ppm selenium do produce toxic seleniferous plants. This apparent dichotomy is resolved when the availability of selenium is examined. Elemental selenium is relatively stable in soils and not available to plants. The availability of the element is controlled primarily by whether the soil is alkaline or acidic. In acid soils (as in Hawaii), the selenium is in an unsoluble form, whereas in alkaline soils, selenium may be oxidized to a form that is very soluble in water and thus readily available to plants. This explains how forage crops grown for animals in soil containing soluble selenium are toxic, whereas forage crops grown in soils containing insoluble selenium have selenium deficiencies (17).

Selenium tends to be somewhat concentrated in living organisms. For example, some plants called *selenium accumulators* may contain over 2000 ppm selenium, whereas other plants in the same area may have less than 10 ppm total selenium (18). It is also interesting to note that selenium levels in human blood range from 0.1 to 0.34 ppm, which is about 1000 times that found in river water and several thousand times that found in seawater. Marine fish, on the other hand, contain selenium in amounts of about 2 ppm which is many thousand times that found in seawater (17). Because selenium accumulates in organic material, it may be concentrated in organic sediment and soil. Therefore, fossil fuels such as coal that are developed from organic material also contain selenium. It has been estimated that the annual release of selenium by combustion of coal and oil in the United States is about 4000 tons (17).

Little is known about selenium deficiency and toxicity in humans. There are a few cases of selenium poisoning of people living on food grown in highly seleniferous soils or drinking seleniferous water in undeveloped countries. However, in economically developed countries where interregional shipments of food are standard and diets are high in protein, there is probably little problem with selenium (18).

Geographic patterns of selenium distribution are valuable in assessing the possibility of forage and other feed crops containing relatively high or low selenium concentrations. Figure 11.14 and Table 11.6 show the selenium concentrations in surficial materials of the United States. Figure 11.15 shows the general distribution of selenium in plants in the United States. This map can serve as a general guide to locating areas where the addition of selenium to animal feed might be beneficial. However, it is emphasized that in some areas, selenium content cannot be accurately predicted, and therefore detailed analysis and mapping remain necessary (18).

Human Use, Trace Substances, and Health   Agricultural, industrial, and mining activities have all been responsible for releasing potentially hazardous and toxic

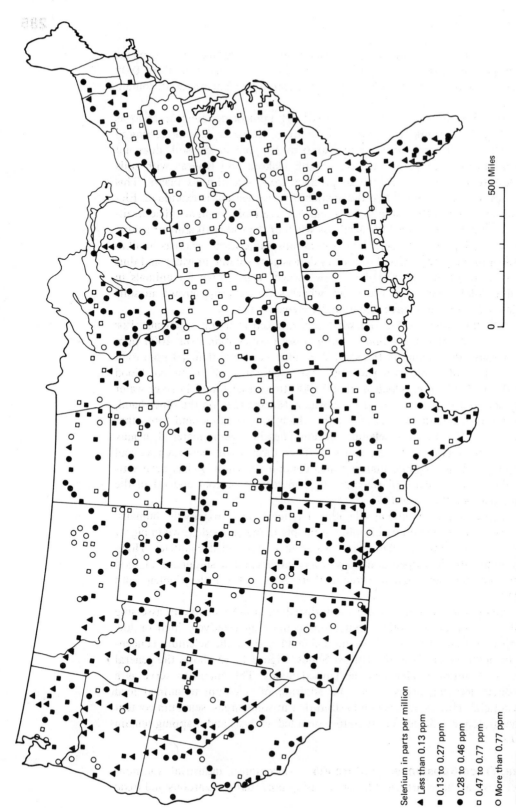

Selenium in parts per million

▲ Less than 0.13 ppm

■ 0.13 to 0.27 ppm

● 0.28 to 0.46 ppm

□ 0.47 to 0.77 ppm

○ More than 0.77 ppm

FIGURE 11.14 *Selenium concentrations in surficial earth material of the conterminous United States. (Source: H. T. Shacklette et al., U.S. Geological Survey Professional Circular, 1974.)*

Area	Range	Geometric Mean	Geometric Deviation	Arithmetic Mean [a]
Entire contermi-nous United States (912)	0.1–4.32	0.31	2.42	0.45
Western United States, west of the 97th meridian (492)	0.1–4.32	0.25	2.53	0.38
Eastern United States, east of the 97th meridian (420)	0.1–3.88	0.39	2.17	0.52

TABLE 11.6 *Concentration of selenium in parts per million in samples of soil and other surficial earth materials from the conterminous United States.*

SOURCE: U.S. Geological Survey Circular 692, 1974.
NOTE: Number of samples is given in parentheses after area.
[a] Estimated by method of Sichel (1952).

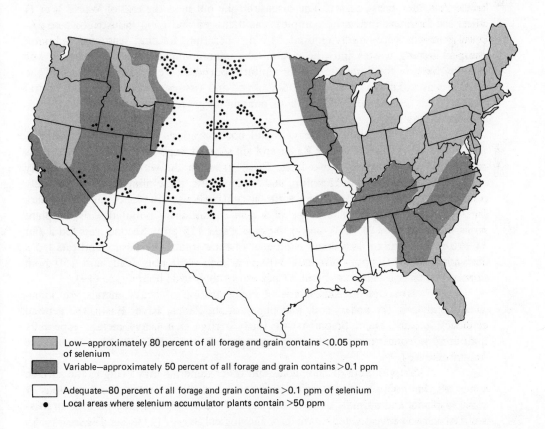

Low—approximately 80 percent of all forage and grain contains <0.05 ppm of selenium

Variable—approximately 50 percent of all forage and grain contains >0.1 ppm

Adequate—80 percent of all forage and grain contains >0.1 ppm of selenium
• Local areas where selenium accumulator plants contain >50 ppm

FIGURE 11.15 *Map showing the concentration of selenium in plants in the conterminous United States. (Reproduced from* Micronutrients In Agriculture, *1972, by permission of the Soil Science Society of America, Inc.)*

materials into the environment. This is part of the price we pay for our life-style. For example, unexpected and unanticipated problems have arisen from seemingly beneficial chemicals that control pests and disease. As we become more sophisticated at anticipating and correcting problems, it is expected that environmental disruption due to release of trace substances will be reduced.

Geologically significant examples of unexpected problems with trace substances are found in mining processes. Ironically, man expends time, energy, and money to extract resources concentrated by the geological cycle and in doing so may concentrate and release potentially harmful and toxic trace elements into the environment. Two rather different examples will illustrate this: first, the occurrence of a serious bone disease in Japan related to mining of zinc, lead, and cadmium; and second, a metabolic imbalance of cattle associated with mining of clay for the ceramic industry in Missouri.

A serious chronic disease known as *Itaiitai* in the Zintsu River Basin, Japan, has claimed many lives. The disease is extremely painful; the name *Itaiitai* literally means, "ouch, ouch." It attacks bones, causing them to become so thin and brittle that they break easily. The disease broke out near the end of World War II when the Japanese industrial complex was damaged and good industrial-waste disposal practices were largely ignored. Mining operations for zinc, lead, and cadmium dumped mining wastes into the rivers. The contaminated water downstream was used by farmers for domestic and agricultural purposes. For years, the cause of the disease was unknown. Then, in 1960, bones and tissues of the victims of Itaiitai disease were examined and found to contain large concentrations of zinc, lead, and cadmium (19).

Measurement of heavy-metal concentrations in the water, sediment, and plants of the Zintsu River Basin and subsequent experiments on rats fed diets of heavy metal established two facts. First, although the water samples generally contained less than 1 ppm cadmium and 50 ppm zinc, these minerals are selectively concentrated at higher rates in the sediment and higher yet in plants. One set of data for five samples shows an average of 6 ppm cadmium in polluted soils. In plant roots, this increased to 1250, and in the rice it was 125 ppm. Second, rats fed a diet of 100 ppm cadmium lost about 3 percent of their total bone tissue, and rats fed a diet containing 30 ppm cadmium, 300 ppm zinc, 150 ppm lead, and 150 ppm copper lost an equivalent of about 33 percent of their total bone tissue (19).

Although measurements of concentration of heavy metals are somewhat variable in the water, soil, and plants of the Zintsu River Basin, the general tendency is quite clear: Scientists are fairly certain that heavy metals, especially cadmium, in concentrations of a few parts per million in the soil and rice produce Itaiitai disease (20).

Strip mining of clay, coal, and other minerals may bring to the surface materials that contain anomalous concentrations of elements that are hazardous or toxic to plants and animals living in the area. An example is a clay pit area in Missouri which was investigated by the U.S. Geological Survey (21). (See Figure 11.16.) Mining of clay for use in the ceramic industry in the area has resulted in severe disturbances in the metabolism of beef cattle; these disturbances have interfered with growth, nutrition, and reproduction of the cattle.

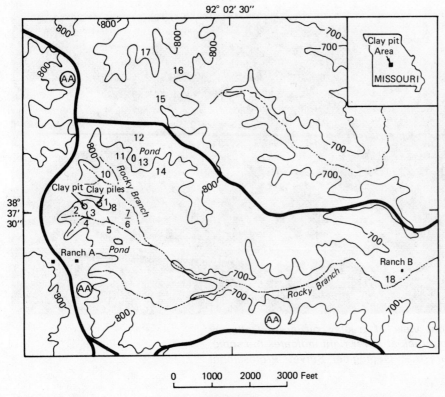

FIGURE 11.16 *Map of a clay pit area in Callaway County, Missouri. The numbers correspond to sample sites. (Source: Ebens et al., U.S. Geological Survey Professional Paper 807, 1971.)*

The topography in the western part of the area is characterized by a flat upland surface generally above 800 feet in altitude. The underlying rocks are interbedded clay, shale, sandstone, and coal covered with a variable thickness of windblown silt deposits. The eastern part of the area is generally below 800 feet in altitude, and the underlying rocks are limestones. Sinkholes in the limestone often contain deposits of fire clay which is mined. Figure 11.17 shows an abandoned clay pit that is several hundred feet wide, about 60 feet deep, and nearly filled with water. The large clay pile consists of clay with numerous fragments of shale, limestone, and pyritic material (iron sulfide). The smaller pit is almost entirely clay. Part of the surface runoff drains into the clay pit, and the remainder drains into the Rocky Branch. Only those cattle that were located on land downstream with direct access to the Rocky Branch suffered from metabolic disorders.

Although the rock types contain normal concentrations of elements, they are relatively rich in some elements compared to the soils. The occurrence of pyrite in the rocks is especially a problem because when pyrite is exposed to water, it weathers to sulfuric acid, which increases the solubility of other compounds. This

FIGURE 11.17 *Aerial view of a clay pit area, Callaway County, Missouri. Dumptruck at lower right indicates the scale. (Source: Ebens et al., U.S. Geological Survey Professional Paper 807, 1973.)*

is essentially the same problem that is prevalent in coal-mining areas where acidic water from mines contaminates streams. Water that drains from clay piles into the pit and streams is very acidic and contains high concentrations of some trace elements. This material is subsequently deposited in the bed of the stream and on the floodplains. Plants that grow on the clay piles along with the floodplains deposit contain some trace elements in concentrations toxic to animals that eat the plants (21).

Extensive analysis of over 30 different elements in the clay, sediment, and plants of the clay pit area revealed that about 20 were present in anomalous concentrations. These fit into one of five general groups (21): First, elements that occur in anomalous amounts in the clay and sediment, or both, and also are found in anomalous amounts in many plants; examples include aluminum, copper, molybdenum, nickel, and sodium. Second, elements that occur in anomalous concentrations in the clay and sediment but that are seldom found in anomalous amounts in plants; examples include barium, beryllium, chromium, and vanadium. Third, elements that occur in relatively low amounts in the clay and sediment but that are concentrated in some plants; examples include boron, cadmium, calcium, and zinc. Fourth, elements that are anomalous in clay or sediment but that are not easily evaluated in plants; examples are carbon, selenium, and silicon. Fifth, elements whose concentrations and rates of movement through the local environment are not easily categorized at this time; examples include iron, lead, magnesium, manganese, and strontium. The above grouping suggests that some elements might influence metabolic imbalances in cattle because the animals eat plants with a toxic level of a particular element as in the first and third groups. On the other hand, some elements might influence the metabolism of grazing animals because the animals directly ingest clay and sediment or drink water containing particularly harmful elements in solution or suspension.

Evaluation of the entire geochemical environment established two items. First, four elements (beryllium, copper, molybdenum, and nickel) are conspicuously anomalous in the clay, sediment, and plants. Second, other elements are highly mobile (beryllium, cobalt, copper, and nickel, among others). Of these elements, three (cobalt, copper, and molybdenum) are known to be significant in metabolic processes of animals. In trace amounts they are essential, but in high concentrations they are likely to be toxic (21).

The geologic, hydrologic, and biologic processes responsible for providing, transporting, and concentrating toxic trace elements are indeed complex. Figure 11.18 lists and describes the movement of elements through the geochemical system of the clay pit area. This particular example is especially valuable because it indicates the interaction of the complex, multidimensional aspects of geochemical

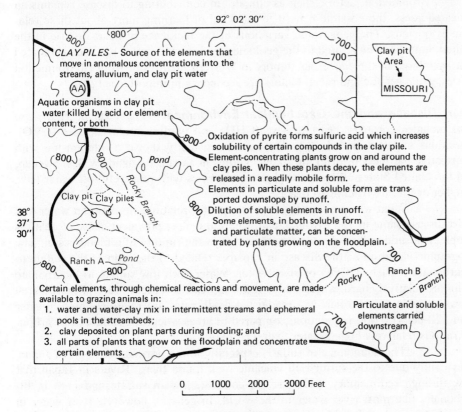

FIGURE 11.18 *Movement of elements through the geochemical system of a clay pit area, Callaway County, Missouri. (Source: Ebens et al., U.S. Geological Survey Professional Paper 807, 1973.)*

systems. Indeed, complexity is much more common than simplicity in the study of anomalous concentrations of trace elements and their effects on the biosphere.

Chronic Disease and Geologic Environment

Human health is a state of adjustment by an organism to its own internal environment and to its external environment. The relations between the geologic environment and regional or local variations in chronic disease such as cancer and heart disease have been observed for many years. Although evidence suggesting associations between the geochemical environment and chronic disease continues to accumulate, the real significance remains to be discovered. Reasons for the lack of conclusive results are twofold. First, hypotheses to test relations between the geologic environment and disease have not been specific enough. Basic research and field verification need to be better coordinated. Second, there remain many difficulties in obtaining reliable and comparable data for medical-geologic studies (22).

Although the significance of geologic variations compared to that of other environmental factors, such as climate, in contributing to disease remains an educated guess, the possible benefit to mankind of learning more about these relations is obvious. The geographic variations of the incidence of heart disease in the United States may be related to the geologic environment; and it has been estimated that two-thirds of the cancerous tumors in Western man result from environmental causes, while genetic and racial factors are secondary in importance (22).

Heart Disease and the Geochemical Environment   The term *heart disease* here includes coronary heart disease (CHD) and cardiovascular disease (CVD). Variations of heart disease mortality have consistently shown a relationship with the relative hardness of drinking water. Studies in Japan, England, Wales, Sweden, and the United States all conclude that communities with relatively soft water have a higher death rate from heart disease.

What was perhaps the first report of a relationship between water and heart disease came in 1957 from Japan, where the most prevalent cause of death is apoplexy, a sudden loss of body functions due to the rupture of a blood vessel. The geographic variation of the disease in Japan is related to the ratio of the sulfate to bicarbonate (SO_4 to HCO_3) of river water. Water with low sulfate-to-bicarbonate ratios are relatively hard, whereas high ratios indicate soft (acid) water. Figure 11.19 shows areas in Japan where the readjusted death rates are relatively high, greater than 120 per 100,000 population, compared to areas where the sulfate to bicarbonate ratio is relatively high, greater than 0.6 (23).

The abundance of sulfate, especially in the northeastern part of Japan, is evidently due to the sulfur-rich volcanic rock found there. Rivers in Japan that flow through sedimentary rocks are, in contrast, low in sulfate and high in bicarbonate, like most river water in the world. In general, however, river water in Japan is relatively soft, with a hardness less than 40 ppm, compared to in the United States where the mean hardness of raw municipal water is 139 ppm (24).

The consistent relationship between soft water and high death rates from heart disease is present in the United States. Figure 11.20 shows the adjusted

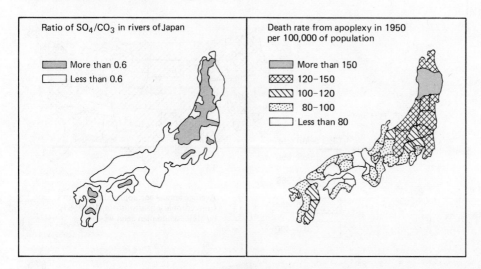

FIGURE 11.19 *Maps of Japan comparing the ratio of SO_4/CO_3 in rivers to the death rate of apoplexy in 1950. (Data from T. Kobayashi, 1957.)*

death rate from heart disease from 1949 to 1951 and the average hardness of the water (25). The figures are consistent with a more recent study in the United States which also found a significant negative correlation between hardness of water and death rates from heart disease (24).

The meaning of the relationship between hardness of water and death rates from heart disease is difficult to surmise. Several possibilities exist. First, the relationship might have nothing to do with heart disease. Second, the soft water is acidic and might, through corrosion, attack pipes and release into the water trace elements that cause heart disease. Third, there are some substances dissolved in hard water that might retard disease. Or fourth, some characteristics of soft water might enhance heart disease. Of course, some combination of the second, third, and fourth factors, along with others, is also likely. This is consistent with our observation that a disease may have several causes. Additional research is needed before we can prove the benefit of hard water and perhaps treat soft water to reduce heart disease. At this time, all we can say is that heart disease may indeed be related to hardness of drinking water, but the explanation remains obscured.

An interesting study of patterns of heart disease mortality and possible relations to the geologic environment was conducted in Georgia by the U.S. Geological Survey (26, 27). The contrast in heart disease mortality among the 159 counties in Georgia is such that the differences between the highest and lowest rates are nearly as great as can be found in any two counties in the United States. Nine counties in northern Georgia with low rates of death due to heart disease and nine counties in central and south central Georgia with high death rates due to heart disease were selected for geochemical analysis.

The locations of these counties are shown in Figure 11.21. Table 11.7 lists the death rates from heart disease and other causes in the selected counties. In the nine selected counties with low rates of death from heart disease, the range is

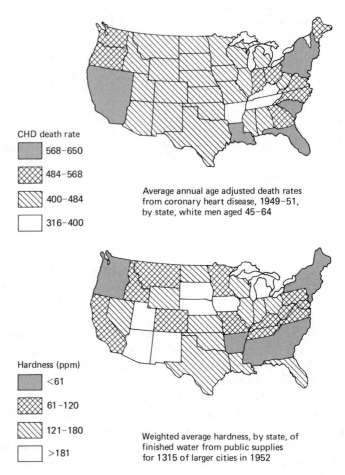

CHD death rate

▓	568–650
▨	484–568
▧	400–484
☐	316–400

Average annual age adjusted death rates
from coronary heart disease, 1949–51,
by state, white men aged 45–64

Hardness (ppm)

▓	<61
▨	61–120
▧	121–180
☐	>181

Weighted average hardness, by state, of
finished water from public supplies
for 1315 of larger cities in 1952

FIGURE 11.20   *Maps of the conterminous United States comparing the average annual adjusted death rate from coronary heart disease (CHD) to the weighted average hardness of the public water supplies. (Reprinted from E. F. Winton and L. J. McCabe,* Journal of American Water Works Association, *Volume 62, by permission of the Association. Copyrighted 1971 by the American Water Works Association, Inc., 6666 West Quincy Avenue, Denver, Colorado 80235.)*

from 560 to 682 deaths per 100,000 population for white males 35 to 74 years of age during the period of 1950 to 1959. In the nine counties selected with high mortality rates, the range is from 1151 to 1446 deaths per 100,000 population for the same age group and time period (26). Therefore, mortality rates for the south central part of the state are about twice as high as for the northern section.

      The nine counties with high heart-disease rates are primarily in southern Georgia on the Atlantic Coastal Plain (Figure 11.21). The landscape characteristically has low relief, sluggish drainage, and swamps. In general, sandy soils overlie Cenozoic marine sedimentary rocks that have experienced intensive weather-

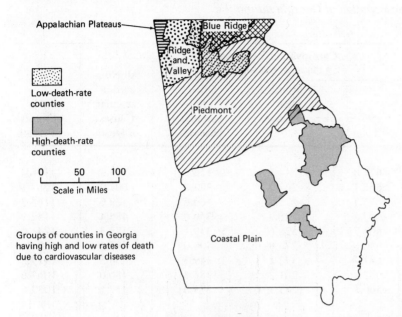

Appalachian Plateaus

Blue Ridge

Ridge and Valley

Low-death-rate counties

High-death-rate counties

0   50   100
Scale in Miles

Piedmont

Groups of counties in Georgia having high and low rates of death due to cardiovascular diseases

Coastal Plain

FIGURE 11.21 *Map showing physiographic regions of Georgia counties having high and low death rates due to cardiovascular disease. (Source: Shacklette, Sauer, and Miesch, U.S. Geological Survey Professional Paper 574C, 1970.)*

ing. Small-scale agricultural activity is prevalent where relief and drainage are sufficient.

The nine counties with low heart-disease rates are located in northern Georgia in the Appalachian Highlands, known as the Piedmont, the Blue Ridge, and the Valley and Ridge (Figure 11.21). The *Piedmont* extends from the Blue Ridge Mountains to the Coastal Plain. Rocks here are a mixture of Precambrian and Paleozoic metamorphic rocks and intrusive igneous rocks of varying ages. Soils are generally mixtures of sand, silt, and clay (loam) with clay subsoils. The *Blue Ridge* topography is mountainous with narrow valleys and turbulent streams. The rocks include metamorphic and sedimentary rocks of Precambrian to early Cambrian age and Paleozoic igneous rocks. Soils are acid with relatively high organic content; high relief is not favorable to agricultural activity. The *Valley and Ridge* is characterized by linear ridges and parallel valleys oriented northeast to southwest; rocks are folded and faulted sedimentary rock of Paleozoic age. Soils vary from rich loam soils in valleys with limestone bed rock to less fertile clay soils on shales and relatively infertile soils derived from sandstone bed rock (26).

A detailed geochemical investigation of the soils and plants of the low-death-rate area in northern Georgia and high-death-rate area in south central Georgia was conducted (26, 27). It was concluded that the geochemical variations in the soil reflected differences in the bed rock. Thirty elements were analyzed, and significant concentrations of aluminum, barium, calcium, chromium, copper, iron, potassium, magnesium, manganese, niobium, phosphorus, titanium, and vanadium were found. Zirconium was the only element that occurred in significantly larger

TABLE 11.7 *The death rates from selected causes, white males age 35 to 74, for selected counties of Georgia during the period 1950 to 1959.*

| County | Cardiovascular diseases (330–334, 400–468 [a]) | | | All Non-cardio-vascular Causes of Death | All Causes of Death |
	Total	Coronary Heart Disease (420 [a])	Other Cardio-vascular Diseases		
Low-rate counties					
Cherokee	671.0	344.1	326.9	525.4	1196.4
Fannin	655.2	262.2	393.0	607.9	1263.1
Forsyth	635.7	275.1	360.6	529.6	1165.3
Gilmer	647.5	296.9	350.6	485.4	1132.9
Hall	667.7	328.6	339.1	573.3	1241.0
Murray	680.8	274.8	406.0	593.6	1274.4
Pickens	681.7	332.2	349.5	735.5	1417.2
Towns	587.6	201.2	386.4	480.0	1067.6
Union	560.1	287.1	273.0	477.5	1037.6
High-rate counties that were studied					
Bacon	1302.2	743.1	559.1	789.9	2092.1
Bleckley	1205.5	650.6	554.9	546.0	1751.5
Burke	1167.0	603.7	563.3	834.3	2001.3
Dodge	1206.3	569.4	636.9	736.1	1942.4
Emanuel	1167.8	467.4	700.4	731.5	1899.3
Jeff Davis	1446.4	701.1	745.3	568.6	2015.0
Jefferson	1151.2	679.3	471.9	677.9	1829.1
Jenkins	1330.8	530.1	800.7	802.5	2133.3
Warren	1321.7	915.6	406.1	760.2	2081.9
Other high-rate counties					
Baldwin [b]	1218.8	781.5	437.3	744.7	1963.5
Chatham	1177.2	753.7	423.5	757.2	1934.4
Johnson	1170.3	644.1	526.2	700.1	1870.4
Lee	1170.6	754.5	416.1	927.8	2098.4
Long	1162.3	614.6	547.7	638.1	1800.4
Marion	1155.8	667.2	488.6	686.6	1842.4
Randolph	1252.1	572.2	679.9	618.1	1870.2
Richmond [b]	1203.0	773.5	429.5	771.3	1974.3
Treutlen	1171.7	533.2	638.5	766.2	1937.9
All counties of Georgia	925.2	520.4	404.8	649.0	1574.2

SOURCE: Shacklette, Sauer, and Miesch, U.S. Geological Survey Professional Paper 574C, 1970.
NOTE: Average annual death rates per 100,000 population, age adjusted by 10-year age groups by the direct method to the entire United States population age 35–74 in 1950. Deaths tabulated by Georgia State Department of Health, by county of usual residence.
[a] Code number of International Statistical Classification of Diseases, 7th Revision (World Health Organization, 1957).
[b] With adjustments for resident institution populations.

concentrations in soils of the high-death-rate counties (27). The differences probably reflect the fact that soils in the high-death-rate counties of the Coastal Plain are derived from unconsolidated sediments (sands and clays) that were previously weathered and leached during deposition.

Several trace elements are known to have beneficial effects on heart disease. Of these, manganese, chromium, vanadium, and copper are more highly concentrated in the low-death-rate areas of Georgia. Therefore, the low death rate may be due to an abundance of beneficial trace elements in the soils rather than concentrations of harmful elements. In other words, the tendency for higher mortality due to heart disease in Georgia may be a causal relationship facilitated by a deficiency rather than an excess of certain trace elements (27).

Additional information is needed about other elements and their relationship to heart disease. Cadmium, fluorine, and selenium, among other elements, require further study to increase our understanding of how trace elements affect the heart and the circulatory system. With respect to cadmium, we know that individuals dying from hypertensive complications generally have increased concentrations of cadmium or increased ratios of cadmium to zinc in their kidneys compared to individuals dying from other diseases. However, surprisingly, workers who are exposed to cadmium dust and accumulate the element in their lungs do not have abnormal rates of hypertension (28).

Cancer and the Geochemical Environment   Cancer, one of the most serious of diseases, tends to be strongly related to environmental conditions. However, as with heart disease, relations between the geochemical environment and cancer cannot be proven, and undoubtedly the causes of the various types of cancers are complex and involve many variables, some of which may be the presence or absence of certain earth materials which contribute to, or help protect people from, disease.

Relations between cancer and environment have two aspects to consider: cancer-causing (carcinogenic) substances released into the environment by *human use* of resources; and cancer-causing substances that occur *naturally* in earth materials, such as soil and water.

Recent information released to the news media suggests that the presence of cancer-causing substances in drinking water in the United States is ubiquitous. This may or may not be true, but certainly water polluted with industrial waste containing toxic chemicals, some of which are possible carcinogens, is being released into our surficial water supplies. The Mississippi River, particularly, has pollution problems, and ironically even present water treatment causes problems. Some industrial waste when combined with chlorination in water treatment turns into cancer-producing material. Other carcinogens escape water treatment because antiquated treatment procedures cannot remove toxic substances.

The occurrence of certain cancers has been related to the natural environment. Examples include iodine deficiency and its relation to breast cancer (14); mineralized drinking water and its relation to stomach cancer (29); organic matter, zinc, and cobalt in soil and their relation to stomach cancer (30, 31); and soil salinity, climate, vegetation, and agriculture and their relation to esophageal cancer (32). It is emphasized that these studies are not proof that there is a cause-and-effect relationship between certain earth materials and cancer. They do, however, indicate an area for future research that promises to be significant in solving environmental health problems.

The leading cause of death for women between 40 and 44 years of age is breast cancer. It is also the leading cause of death for all types of cancer for all women between the ages 35 and 55 (14).

The occurrence of breast cancer in the United States has been related to areas with an iodine deficiency (the goiter belt). Figure 11.22 shows areas with a relatively high incidence of breast cancer superposed on areas with an iodine deficiency. It has also been observed that countries with iodine deficiencies and resulting high incidence of goiter also have high incidences of breast cancer. The converse also holds; that is, countries with sufficient iodine have low incidence of goiter and breast cancer (14). The significance of this observation is not well understood, but it apparently is real.

An interesting study into the relations between the occurrence of cancer, especially stomach cancer, in West Devon, England, established the hypothesis that cancer in that area is connected with water supplies derived from a particular rock type. Figure 11.23 shows the geology of the area and the high incidence of cancer where drinking water is pumped from the rocks of Devonian age (345 to 400 million years old). The sedimentary rocks are highly mineralized compared to the sedimentary rocks of Carboniferous age (280 to 345 million years old) or the granite which is mineralized but to a lesser extent than the Devonian rocks. The research suggests that the cancer occurrence is associated with the mineralization of the Devonian rocks, but no specific cancer-causing substance was isolated (29).

Two studies in Wales and England have established relationships between stomach cancer and soil characteristics. An earlier study concluded that areas with a high incidence of stomach cancer had soils with a relatively high amount of organic matter (30). The amount of organic material is measured by the percentage of weight loss of a dry soil after burning. Figure 11.24 shows the correlation of areas with high mortality rates of stomach cancer with organic content of cultivated soils.

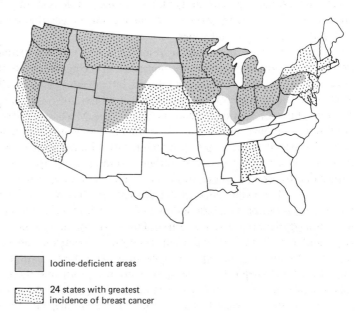

FIGURE 11.22 *Map of conterminous United States comparing iodine-deficient areas ("goiter belt") with the 24 states with the greatest incidence of breast cancer. (Data from Bogardus and Finley, 1961, and Spencer,* The Texas Journal of Science, *1970.)*

Iodine-deficient areas

24 states with greatest
incidence of breast cancer

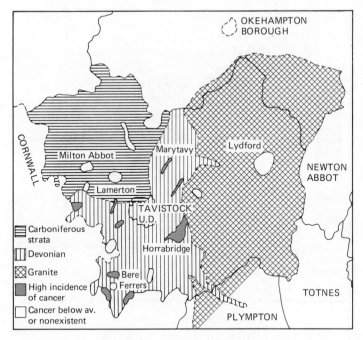

FIGURE 11.23 *Map of West Devin, England, showing the geology and areas where the incidence of cancer is high or low. (Source: Allen-Price,* The Lancet, *vol. 1, 1960.)*

However, this study was not able to associate disease rates with specific cancer-producing substances in the soils.

A more recent, detailed study of soils in northern Wales and Cheshire, England, combined geochemical analysis of the soil and measurement of organic content (31). It concluded that the abnormal rates of stomach cancer were associated with the amount of organic material in the soil such that abnormally high rates of the disease were related to long residence on soils with an organic content between certain limits. In addition, the study found a positive relation between the concentration of zinc, cobalt, and chromium and cancer of the stomach. Zinc and cobalt may be especially significant for two reasons. First, zinc is an active part of some essential enzyme systems in the human body and is also active in the processes of gastric digestion. Second, cobalt is known to have carcinogenic properties and is important in plant and animal activity. However, it is emphasized that although the high rates of cancer in this study are related to excesses of trace elements in the soil, the geographic distribution of such soils is apparently unrelated to that of stomach cancer (31).

Incident rates of esophageal cancer in northern Iran, near the Caspian Sea, offer another opportunity to examine relations between environment and chronic disease. In this area, the rates of esophageal cancer are tremendously variable, and marked differences occur over short distances (32). Research in the area indicates a relationship between cancer and the climate, soils, vegetation, and agricultural practices. However, there is a great deal of interrelationship between these factors as the climate obviously affects the soils, vegetation, and agricultural practices. The best relation between rates of esophageal cancer and environment is for soil types. Figure 11.25 shows the relative rates of cancer and the general soil types. It is obvious that the highest rates of the disease are associated with the saline

FIGURE 11.24 *Maps of North Wales rural districts comparing cancer mortality ratios and ignition loss of cultivated soils. (Source: C. D. Legon,* British Medical Journal, *September 27, 1952.)*

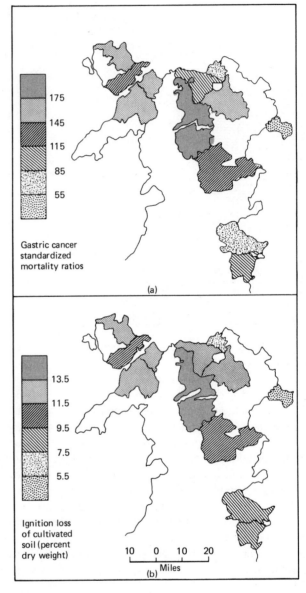

(salty) soils in the eastern section. These soils are in an area of little rainfall. Rainfall increases to the west by a factor of 4, and the soils become more and more leached. The lowest incidence of cancer is in the rain belt in the western section (Figure 11.25). The vegetation and agricultural practices of the east systematically change from grazing on sparse cover of specially adapted plants to lush forests and dry farming of rice, fruit, and tea in the west. Paralleling this change from east to west is the continuous lessening of incidence of esophageal cancer (32).

The examples of relations between environment and incidence of cancer further establish the importance of multiple causes of chronic disease. Re-

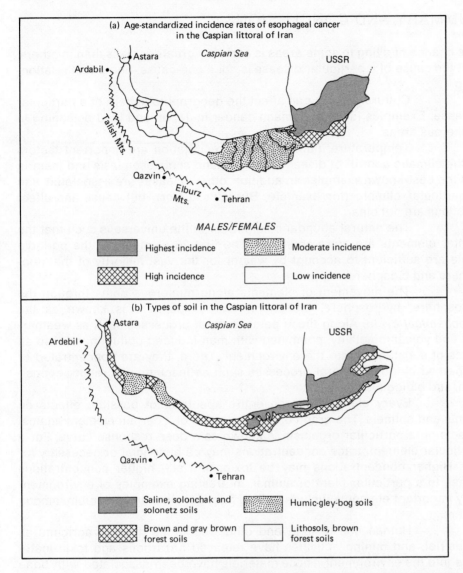

(a) Age-standardized incidence rates of esophageal cancer in the Caspian littoral of Iran

Caspian Sea

Astara

Ardabil

USSR

Talish Mts.

Qazvin

Elburz Mts.

Tehran

MALES/FEMALES

Highest incidence   Moderate incidence

High incidence   Low incidence

(b) Types of soil in the Caspian littoral of Iran

Caspian Sea

Astara

Ardebil

USSR

Qazvin

Tehran

Saline, solonchak and solonetz soils   Humic-gley-bog soils

Brown and gray brown forest soils   Lithosols, brown forest soils

FIGURE 11.25 *Maps of the esophageal cancer belt in Asia comparing age and standardized incidence rates of the cancer (top) with soil types (lower). (Source: Kmet and Mahboubi, Science, vol. 175, pp. 846–53, February 1972. Copyright 1972 by the American Association for the Advancement of Science.)*

gardless of the numerous contradictions and exceptions that can be found, sufficient data are now available to establish beyond a reasonable doubt that the hypothesis of a single cause for cancer is invalid. Rather, we must pursue a multifactorial approach, and variations in the natural and man-induced environment are certainly one of the significant factors.

SUMMARY AND CONCLUSIONS

The chance of dying in some areas is certainly greater or less than in others, and the cause of a particular disease is not a one-cause, one-effect relationship.

Cultural factors can affect the geographic pattern of a particular disease. Examples include stomach cancer in Japan and lead poisoning in numerous areas.

Temperature, humidity, and precipitation are important factors in the climatic control of disease patterns, and schistosomiasis and malaria are the best-known examples. In addition, other diseases are associated with a particular climate (for example, Burkett's tumor), but cause-and-effect relations are not clear.

The natural abundance of matter in the universe is such that the lighter elements are more abundant. The first 26 elements of the periodic table are sufficient to account by weight for the vast majority of the lithosphere and biosphere.

The movement of elements along numerous paths through the lithosphere, hydrosphere, biosphere, and atmosphere is known as the *geochemical cycle*. Along these paths, natural processes such as weathering and volcanic activity, combined with man-induced pollution, release all types of substances into the environment. There, they are concentrated or dispersed by other natural processes such as leaching, accretion, deposition, and biologic activity.

Every element has an entire spectrum of possible effects on plants and animals. The graph of concentration of a certain element against effects on a particular organism is known as a dose-response curve. For a particular element, trace concentrations may be beneficial or necessary for life. Higher concentrations may be toxic, and still higher concentrations lethal, to a particular plant or animal. Interesting examples of environmentally important elements include iodine, fluorine, zinc, and selenium, among others.

Human use of land and other resources through agricultural, industrial, and mining activities have released hazardous and toxic materials into the environment. These materials have been associated with bone disease in people, metabolic disorders in cattle, and other biological problems.

Relations between chronic disease and the geologic environment are complex and difficult to analyze. Furthermore, relationships are not proof of a cause-and-effect relation. Nevertheless, considerable evidence is being gathered and studied, and preliminary results suggest that the geochemical environment is indeed a significant factor in the incidence of serious health problems such as heart disease and cancer.

Heart disease rates are apparently inversely related to the hardness of water. In addition, a study in Georgia suggests that deficiencies of trace elements may also be related to abnormal rates of heart disease.

Studies of cancer rates suggest that abnormal incidence in specific areas may be related to such factors as organic content of the soil and abundance of certain trace elements. In other areas, saline soils may be related to cancer. Although most tumors in Western man are assumed to be caused by environmental factors, the evidence of simple cause-and-effect relations are lacking, and multiple causation of disease remains more likely.

References

1. Sauer, H. I., and Brand, F. R. 1971. Geographic patterns in the risk of dying. In *Environmental geochemistry in health,* eds. H. L. Cannon and H. C. Hopps, pp. 131–50. Geological Society of America Memoir 123.

2. Hopps, H. C. 1971. Geographic pathology and the medical implications of environmental geochemistry. In *Environmental geochemistry in health,* eds. H. L. Cannon and H. C. Hopps, pp. 1–11. Geological Society of America Memoir 123.

3. Young, K. 1975. *Geology: the paradox of earth and man.* Boston: Houghton Mifflin.

4. Bylinsky, G. 1972. *Metallic menaces.* In *Man, health and environment,* ed. B. Hafen, pp. 174–85. Minneapolis: Burgess Publishing.

5. Warren, H. V., and Delavault, R. E. 1967. A geologist looks at pollution: mineral variety. *Western Mines* 40: 23–32.

6. Furst, A. 1971. Trace elements related to specific chronic diseases: cancer. In *Environmental geochemistry in health,* eds. H. L. Cannon and H. C. Hopps, pp. 109–30. Geological Society of America Memoir 123.

7. Cargo, D. N., and Mallory, B. F. 1974. *Man and his geologic environment.* Reading, Massachusetts: Addison-Wesley.

8. Underwood, E. J. 1971. Geographical and geochemical relationships of trace elements to health and disease. In *Trace substances in environmental health—IV,* ed. D. D. Hemphill, pp. 3–11. Columbia, Missouri: University of Missouri Press.

9. Mertz, W. 1968. Problems in trace element research. In *Trace substances in environmental health—II,* ed. D. D. Hemphill, pp. 163–69. Columbia, Missouri: University of Missouri Press.

10. Horne, R. A. Biological effects of chemical agents. *Science* 177: 1152–53.

11. Maier, F. J. 1971. Fluorides in water. In *Water quality and treatment,* 3rd ed., pp. 397–440. New York: McGraw-Hill.

12. Bernstein, D. S.; Sadowsky, N.; Hegsted, D. M.; Guri, C. D.; and Stare, F. J. 1966. Prevalence of osteoporosis in high- and low-fluoride areas in North Dakota. *The Journal of the American Medical Association* 198: 499–504.

13. Gilbert, F. A. 1947. *Mineral nutrition and the balance of life.* Norman, Oklahoma: University of Oklahoma Press.

14. Spencer, J. M. 1970. Geologic influence on regional health problems. *The Texas Journal of Science* 21: 459–69.

15. Shacklette, H. T., and Cuthbert, M. E. 1967. Iodine content of plant groups as influenced by variations in rock and soil type. In *Relations of geology and trace elements to nutrition,* eds. H. L. Cannon and D. F. Davidson, pp. 31–45. Geological Society of America Special Paper 90.

16. Pories, W. J.; Strain, W. H.; and Rob, C. G. 1971. Zinc deficiency in delayed health and chronic disease. In *Environmental geochemistry in health and disease,* eds. H. L. Cannon and H. C. Hopps, pp. 73–95. Geological Society of America Memoir 123.

17. Lakin, H. W. 1973. Selenium in our environment. In *Trace elements in the environment,* ed. E. L. Kothny, pp. 96–111. Advances in Chemistry Series 123, American Chemical Society.

18. Allaway, W. H. 1969. Control of environmental levels of selenium. In *Trace substances in environmental health—II,* ed. D. D. Hemphill, pp. 181–206. Columbia, Missouri: University of Missouri Press.

19. Pettyjohn, W. A. 1972. No thing is without poison. In *Man and his physical environment,* eds. G. D. McKenzie and R. O. Utgard, pp. 109–10. Minneapolis: Burgess Publishing.

20. Takahisa, H. 1971. Discussion in *Environmental geochemistry in health and disease,* eds. H. L. Cannon and H. C. Hopps, pp. 221–22. Geological Society of America Memoir 123.

21. Ebens, R. J.; Erdman, J. A.; Feder, G. L.; Case, A. A.; and Selby, L. A. 1973. *Geochemical anomalies of a claypit area, Callaway County, Missouri, and related metabolic imbalance in beef cattle.* U.S. Geological Survey Professional Paper 807.

22. Armstrong, R. W. 1971. Medical geography and its geologic substrate. In *Environmental geochemistry in health and disease,* eds. H. L. Cannon and H. C. Hopps, pp. 211–19. Geological Society of America Memoir 123.

23. Kobayashi, J. 1957. On geographical relationship between the chemical nature of river water and death-rate of apoplexy. *Berichte des Ohara Institute fur landwirtshaftliche biologie* 11: 12–21.

24. Schroeder, H. A. 1966. Municipal drinking water and cardiovascular death-rates. *Journal of the American Medical Association* 195: 125–29.

25. Winton, E. F., and McCabe, L. J. 1970. Studies relating to water mineralization and health. *Journal of American Water Works Association* 62: 26–30.

26. Shacklette, H. T.; Sauer, H. I.; and Miesch, A. T. 1970. *Geochemical environments and cardiovascular mortality rates in Georgia.* U.S. Geological Survey Professional Paper 574C.

27. Shacklette, H. T.; Sauer, H. I.; and Miesch, A. T. 1972. Distribution of trace elements in the occurrence of heart disease in Georgia. *Geological Society of America Bulletin* 83: 1077–82.

28. Schroeder, H. A. 1965. Cadmium as a factor in hypertension. *Journal of Chronic Disease* 18: 647–56.

29. Allen-Price, E. D. 1960. Uneven distribution of cancer in West Devon. *Lancet* 1: 1235–38.

30. Legon, C. D. 1952. The Aetiological significance of geographical variations in cancer mortality. *British Medical Journal,* 700–702.

31. Stocks, P., and Davies, R. I. 1960. Epidemiological evidence from chemical and spectrographic analysis that soil is concerned in the causation of cancer. *British Journal of Cancer* 14: 8–22.

32. Kmet, J., and Mahboubi, E. 1972. Esophageal cancer in the Caspian littoral of Iran. *Science* 175: 846–53.

Part Four

MINERALS, ENERGY, AND ENVIRONMENT

One of the most fundamental concepts of environmental geology is that this earth is our only suitable habitat and its resources are limited. This belief holds even though some resources such as timber, water, air, and food are renewable, whereas other resources such as oil, gas, and minerals are recycled so slowly in the geologic cycle that they are essentially nonrenewable. It is important to recognize, however, that renewable resources are renewable only as long as environmental conditions remain favorable for their natural or man-induced reproduction. Careless use of renewable resources, such as air, water, vegetation, and, to a lesser extent, soils, may render these renewable resources less renewable than desired.

Because resources are indeed limited, important questions arise: How long will a particular resource last? How much short- or long-term environmental deterioration are we willing to concede in order to insure that resources are developed in a particular area? How can we make the best use of available resources? These questions have no easy answers. We are now struggling with ways to better estimate the quality and quantity of resources. Unfortunately, it is extremely hard to address the second and third questions without a satisfactory answer to the first question, but how long a particular resource will last is further complicated because availability will change as our technological skills and discovery techniques become more sophisticated. Furthermore, the nature and extent of people's and institutions' willingness to conserve resources and protect sensitive environments from mineral exploitation and development depends on factors such as national security, desired standard of living, environmental awareness, and economic incentives, as well as many other factors, all of which are difficult to evaluate and which change with time.

Discussion of mineral resources often center on elements such as aluminum, copper, iron, lead, and zinc, and many people are surprised to learn that with the exception of iron, none of these elements are used as much as carbon, sodium, nitrogen, oxygen, sulfur, potassium, and calcium. The annual world consumption of carbon is on the order of 10 billion tons, reflecting our dependence on coal, oil, and gas. Sodium and iron are consumed at a rate of about 1 billion tons per year. Iron is the structural metal our civilization most depends on, and sodium is used primarily in chemical industries (chlorine production, glass, soda ash, etc.). Nitrogen, oxygen, sulfur, potassium, and calcium are each consumed at about the rate of 100 million tons per year, and of these five elements, four are used primarily as soil conditioners or fertilizers. Elements such as copper, aluminum, and lead have annual world consumption rates of about 10 million tons, and the much discussed metals such as gold and silver have annual consumption rates less than 10,000 tons. Therefore, a significant conclusion concerning the consumption of elements is that the nonmetallics, with the exception of iron, are consumed at much greater rates than elements used for their metallic properties (1).

Basically, our present resource problem (or crisis) is related to a people problem—too many people, for as stated earlier it is impossible to maintain an ever-growing population on a finite resource base. With this perspective, Chapter 12 will explore geologic and environmental aspects of mineral resources, and Chapter 13 will discuss energy resources. Although there is considerable overlap between these two subjects, for organizational purposes, all resources associated with energy production will be discussed separately. Furthermore, mineral resources will include broadly a wide variety of earth materials, including metallic and nonmetallic minerals as well as sand, gravel, crushed rock, and ornamental rock (dimension stone) which have commercial value. Some earth materials, such as those used for construction material (sand and gravel, etc.), are a low-value resource and have primarily a "place value." That is, they are economically extracted because they are located close to where they are to be used. Long hauls of these low-value, high-bulk materials drastically increase their price and reduce their chance of competing with other closer supplies. On the other hand, materials such as diamonds, copper, gold, and aluminum are high-value resources. These materials are extracted wherever they are found, and they are transported around the world to numerous markets regardless of distances (2).

References

1. Skinner, B. J. 1969. *Earth resources*. Englewood Cliffs, New Jersey: Prentice-Hall.
2. Flawn, P. T. 1970. *Environmental geology*. New York: Harper & Row.

12

MINERAL RESOURCES AND ENVIRONMENT

Minerals and Population

Infinite resources (including space) could support an infinite population. Unfortunately, the earth and its resources are finite while the population grows at an ever-faster pace. Although overpopulation has been a problem in certain areas for at least several hundreds of years, it is now apparent that it is a global problem, sometimes termed the *population bomb* (Figure 12.1). The world population doubled from 1 to 2 billion people from 1830 to 1930, an annual growth rate of less than 1 percent. By 1970, it had nearly doubled again, and by the year 2000, it is expected to double once more. This suggests that in the last 40 years the growth rate was a little less than 2 percent per year, and now it is a little greater than 2 percent per year. The problem is that this is *exponential* growth, which is a very dynamic process. Consider the simple example of a student who upon taking a job for one month requests from his employer that he be paid 1 cent for the first day of work, 2 cents for the second day, 4 for the third day, and so on. In other words, his pay would be doubled each day. What would his pay total? It would take the student one

FIGURE 12.1 *The population bomb.*
(Source: U.S. Department of State.)

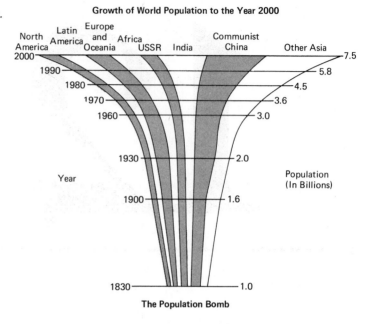

The Population Bomb

week to earn a wage of over $1 per day, and by the tenth day, he would be earning over $10 per day. By the middle of the month he would be receiving over $300 per day, and on the last day of the month (30 days), his earnings for that one day would be over $10 million.

Consider another hypothetical and completely fictitious example of exponential growth with more environmental significance (1). A farmer owns a pond in which a fast growing water plant becomes established. The plant chokes off all other life in the pond as it grows. The farmer is advised that the plant, which doubles its area each day, will cover the entire pond in about 6 weeks, or 42 days. As the plant appears to be growing rather slowly the first few days and weeks, the farmer decides not to worry about it until it covers approximately one-half his pond, at which time he plans to remove it. That time will come on the forty-first day, and the farmer will have one day to save his pond!

A question often asked is, How safe is our way of life? From a resource standpoint, affluence is indeed in jeopardy (1). Consider that approximately 22,000 pounds of new mineral material (excluding energy resources) are required each year for each person in the United States (Figure 12.2). When these data are combined with population data (Figure 12.3), the conclusion is clear—it is impossible in the long-run to match exponential growth of population with production of useful materials based on a finite resource base. Considering that many other countries aspire to become affluent also, while the world population increases, then the picture becomes more bleak. Table 12.1 summarizes for three commonly used metals (iron, copper, and lead) how much production would have to increase (from 1967 to 2000) to bring the rest of the world up to the standard of the United States. While these projections are not precise because of export practices and greater-than-expected increase in per capita use of minerals, they are useful and practical only because they suggest the possibility of a resource crisis that in the future may lower the ex-

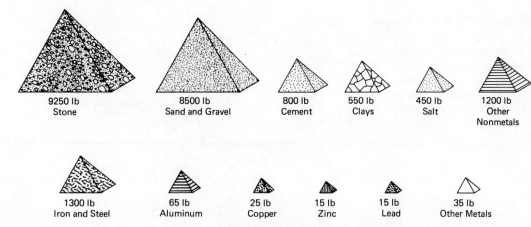

9250 lb
Stone

8500 lb
Sand and Gravel

800 lb
Cement

550 lb
Clays

450 lb
Salt

1200 lb
Other
Nonmetals

1300 lb
Iron and Steel

65 lb
Aluminum

25 lb
Copper

15 lb
Zinc

15 lb
Lead

35 lb
Other Metals

FIGURE 12.2 *Amount of new mineral materials required annually by each U.S. citizen. (Source: U.S. Bureau of Mines, Mining and Mineral Policy, 1975.)*

pected standard of living in affluent countries. However, historically, many expected crises have not developed, and hopefully there is time to insure through conservation, planning, and scientific advancements that the looming resource crisis also fails to bloom.

The United States and other affluent societies today are completely dependent on minerals and in fact could not survive without them (2). Table 12.2 lists a few of the mineral products, including petroleum, in a typical American home. This list shows that minerals are extensively used in almost every activity in our homes, offices, and industries.

A rather simple relation summarizes the role of minerals in influencing our expected standard of living (3):

$$L \approx \frac{R \times E \times I}{P}$$

where L is the average standard of living as measured by useful consumption of goods and services and is a function of R, the useful consumption of resources such as metals, nonmetals, soil, water, etc.; E is the useful consumption of energy; I is the

TABLE 12.1 *Comparison of the use of selected metals in 1967 and the predicted use by the year 2000.*

| Metal | Year 1967 | | | | Year 2000 | |
	United States per Capita	World per Capita	World Production	Bring World to U.S. Standard	Double Population as Today	Double Population World to U.S. Standard
Iron	1 ton	.17 ton	550×10^6 tons	$\times 6$	$\times 2$	$\times 12$
Copper	18 lb	3.2 lb	5.25×10^6 tons	$\times 5.5$	$\times 2$	$\times 11$
Lead	12 lb	1.5 lb	2.44×10^6 tons	$\times 8$	$\times 2$	$\times 16$

SOURCE: Data from C. F. Park, Jr., *Affluence in Jeopardy* (San Francisco: Freeman, Cooper & Company, 1968).

FIGURE 12.3 *Predicted increase in use of copper, lead, and iron compared to predicted increase in population. (From* Affluence in Jeopardy, *by Charles F. Park, Jr., published by Freeman, Cooper & Company, San Francisco, 1968.)*

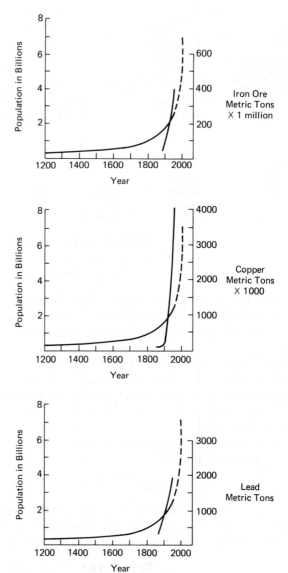

useful consumption of all types of ingenuity, including technical, social, economic, and political; and P is the total number of people involved in the system. This relationship clearly shows that any increase in R, E, or I will increase the standard of living, while an increase in P will decrease the standard of living. Of course, an even worse condition exists where R and E decrease while P increases and I cannot increase fast enough to compensate.

A problem facing the United States and many other affluent nations is that domestic supplies of many mineral resources must be supplemented to a lesser or greater extent by importing them from other nations. Figure 12.4 shows that imports in 1974 supplied a significant percentage of the total United States resource demands. However, that a particular material is imported does not necessarily suggest that it does not exist in quantities that could be mined in the United States.

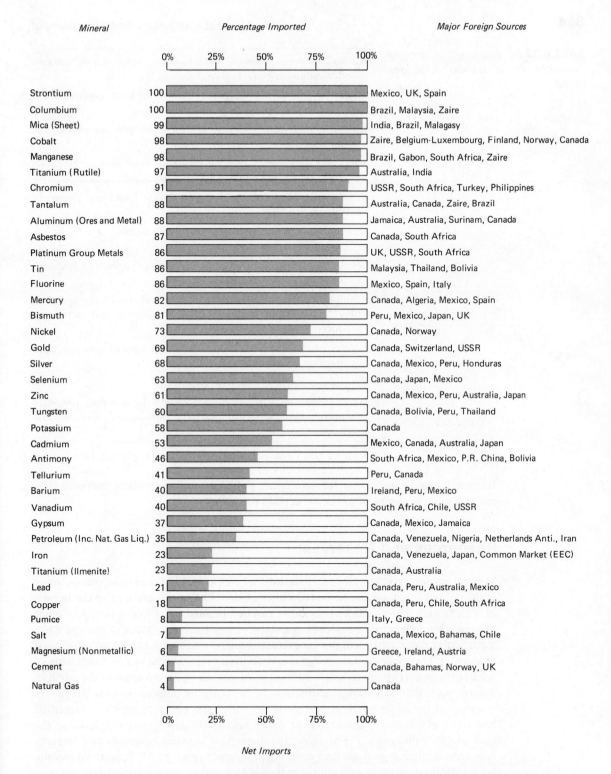

Mineral	Percentage Imported					Major Foreign Sources
	0%	25%	50%	75%	100%	
Strontium	100					Mexico, UK, Spain
Columbium	100					Brazil, Malaysia, Zaire
Mica (Sheet)	99					India, Brazil, Malagasy
Cobalt	98					Zaire, Belgium-Luxembourg, Finland, Norway, Canada
Manganese	98					Brazil, Gabon, South Africa, Zaire
Titanium (Rutile)	97					Australia, India
Chromium	91					USSR, South Africa, Turkey, Philippines
Tantalum	88					Australia, Canada, Zaire, Brazil
Aluminum (Ores and Metal)	88					Jamaica, Australia, Surinam, Canada
Asbestos	87					Canada, South Africa
Platinum Group Metals	86					UK, USSR, South Africa
Tin	86					Malaysia, Thailand, Bolivia
Fluorine	86					Mexico, Spain, Italy
Mercury	82					Canada, Algeria, Mexico, Spain
Bismuth	81					Peru, Mexico, Japan, UK
Nickel	73					Canada, Norway
Gold	69					Canada, Switzerland, USSR
Silver	68					Canada, Mexico, Peru, Honduras
Selenium	63					Canada, Japan, Mexico
Zinc	61					Canada, Mexico, Peru, Australia, Japan
Tungsten	60					Canada, Bolivia, Peru, Thailand
Potassium	58					Canada
Cadmium	53					Mexico, Canada, Australia, Japan
Antimony	46					South Africa, Mexico, P.R. China, Bolivia
Tellurium	41					Peru, Canada
Barium	40					Ireland, Peru, Mexico
Vanadium	40					South Africa, Chile, USSR
Gypsum	37					Canada, Mexico, Jamaica
Petroleum (Inc. Nat. Gas Liq.)	35					Canada, Venezuela, Nigeria, Netherlands Anti., Iran
Iron	23					Canada, Venezuela, Japan, Common Market (EEC)
Titanium (Ilmenite)	23					Canada, Australia
Lead	21					Canada, Peru, Australia, Mexico
Copper	18					Canada, Peru, Chile, South Africa
Pumice	8					Italy, Greece
Salt	7					Canada, Mexico, Bahamas, Chile
Magnesium (Nonmetallic)	6					Greece, Ireland, Austria
Cement	4					Canada, Bahamas, Norway, UK
Natural Gas	4					Canada
	0%	25%	50%	75%	100%	

Net Imports

FIGURE 12.4 *Imports supplied a significant percentage of total U.S. demand for minerals in 1974. (Source: U.S. Bureau of Mines,* Mining and Mineral Policy, *1975.)*

TABLE 12.2 *A few of the mineral products in a typical American home.*

Building materials: sand, gravel, stone, brick (clay), cement, steel, aluminum, asphalt, glass.

Plumbing and wiring materials: iron and steel, copper, brass, lead, cement, asbestos, glass, tile, plastic.

Insulating materials: rock wool, fiberglass, gypsum (plaster and wallboard).

Paint and wallpaper: mineral pigments (such as iron, zinc, and titanium) and fillers (such as talc and asbestos).

Plastic floor tiles, other plastics: mineral fillers and pigments, petroleum products.

Appliances: iron, copper, and many rare metals.

Furniture: synthetic fibers made from minerals (principally coal and petroleum products); steel springs; wood finished with rotten-stone polish and mineral varnish.

Clothing: natural fibers grown with mineral fertilizers; synthetic fibers made from minerals (principally coal and petroleum products).

Food: grown with mineral fertilizers; processed and packaged by machines made of metals.

Drug and cosmetics: mineral chemicals.

Other items, such as windows, screens, light bulbs, porcelain fixtures, china, utensils, jewelry: all made from mineral products.

SOURCES: U.S. Geological Survey Professional Paper 940, 1975.

Rather, it may suggest that for economic, political, or conservation purposes it is easier and more practical to import certain resources.

Resources and Reserves

Mineral resources may be broadly defined as elements, chemical compounds, minerals, or rocks that are concentrated either in the crust of the earth or in the oceans in a form such that a usable commodity can be obtained (4). From a practical point of view, this concept of a resource is unsatisfactory because, except in emergencies, a particular commodity will not be extracted unless extraction can be accomplished at a profit. Therefore, a more pragmatic definition of a *resource* is a concentration of a naturally occurring material (solid, liquid, or gas) in or on the crust of the earth in a form such that economical extraction is currently or potentially feasible (2). A *reserve,* on the other hand, is that portion of the total resource which is identified and from which usable materials can be economically and legally extracted *at the time* of the evaluation (2). Therefore, the distinction between resources and reserves is based on current geologic and economic factors (Figure 12.5). Resources include both identified reserves and other materials that either are identified but not currently economically or legally available or are undiscovered (currently hypothetical or speculative) (2).

The important point about resources and reserves remains: Resources

	Identified	Undiscovered		
		In known districts	In undiscovered districts or forms	
Economic	Reserves	Hypothetical	Speculative	
Subeconomic	Identified Subeconomic Resources	Resources	Resources	

Increasing degree of economic feasibility ↑

← Increasing degree of geologic assurance →

FIGURE 12.5 *Classification of mineral resources used by the U.S. Geological Survey and the U.S. Bureau of Mines. (Source: W. P. Pratt and D. A. Brobst, U.S. Geological Survey Circular 698, 1974.)*

Reserves: Identified resources from which a usable mineral or energy commodity can be economically and legally extracted at the time of determination.

Identified-subeconomic resources: Materials that are not reserves but that may become reserves as a result of changes in economic and legal conditions.

Hypothetical resources: Undiscovered materials that may reasonably be expected to exist in a known mining district under known geologic conditions.

Speculative resources: Undiscovered materials that may occur either in known types of deposits in a favorable geologic setting where no discoveries have been made, or in as-yet-unknown types of deposits that remain to be recognized.

are not reserves! An analogy from personal finances might help clarify this point (2). Reserves (financial) are an individual's liquid assets, such as money in the bank, whereas resources include the total income the individual can be expected to earn during his lifetime. This distinction between resources and reserves may have serious consequences because, reiterating, potential resources are not reserves. From our financial analogy, resources are "frozen assets" or next year's income and cannot be used to pay this month's bills (4).

Regardless of the inherent danger, it is very important from a long-range planning perspective to compute future resources, and this necessitates a continual reassessment of all components of total resources by considering new technology, probability of geologic discovery, and shifts in economic and political conditions (2).

The approach to resources and reserves outlined above and developed by the U.S. Geological Survey and the U.S. Bureau of Mines is superior to a simple periodical listing of the total amount of material available or likely to become available. Such a list is misleading when used for planning purposes. A superior method is to list or graph mineral resources in terms of the resource classification (Figure 12.5). United States domestic reserves and resources projected through the year 2000 for selected materials are shown in Table 12.3. Figure 12.6 is an example of how data from an identified resource can be tabulated, and it also shows how identified resources can be further subdivided. "Measured-identified" resources refer to those which are well known and measured and for which the total tonnage or grade is known within 20 percent. "Indicated-identified" resources are not so well known and measured and therefore cannot be completely outlined by tonnage or grade. "Inferred-identified" resources have quantitative estimates based on broad geologic knowledge of the deposit. The category that a particular identified resource will fit in is obviously a function of available geologic information which involves testing, drilling, and mapping, all of which become more expensive with greater depth. Therefore, it is not surprising that most of the information for coal is available for the identified resources that are buried less than 1000 feet from the surface (this includes 89 percent of the identified resources; see Figure 12.6). With greater depth, less and less information is available because most detailed mapping is still done from surface exposure (outcrop) of rocks where local relief is seldom more

TABLE 12.3 *A generalized outlook of domestic reserves and resources through the year 2000.*

Group 1: RESERVES in quantities adequate to fulfill projected needs well beyond 25 years.

Coal	Phosphorus
Construction stone	Silicon
Sand and gravel	Molybdenum
Nitrogen	Gypsum
Chlorine	Bromine
Hydrogen	Boron
Titanium (except rutile)	Argon
	Diatomite
Soda	[a]Barite
Calcium	Lightweight aggregates
Clays	Helium
Potash	Peat
Magnesium	[a]Rare earths
Oxygen	[a]Lithium

Group 2: IDENTIFIED SUBECONOMIC RESOURCES in quantities adequate to fulfill projected needs beyond 25 years and in quantities significantly or slightly greater than estimated UNDISCOVERED RESOURCES.

Aluminum	Vanadium
[a]Nickel	[a]Zircon
Uranium	Thorium
Manganese	

Group 3: Estimated UNDISCOVERED (hypothetical and speculative) RESOURCES in quantities adequate to fulfill projected needs beyond 25 years and in quantities significantly greater than IDENTIFIED SUBECONOMIC RESOURCES. Research efforts for these commodities should concentrate on geologic theory and exploration methods aimed at discovering new resources.

Iron	Platinum
[a]Copper	Tungsten
[a]Zinc	[a]Beryllium
Gold	[a]Cobalt
[a]Lead	[a]Cadmium
Sulfur	[a]Bismuth
[a]Silver	Selenium
[a]Fluorine	[a]Niobium

Group 4: IDENTIFIED SUBECONOMIC and UNDISCOVERED RESOURCES together in quantities probably not adequate to fulfill projected needs beyond the end of the century; research on possible new exploration targets, new types of deposits, and substitutes is necessary to relieve ultimate dependence on imports.

Tin	[a]Antimony
Asbestos	[a]Mercury
Chromium	[a]Tantalum

SOURCE: U.S. Geological Survey Professional Paper 940, 1975.
[a]Those commodities that may be in much greater demand than is now predicted because of known or potentially new applications in the production of energy.

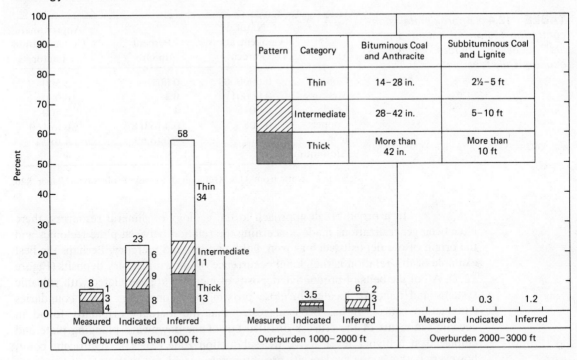

FIGURE 12.6 *Major resource categories of identified U.S. coal resources. (Source: D. A. Brobst and W. P. Pratt, eds., U.S. Geological Survey Professional Paper 820, 1973.)*

than several hundred feet. Sampling of subsurface materials is less certain, and indirect geophysical observations are usually not sufficient to make accurate measurements of identified resources that are well below the surface of the earth.

Geology of Mineral Resources

The geology of mineral resources is intimately related to the entire geologic cycle, and nearly all aspects and processes of the cycle are involved to a lesser or greater extent in producing local concentrations or occurrences of materials useful to man. The term *ore* is sometimes used for those useful metallic minerals that can be mined at a profit, and locations where ore is found are anomalous concentrations of these minerals. The concentration factor of a particular mineral necessary before the mineral can be classified as an ore varies with technology, economics, and politics as well as its natural concentration in the earth's crust. For example, the natural concentration of the element aluminum in the crust is about 8 percent, and it needs to be concentrated only to about 40 percent (concentration factor of five), whereas the natural concentration of mercury in the crust is only a tiny fraction of one percent and must have a concentration factor of about 10,000 to be economically mined. Table 12.4 lists other metallic elements and their current concentration factors necessary before mining is profitable. However, it is emphasized that these concentration factors necessary for mining may vary as demand changes.

TABLE 12.4 *Approximate concentration factors of selected metals necessary before mining is economically feasible.*

Metal	Natural Concentration (Percent)	Percent in Ore	Approximate Concentration Factor
Gold	0.0000004	0.001	2 500
Mercury	0.00001	0.1	10,000
Lead	0.0015	4	2 500
Copper	0.005	0.4 to 0.8	80 to 160
Iron	5	20 to 69	4 to 14
Aluminum	8	35	4

SOURCE: Data from U.S. Geological Survey Professional Paper 820, 1973.

In a broad-brush approach to the geology of mineral resources, there have been generalizations made concerning the relation between plate tectonics and the origin of ore deposits such as iron, gold, copper, and mercury. Perhaps the best example of this relation is the global occurrence of known mercury deposits (Figure 12.7). All of the belts of productive deposits of mercury are associated with volcanic systems and thus are located in close proximity to convergent plate boundaries (subduction zones). It has been suggested that the mercury originally found in oceanic sediments of the crust is distilled out of the downward-plunging plate and emplaced at a higher level above the subduction zone (2). The significant point, however, is that mercury deposits are intimately associated with volcanic and tectonic processes, and therefore convergent plate junctions which are characterized by volcanism and tectonic activities are likely places to find mercury. A similar argument can be made for other ore deposits, but there is danger in oversimplification, as many deposits are not directly associated with plate boundaries (Figure 12.8).

The geology of economic mineral and rock materials is as diversified and complex as the processes responsible for their formation or accumulation in the natural environment. Most deposits, however, can be related to various parts of the rock cycle within the influence of the tectonic, geochemical, and hydrologic cycles.

The genesis of mineral resources with commercial value can be subdivided into several categories: first, igneous processes, including crystal settling and late magmatic and hydrothermal activities; second, metamorphic processes such as contact metamorphism; third, sedimentary processes, including accumulation in oceanic, lake, stream, wind, and glacial environments; and fourth, weathering processes such as soil formations and insitu (in-place) concentration of insoluble minerals in weathered rock debris. Table 12.5 lists several examples of ore deposits from the above categories.

Igneous Processes   Most ore deposits resulting from igneous processes are due to an enrichment process that concentrates an economically desirable ore of metals such as copper, nickel, or gold. However, in some cases an entire igneous rock mass contains disseminated crystals that may be economically recovered. Perhaps the best-known example of this is the occurrence of diamond crystals found in a coarse-grained igneous rock called *kimberlite,* which characteristically is found as a pipe-shaped body of rock that decreases in diameter with depth. Almost the entire

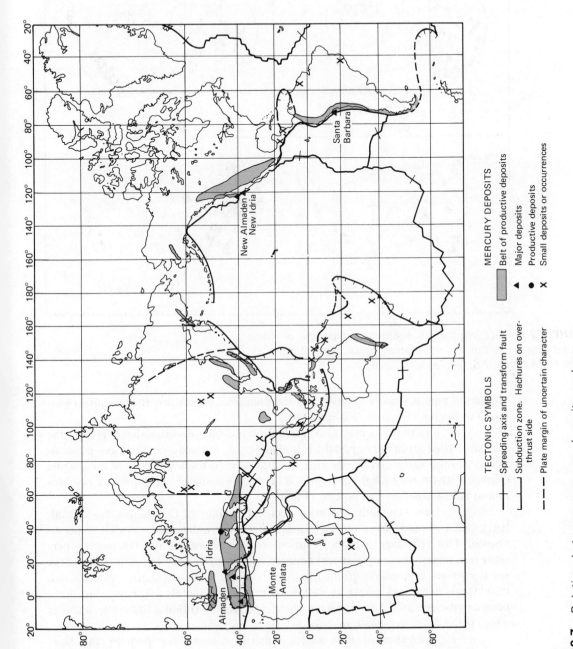

TECTONIC SYMBOLS

Spreading axis and transform fault

Subduction zone. Hachures on over-
thrust side

Plate margin of uncertain character

MERCURY DEPOSITS

Belt of productive deposits

▲ Major deposits

● Productive deposits

✕ Small deposits or occurrences

FIGURE 12.7 *Relation between mercury deposits and recently active subduction zones. (Source: D. A. Brobst and
W. P. Pratt, eds., U.S. Geological Survey Professional Paper
820, 1973.)*

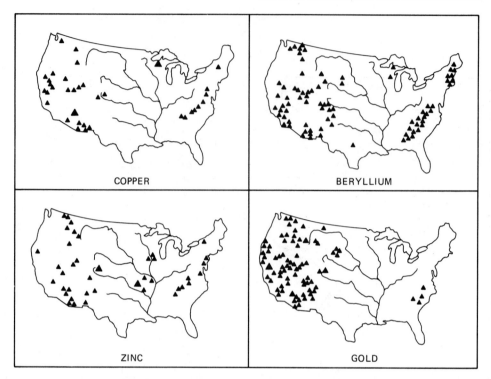

FIGURE 12.8 *Occurrence of economic deposits of some selected metals in the United States. (Data from mineral resource maps of the U.S. Geological Survey.)*

kimberlite pipe is the ore deposit, and the diamond crystals are disseminated (scattered) throughout the rock (Figure 12.9).

Ore deposits may also result from igneous processes that segregate the deposits earlier from later crystals. For example, as a magma cools, heavy minerals that crystallize early may slowly sink or settle toward the lower part of the magma chamber, where they form concentrated layers. Deposits of chromite (ore of chromium) have formed by this process (Figure 12.10).

Late magmatic processes occur after most of the magma has crystallized, and rare and heavy metalliferous material in water- and gas-rich solutions remain. This late-stage metallic solution may be squeezed into fracture or may settle into interstices (empty space) between earlier formed crystals. Other late-stage solutions form very coarse-grained igneous rock known as *pegmatite,* which is rich in feldspar, mica, and quartz as well as certain rare minerals (5). Pegmatites have been extensively mined for feldspar, mica, spodumene (lithium mineral), and clay which forms from weathered feldspar.

Hydrothermal (hot-water) mineral deposits are perhaps the most common type of ore deposit. They originate from late-stage magmatic processes and give rise to a variety of deposits, including gold, silver, copper, mercury, lead, zinc, and other metals as well as many nonmetallic minerals. The mineral material either is directly produced from the igneous parent rock, as ideally shown in Figure 12.11,

Type	Example
Igneous	
Disseminated	Diamonds—South Africa
Crystal settling	Chromite—Stillwater, Montana
Late magmatic	Magnetite—Adirondack Mountains, New York
Pegmatite	Beryl and lithium—Black Hills, South Dakota
Hydrothermal	Copper—Butte, Montana
Metamorphic	
Contact metamorphism	Lead and silver—Leadville, Colorado
Regional metamorphism	Asbestos—Quebec, Canada
Sedimentary	
Evaporite (lake or ocean)	Potassium—Carlsbad, New Mexico
Placer (stream)	Gold—Sierra Nevada foothills, California
Glacial	Sand and gravel—northern Indiana
Deep-ocean	Manganese oxide nodules—central and southern Pacific Ocean
Weathering	
Residual soil	Bauxite—Arkansas
Secondary enrichment	Copper—Utah

TABLE 12.5 *Examples of different types of ore deposits. (Modified from Robert J. Foster, Physical Geology, 2nd ed. Columbus: Charles E. Merrill, 1975.)*

or else is altered by metamorphic processes as it is intruded into the surrounding rock as veins or small dikes. However, many hydrothermal deposits cannot be traced to a parent igneous rock mass, and for these cases the origin remains unknown. However, it is speculated that circulating groundwater, heated and enriched with minerals after contact with deeply buried magma, might be responsible for some of these deposits (5, 6).

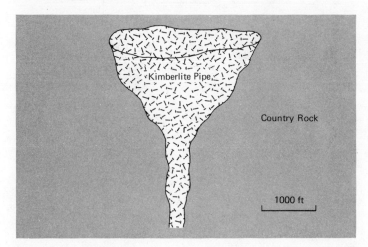

FIGURE 12.9 *Idealized diagram showing a typical South African diamond pipe. The diamonds are found scattered throughout the cylindrical body of igneous rock known as kimberlite. (Source: Foster, Physical Geology, 2nd ed., Merrill, 1975.)*

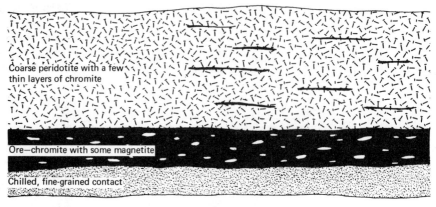

FIGURE 12.10 *Idealized diagram showing how chromite layers might form. The chromite crystallizes early, and the heavy crystals sink to the bottom and accumulate in layers. (Source: Foster,* Physical Geology, *2nd ed., Merrill, 1975.)*

Hydrothermal solutions that form ore deposits are basically mineralizing fluids that migrate through a host rock. Two types of deposits are recognized: *cavity-filling* and *replacement*. Cavity-filling deposits are formed when hydrothermal solutions migrate along openings in rocks such as fracture systems, pore spaces, or bedding planes and precipitate ore minerals. Replacement deposits, on the other hand, form as hydrothermal solutions react with the host rock, forming a zone in which ore minerals precipitate from the mineralizing fluids and replace part of the host rock. Although the replacement deposits are generally believed to dominate at higher temperatures and pressures than the cavity-filling deposits, actually both may be found closely associated as one grades into the other. That is, filling of an open

FIGURE 12.11 *Idealized diagram showing how hydrothermal and contact metamorphic ore deposits might form.*

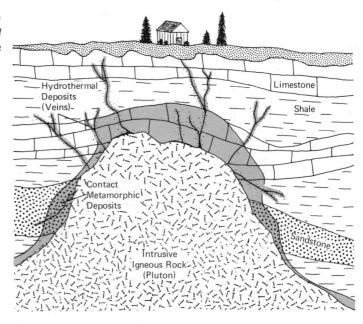

fracture by precipitation from hydrothermal solutions may be simultaneous with replacement of the rock lining the fracture (6).

Hydrothermal replacement processes are very significant because, excluding some iron and nonmetallic deposits, they have produced some of the world's largest and most important deposits of minerals. Some of these deposits result from a massive, nearly complete replacement of host rock with ore minerals that terminate abruptly; others form thin replacement zones along fissures; and still others form disseminated replacement deposits that may involve huge amounts of relatively low-grade ore (6).

The actual sequence of geologic events leading to the development of a hydrothermal ore deposit is usually complex. Consider, for example, the tremendous, disseminated copper deposits of northern Chile. The actual mineralization is thought to be related to igneous activity, faulting, and folding that occurred 60 to 70 million years ago. The ore deposit is an elongated, tabular mass along a highly sheared (fractured) zone separating two types of granitic rock. The concentration of copper results from a number of factors. First, there was a source igneous rock that supplied the copper. Second, the fissure zone tapped the copper supply and facilitated the movement of the mineralizing fluids. Third, the host rock was altered and fractured, preparing it for deposition and replacement processes which produced the ore. Fourth, the copper was leached and redeposited again by meteoric water, which further concentrated the ore (7).

Metamorphic Processes   Ore deposits are often found along the contact between igneous rocks and the country rocks they intrude. This area is characterized by contact metamorphism, which is caused by interactions between heat, pressure, and chemically active fluids of the cooling magma in contact with the surrounding (country) rock. The width of the contact metamorphic zone varies with the type of country rock. Generally, the zone for limestone is thickest because limestone is more reactive due to an increased mobility of reactants facilitated by the release of carbon dioxide (CO_2). The zone is generally thinnest for shale because the fine-grained texture retards the movement of hot, chemically active solutions; and it is intermediate for sandstone, as shown in Figure 12.11. Some of the mineral deposits that form in contact areas originate from the magmatic fluids and some from reactions of these fluids with the country rock.

Metamorphism may also result from regional increase of temperature and pressure associated with deep burial of rocks or tectonic activity. Regardless of its cause, regional metamorphism can change the mineralogy and texture of the preexisting rocks, producing ore deposits of asbestos, talc, graphite, and other valuable nonmetallic deposits (6).

Metamorphism has been suggested as a possible origin of some hydrothermal fluids. It is a particularly likely cause in high-temperature, high-pressure zones where fluids might be produced and forced out into the surrounding rocks to form replacement or cavity-filling deposits. For example, the native copper found along the top of ancient basalt flows in the Michigan copper district was apparently produced by metamorphism and alteration of the basalt which released the copper and other materials that produced the native copper deposits (7).

Our discussion of igneous and metamorphic processes has been concerned primarily with ore deposits. It is interesting and informative to learn also that igneous and metamorphic processes are responsible for producing a good deal of stone used in the construction industry. Granite, basalt, marble (metamorphosed limestone), slate (metamorphosed shale), and quartzite (metamorphosed sandstone), along with other rocks, are quarried to produce crushed rock and dimension stone in the United States (Figures 12.12 and 12.13). Stone is used in many aspects of construction work. However, it is a surprise to many people to learn that the stone industry in total value is the largest nonfuel, nonmetallic mineral industry in the United States (8).

Sedimentary Processes   In all aspects of the sedimentary cycle, from transportation to deposition of sediment, sedimentary processes are often significant in concentrating economically valuable earth materials in sufficient amounts that it is advantageous to recover the materials from depositional areas such as floodplains, lakes, or oceans. Throughout all aspects of the transport of sediments, processes of running water and wind help segregate the sediment by size and shape. Thus, the best sand or sand and gravel deposits for construction purposes are those where water or wind have effectively removed the finer materials, leaving the larger particles behind. Sand dunes, beach deposits, and deposits in stream channels are good examples.

The sand and gravel industry amounts to over $1 billion per year, and by volume mined it is the largest nonfuel mineral industry in the United States. Figure 12.14 shows the production data of sand and gravel compared to those of other construction material. Currently, most sand and gravel is obtained from river channels and water-worked glacial deposits. The United States now produces more sand and gravel than is needed for its requirements, but supply is likely to meet demand by the year 2000 (9).

Stream processes transport and sort all types of materials according to size and density. Therefore, if the bed rock in a river basin contains heavy metals such as gold, streams draining the basin may concentrate heavy metals to form "placer deposits" in areas where the turbulence and velocity of the water are minimized. These areas tend to be found in open crevices or fractures at the bottom of pools, on the inside of bends, or on riffles (Figure 12.15).

The settlement of California, Alaska, and other areas in the United States was facilitated by placer mining of gold, which is known as a "poor man's method" because a miner needed only a shovel, a pan, and a strong back to work his stream-side claim. Furthermore, the gold in California attracted miners who acquired the expertise necessary to locate and develop other resources in the western United States and Alaska.

Placer deposits of gold and diamonds have also been concentrated by coastal processes, primarily wave action; and beach sands and near-shore deposits are mined in Africa and other places.

Rivers and streams that empty into the oceans and lakes carry tremendous quantities of dissolved material derived from the weathering of rocks. From time to time (geologically), shallow marine basins may be isolated by tectonic activity (uplift) which restricts circulation and facilitates evaporation. In other cases, large inland lakes with no outlets essentially dry up. As the evaporation progresses, the dissolved materials precipitate out, forming a wide variety of compounds,

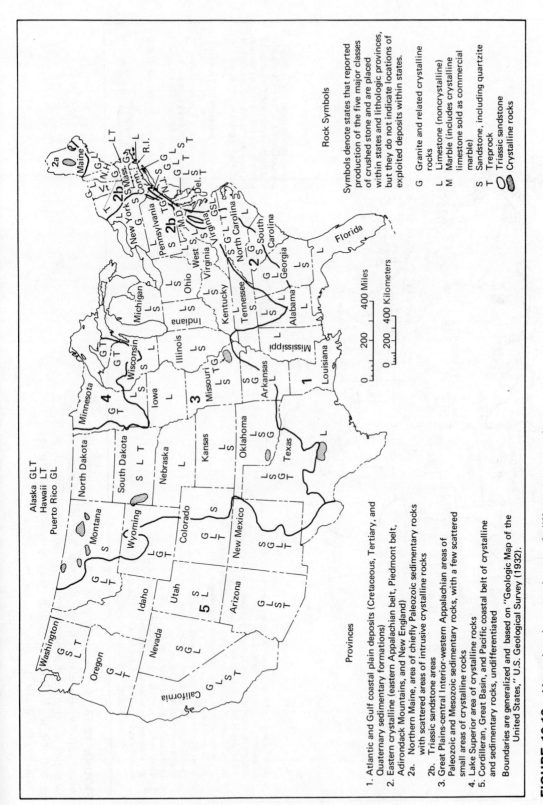

FIGURE 12.12 *Map showing provinces of different types of rocks in the United States and related occurrence and production of principal kinds of crushed stone for 1969. (Source: D. A. Brobst and W. P. Pratt, eds., U.S. Geological Survey Professional Paper 820, 1973.)*

Rock Symbols

Symbols denote states that reported production of the five major classes of crushed stone and are placed within states and lithologic provinces, but they do not indicate locations of exploited deposits within states.

G   Granite and related crystalline rocks
L   Limestone (noncrystalline)
M   Marble (includes crystalline limestone sold as commercial marble)
S   Sandstone, including quartzite
T   Triassic sandstone
⬭   Crystalline rocks

Provinces

1. Atlantic and Gulf coastal plain deposits (Cretaceous, Tertiary, and Quaternary sedimentary formations)
2. Eastern crystalline (eastern Appalachian belt, Piedmont belt, Adirondack Mountains, and New England)
2a. Northern Maine, area of chiefly Paleozoic sedimentary rocks with scattered areas of intrusive crystalline rocks
2b. Triassic sandstone areas
3. Great Plains-central Interior-western Appalachian areas of Paleozoic and Mesozoic sedimentary rocks, with a few scattered small areas of crystalline rocks
4. Lake Superior area of crystalline rocks
5. Cordilleran, Great Basin, and Pacific coastal belt of crystalline and sedimentary rocks, undifferentiated

Boundaries are generalized and based on "Geologic Map of the United States," U.S. Geological Survey (1932).

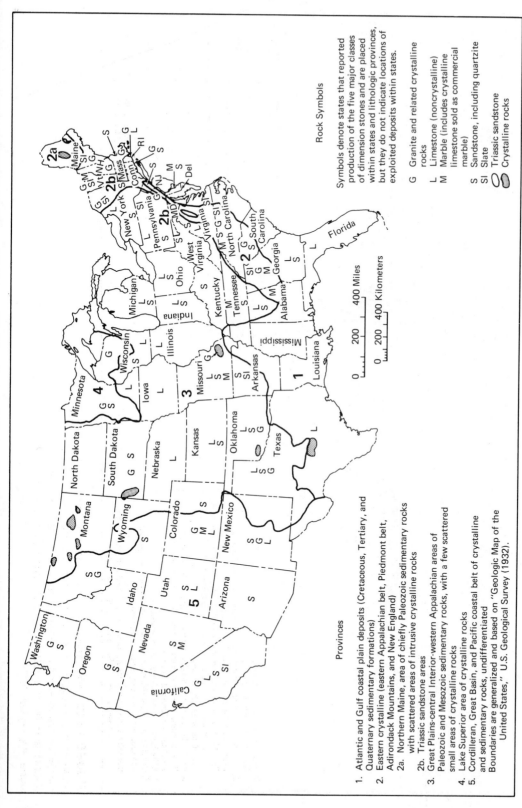

FIGURE 12.13 *Provinces of different types of rocks in the United States as related to occurrence and production of principal kinds of dimension stone for 1969. (Source: D. A. Brobst and W. P. Pratt, eds., U.S. Geological Survey Professional Paper 820, 1973.)*

Rock Symbols

Symbols denote states that reported production of the five major classes of dimension stones and are placed within states and lithologic provinces, but they do not indicate locations of exploited deposits within states.

G   Granite and related crystalline rocks
L   Limestone (noncrystalline)
M   Marble (includes crystalline limestone sold as commercial marble)
S   Sandstone, including quartzite
Sl Slate
▬   Triassic sandstone
⬭   Crystalline rocks

Provinces

1. Atlantic and Gulf coastal plain deposits (Cretaceous, Tertiary, and Quaternary sedimentary formations)
2. Eastern crystalline (eastern Appalachian belt, Piedmont belt, Adirondack Mountains, and New England)
   2a. Northern Maine, area of chiefly Paleozoic sedimentary rocks with scattered areas of intrusive crystalline rocks
   2b. Triassic sandstone areas
3. Great Plains-central Interior-western Appalachian areas of Paleozoic and Mesozoic sedimentary rocks, with a few scattered small areas of crystalline rocks
4. Lake Superior area of crystalline rocks
5. Cordilleran, Great Basin, and Pacific coastal belt of crystalline and sedimentary rocks, undifferentiated
   Boundaries are generalized and based on "Geologic Map of the United States," U.S. Geological Survey (1932).

326

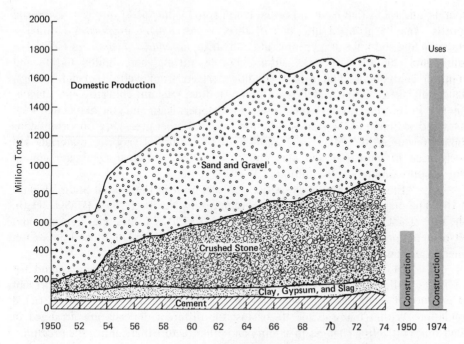

FIGURE 12.14 *Supplies and uses of major nonmetallic construction materials in the United States from 1950 to 1974. (Source: U.S. Bureau of Mines,* Mining and Mineral Policy, *1975.)*

FIGURE 12.15 *Idealized diagram of a stream channel and bottom profile showing areas where placer deposits of gold are likely to occur.*

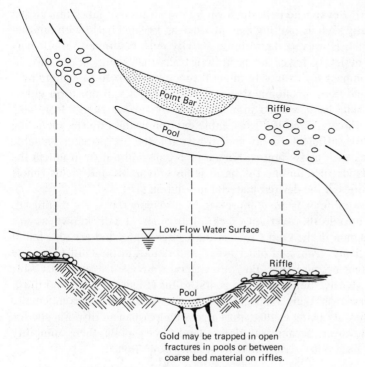

minerals, and rocks that have important commercial value. Most of these "evaporite deposits" can be grouped into one of three types: *marine evaporites* (solids)—potassium and sodium salt, gypsum, and anhydrite; *nonmarine evaporites* (solids)—sodium and calcium carbonate, sulfate, borate, nitrate, and limited iodine and strontium compounds; and *brines* (liquids derived from wells, thermal springs, inland salt lakes, and seawaters)—bromine, iodine, calcium chloride, and magnesium (9). Furthermore, heavy metals (such as copper, lead, and zinc associated with brines and sediment in the Red Sea, Salton Sea, and other areas) are important conditional resources that may be exploited in the future. Evaporite materials are widely used in industrial and agricultural activities, and their aggregate annual value is approximately $600 million (10).

Extensive marine evaporite deposits exist in the United States (Figure 12.16). The major deposits are halite (common salt, NaCl), gypsum ($CaSO_42H_2O$), anhydrite ($CaSO_4$), and interbedded limestone ($CaCO_3$). Limestone, gypsum, and anhydrite are present in nearly all marine evaporite basins, and halite and potassium minerals are found in a few (10).

Marine evaporites may form stratified deposits that may extend for hundreds of miles with a thickness of several thousand feet. The evaporites represent the product of evaporation of seawater in isolated shallow basins with restricted circulation. Within many evaporite basins, the different deposits are arranged in broad zones reflecting changes in salinity and other factors that control the precipitation of evaporites. That is, different materials may be precipitated at the same time in different parts of the evaporite basin. For example, halite is precipitated in areas where the brine is more saline and gypsum where it is less saline. Economic deposits of potassium evaporite minerals are relatively rare but may form from highly concentrated brines.

Nonmarine evaporite deposits form by evaporation of lakes in a closed basin. Tectonic activity such as faulting may produce an isolated basin with internal drainage and no outlet. However, the tectonic activity must continue to uplift barriers across possible outlets or lower the basin floor faster than sediment can raise it, if a favorable environment for evaporite mineral precipitation is to be maintained. Even under these conditions, economic deposits of evaporites will not form unless sufficient dissolved salts have washed into the basin by surface runoff from surrounding highlands. Finally, even if all favorable environmental criteria, including an isolated basin with sufficient runoff and dissolved salts, are present, valuable nonmarine evaporites such as sodium carbonate or borate will not form unless the geology of the highlands surrounding the basin is also favorable and yields runoff with sufficient quantities of the desired material in solution (10).

Some evaporite beds are compressed by overlying rocks, are mobilized, and then pierce or intrude the overlying rocks. Intrusions of salt, known as *salt domes,* are quite common in the Gulf Coast of the United States and are also found in northwestern Germany, Iran, and other areas. Salt domes in the Gulf Coast are economically important because first, they are a good source for nearly pure salt; second, some have extensive deposits of elemental sulfur (Figure 12.17); and third, some have oil reserves on their flanks. Furthermore, they are environmentally important because they are being evaluated as a possible permanent disposal site for radioactive waste. However, because salt domes tend to be mobile, their suitability as a disposal site for hazardous wastes must be seriously questioned.

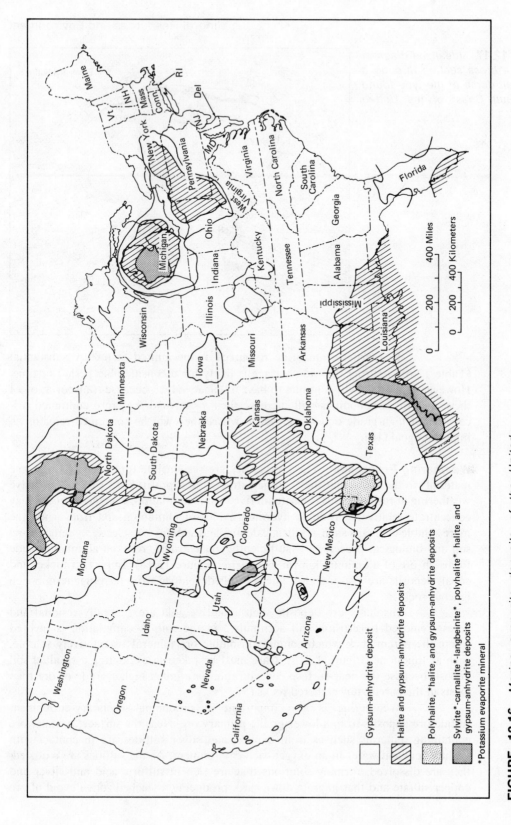

FIGURE 12.16 *Marine evaporite deposits of the United States. (Source: D. A. Brobst and W. P. Pratt, eds., U.S. Geological Survey Professional Paper 820, 1973.)*

Gypsum-anhydrite deposit

Halite and gypsum-anhydrite deposits

Polyhalite,*halite, and gypsum-anhydrite deposits

Sylvite*-carnallite*-langbeinite*, polyhalite*, halite, and gypsum-anhydrite deposits

*Potassium evaporite mineral

FIGURE 12.17 *Idealized diagram showing a cross section through a typical salt dome of the type found in the Gulf Coast of the United States.*

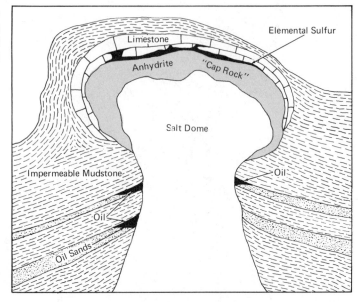

Evaporites from brine resources of the United States are substantial (Table 12.6), assuring that no shortage is likely for a considerable period of time. However evaporites will continue to have a "place value" because transportation of these mineral commodities will increase their price, and therefore continued discoveries of high-grade deposits closer to where they will be consumed remains an important goal (10).

Weathering Processes   These processes are responsible for concentrating some materials to the point that they can be extracted at a profit. For example, intensive weathering of residual soils (laterite) derived from aluminum-rich igneous rocks may concentrate relatively insoluble hydrated oxides of aluminum and iron, while the more soluble elements such as silica, calcium, and sodium are selectively removed by soil and biological processes. If sufficiently concentrated, residual aluminum oxide forms an ore of aluminum known as *bauxite* (Figure 12.18). Important nickel and cobalt deposits are also found in laterite soils developed from ferromagnesium-rich igneous rocks.

Insoluble ore deposits such as native gold are generally residual and unless removed by erosion will accumulate in the soil. Accumulation is favored where the parent rock which contains insoluble ore minerals is relatively soluble, such as limestone (Figure 12.19). Care must be taken in evaluating a residual soil deposit because the near-surface concentration is a much higher-grade of ore than exists in the parent, unweathered rocks (5).

Weathering is also important in producing secondary-enrichment sulfide-ore deposits from a lower-grade, primary ore. Near the surface, primary ore containing minerals such as iron, copper, and silver sulfides are in contact with slightly acid soil water in an oxygen-rich environment. As the sulfides are oxidized, they are dissolved, forming solutions that are rich in sulfuric acid and silver and copper sulfate and that migrate downward, producing a leached zone devoid of ore

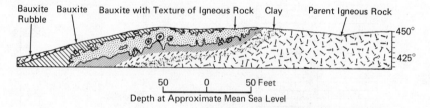

FIGURE 12.18 *Cross section of the Pruden bauxite mine, Arkansas. The bauxite has formed by intensive weathering of the aluminum-rich igneous rocks. (Source: G. Mackenzie, Jr., et. al., U.S. Geological Survey Professional Paper 299, 1958.)*

TABLE 12.6 *Evaporite and brine resources of the United States expressed as year's supply at current rates of domestic consumption.*

Commodity	Identified Resources[a] (Reserves[b] and Subeconomic Deposits)	Undiscovered Resources (Hypothetical[c] and Speculative[d] Resources)
Potassium compounds	100 years	Virtually inexhaustible
Salt	1000 + years	Do
Gypsum and anhydrite	500 + years	Do
Sodium carbonate	6000 years	5000 years
Sodium sulfate	700 years	2000 years
Borates	300 years	1000 years
Nitrates	Unlimited (air)	Unlimited (air)
Strontium	500 years	2000 years
Bromine	Unlimited (seawater)	Unlimited (seawater)
Iodine	100 years	500 years
Calcium chloride	100 + years	1000 + years
Magnesium	Unlimited (seawater)	Unlimited (seawater)

Source: G. I. Smith et al., U.S. Geological Survey Professional Paper 820, 1973.

[a]Identified resources: Specific, identified mineral deposits that may or may not be evaluated as to extent and grade, and whose contained minerals may or may not be profitably recovered with existing technology and economic conditions.

[b]Reserves: Identified deposits from which minerals can be extracted profitably with existing technology and under present economic conditions.

[c]Hypothetical resources: Undiscovered mineral deposits, whether of recoverable or subeconomic grade, that are geologically predictable as existing in known districts.

[d]Speculative resources: Undiscovered mineral deposits, whether of recoverable or subeconomic grade, that may exist in unknown districts or in unrecognized or unconventional form.

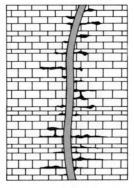 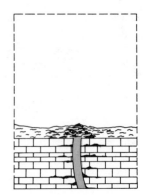

FIGURE 12.19 *Idealized diagram showing how an ore deposit of insoluble minerals might form due to weathering and formation of a residual soil. As the limestone that contained the deposit weathered, the ore minerals become concentrated in the residual soil. (Source: Foster,* Physical Geology, *2nd ed., Merrill, 1975.)*

minerals (Figure 12.20). Below the leached zone and above the groundwater table, oxidation continues, and sulfate solutions continue their downward migration. Below the water table, if oxygen is no longer available, the solutions are deposited as sulfides, enriching the metal content of the primary ore by as much as ten times. In this way, low-grade primary ore is rendered more valuable, and high-grade primary ore is made even more attractive (6, 7).

The presence of a residual iron oxide cap at the surface indicates the possibility of an enriched ore below, but it is not always conclusive. Of particular importance as to whether a zone of secondary enrichment forms is the presence of iron sulfide (for example, pyrite) in the primary ore, without which secondary enrichment seldom takes place because iron sulfide in the presence of oxygen and water forms sulfuric acid, which is a necessary solvent. Another factor that favors the development of a secondary-enrichment ore deposit is that the primary ore is sufficiently permeable to allow water and solutions to migrate freely downward.

FIGURE 12.20 *Idealized diagram showing the typical zones that form during secondary enrichment processes. Sulfide ore minerals in the primary ore vein are oxidized and altered, and then are leached from the oxidized zone and redeposited in the enriched zone. The iron oxide cap is generally a reddish color and may be helpful in locating ore deposits that have been enriched. (Source: Foster,* Physical Geology, *2nd ed., Merrill, 1975.)*

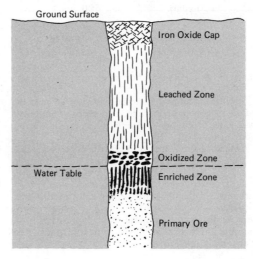

Given a primary ore that meets the above criteria, the reddish iron oxide cap probably does indicate that secondary enrichment has taken place (6).

Secondary enrichment is important in concentrating dispersed metals, and the economic success of several disseminated copper deposits has resulted from this process. For example, secondary enrichment of a disseminated copper deposit at Miami, Arizona, increased the grade of the ore from less than 1 percent copper in the primary ore to as much as 5 percent in some localized zones of enrichment (7).

Minerals from the Sea   Mineral resources in seawater or on the bottom of the ocean are vast and in some cases, such as magnesium, are nearly unlimited. In the United States, magnesium was first extracted from seawater in 1940, and by 1972, 80 percent of the domestic production capacity of magnesium was from one company in Texas utilizing seawater as its raw material source. In addition, companies in Alabama, California, Florida, Mississippi, and New Jersey are also extracting magnesium from seawater.

The deep-ocean floor may be the site of the next big mineral rush. It is now known that manganese oxide nodules (Figure 12.21) which contain on a very rough average dominant manganese (24 percent) and iron (14 percent) with secondary copper (1 percent), nickel (1 percent), and cobalt (0.25 percent) cover vast areas of the deep-ocean floor. The nodules are found in the Atlantic Ocean off Florida, but the richest and most extensive accumulations occur in large areas of the northeastern, central, and southern Pacific where the nodules cover 20 to 50 percent of the ocean floor (11).

The manganese oxide nodules are usually discreet but locally are welded together to form a continuous pavement. Although a few nodules have been found buried in sediment, they are basically a surficial deposit on the sea bed. The average size of the nodules varies from a few millimeters to a few tens of centimeters in diameter. Individual nodules are composed primarily of concentric layers of manganese in iron oxides mixed with a variety of other materials. Deposition is around a nucleus of broken nodules, fragments of volcanic rock, and sometimes fossils. The estimated rate of growth is 1 to 5 millimeters per million years. The

FIGURE 12.21 *Manganese oxide nodules on the floor of the Atlantic Ocean. (Photo courtesy of R. M. Pratt.)*

nodules are most abundant in those parts of the ocean where sediment accumulation is a minimum, generally at depths of 2500 meters to 6000 meters (12).

The origin of the nodules is not well understood, and there probably are several ways they might form. The most likely idea is that they form from material weathered from the continents and transported by rivers to the oceans where ocean currents carry it to the deposition site in the deep-ocean basins. Other ideas are that the minerals from which the nodules form are derived from submarine volcanism or are released during processes and reactions that occur near the water-sediment interface after deposition of the sediments (12).

Expenditures for mining and metallurgical research to recover the nodules have surpassed $190 million, and proposed expenditures through the early 1980s are expected to approach $800 million. At least 20 corporations in several countries are in the race, and metallurgical systems are being examined. Some would produce cobalt, copper, nickel, and manganese, while others would produce combinations of only copper and nickel (13).

Actual mining involves development of a system to lift the nodules off the bottom and up to the mining ship. French and Japanese researchers are examining a system with a continuous line bucket dredge, which involves a continuous rope with buckets attached at prescribed intervals strung out between two ships. The buckets drag along the bottom as the two ships move, and the nodules are dumped into one of the ships as the rope loop is reeled from one ship to the other. Other methods of recovery being examined are hydraulic lifting and use of airlift in conjunction with hydraulic dredging (13). It is a fair statement that the whole manganese oxide nodule industry is in its infancy, and considerable change in methods and technology is likely. Nevertheless, from present data it has been determined that mining of the nodules is technologically feasible and potentially profitable, provided necessary cooperation and management of the resource can be worked out by prospective interest groups.

Environmental Impact of Mineral Development

The impact of mineral exploitation on the environment depends upon such factors as mining procedures, local hydrologic conditions, climate, rock types encountered, size of operation, topography, and many more interrelated factors. Furthermore, the impact varies with the stage of development of the resource. For example, the exploration and testing stage involves considerably less impact than the mining and processing stages.

Exploration activities for mineral deposits vary from collecting and analyzing remote sensing data gathered from airplanes or satellites to field work involving surface mapping, drilling, and gathering of geophysical data. Generally, exploration has a minimal impact on the environment provided care is taken in sensitive areas.

Mining and processing of mineral resources, on the other hand, may have a considerable impact on the land, water, air, and biologic resources. The impact is part of the price we bear for the benefits of mineral consumption, and it is unrealistic to expect that mining of resources can be accomplished without affecting some aspect of the local environment. What must be done is to develop resources with the minimum of adverse impact. However, minimizing environmental degradation caused by mining may be very difficult considering that the demand for minerals is continuing to grow while the volumes of highly concentrated mineral deposits diminish. Therefore, to provide more and more material, larger and larger operations mining ever-poorer grades of ore will be necessary. However, mining of ore deposits that are transitional into the surrounding rocks, allowing recovery of ever-lower grades of ore, is not always possible because some ore deposits terminate abruptly along geologic boundaries, as shown in Figure 12.22 (14).

The trend in recent years has been away from subsurface mining and toward large, open-pit (surface) mines such as the Bingham Canyon copper mine in Utah (Figure 12.23) and the Liberty Pit near Ruth, Nevada (Figure 12.24). The Bingham Canyon mine is one of the world's largest man-made excavations, covering nearly 3 square miles to a maximum depth of nearly 1/2 mile.

Surface mines and quarries today cover less than 1/2 of 1 percent of the total area of the United States, and even though the impact is a local phenomenon, there is growing awareness that eventually numerous local occurrences will constitute a larger problem. The concern is that environmental degradation may tend to extend beyond the excavation and surface plant area. For example, area streams and groundwater may be polluted by particulate or dissolved sediment derived from rapidly eroding, unprotected waste piles exposed to surface erosion and migrating groundwater.

The demand for mineral resources is going to increase. Therefore, the only logical approach to the eventual environmental degradation is to strive to minimize both the on-site and off-site problems by controlling sediment, water, and air

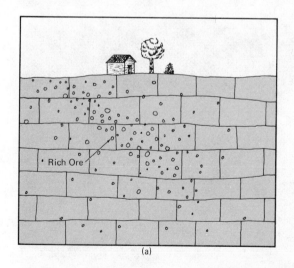
(a)

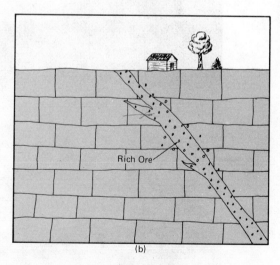
(b)

FIGURE 12.22 *Idealized diagram showing (a) an ore deposit in which the grade of the ore gradually lessens with distance from the greatest concentration, and (b) an ore deposit in which the high-grade ore is concentrated in a definite zone with no outward gradation to lower concentrations of the ore.*

FIGURE 12.23 *The Bingham Canyon copper mine, one of the largest man-made excavations in the world. (Photo courtesy of Kennecott Copper Corporation.)*

FIGURE 12.24 *The Liberty Pit surface mine near Ruth, Nevada. (Photo courtesy of Kennecott Copper Corporation.)*

pollution through good engineering and conservation practice. This will raise the cost of mineral commodities, and hence the price of all products produced from these materials, but it will yield other returns of equal or higher value to future generations.

A serious problem associated with mineral resource development is the possible release of trace elements in hazardous concentrations into the environment. Trace elements such as cadmium, cobalt, copper, lead, molybdenum, and others, when leached from mining wastes and concentrated in water, soil, or plants, may be toxic or may cause diseases in man and other animals who drink the water, eat the plants, or use the soil. Although trace elements in connection with mining activity were discussed in Chapter 11, the problem is mentioned again because it is significant.

Recycling of Resources

Over 200 million tons of household and industrial waste are collected each year in the United States alone. Of this, about 30 million tons are incinerated, and the

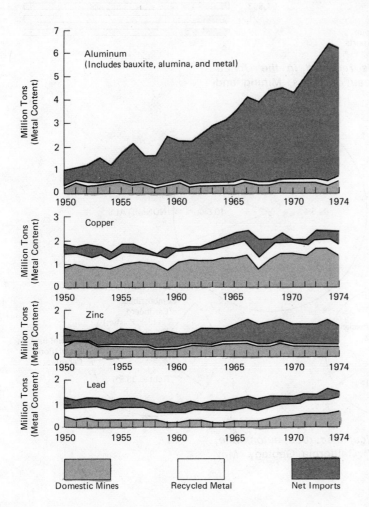

FIGURE 12.25 *United States supplies of major nonferrous metals. (Source: U.S. Bureau of Mines, Mining and Mineral Policy, 1975.)*

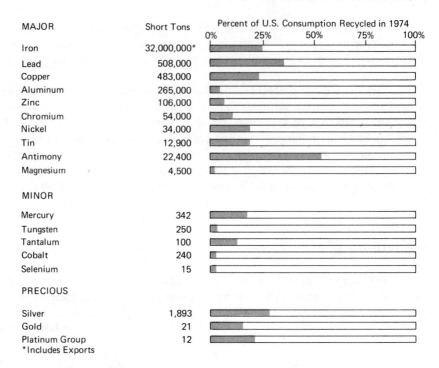

MAJOR	Short Tons	Percent of U.S. Consumption Recycled in 1974
Iron	32,000,000*	
Lead	508,000	
Copper	483,000	
Aluminum	265,000	
Zinc	106,000	
Chromium	54,000	
Nickel	34,000	
Tin	12,900	
Antimony	22,400	
Magnesium	4,500	
MINOR		
Mercury	342	
Tungsten	250	
Tantalum	100	
Cobalt	240	
Selenium	15	
PRECIOUS		
Silver	1,893	
Gold	21	
Platinum Group	12	

*Includes Exports

FIGURE 12.26 *Amount of metals recycled in the United States in 1974. (Source: U.S. Bureau of Mines,* Mining and Mineral Policy, *1975.)*

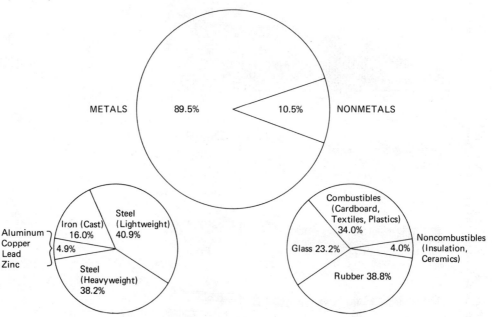

FIGURE 12.27 *Content of a typical discarded automobile. (Source, F. F. Davis, "Urban Ore,"* California Geology, *May 1972.)*

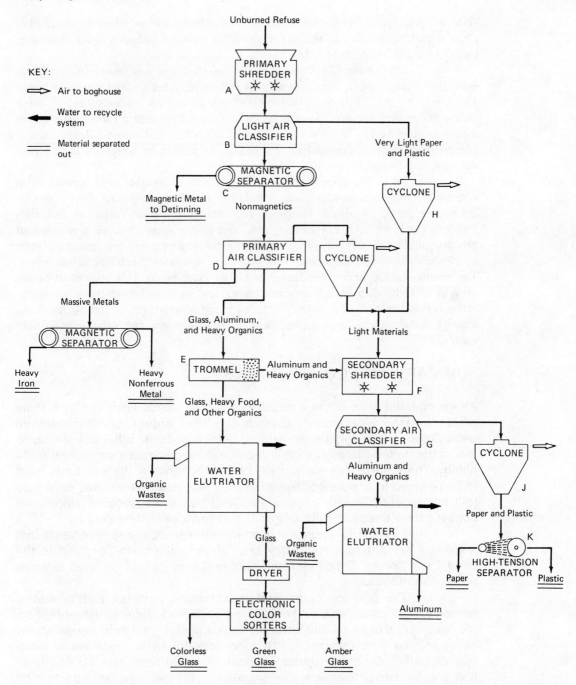

FIGURE 12.28 *Flow chart showing how raw refuse may be separated for recycling of resources. (Source: T. M. Sullivan et al., U.S. Bureau of Mines, R.I. 7760, 1973.)*

process generates 7 million tons of residue. The tremendous tonnage of waste not incinerated and the residues from burning are usually disposed of at sanitary landfill sites or open dumps. These materials are sometimes referred to as *urban ore* because they contain many materials that could be recycled and used again to provide energy or useful products (15, 16).

The notion of reusing waste materials is not new, and metals such as iron, aluminum, copper, zinc, and lead have been recycled for many years (Figures 12.25 and 12.26). Millions of automobiles are discarded annually, and of these nearly 90 percent are now dismantled by auto wreckers and scrap processors for metals to be recycled (16). Recycling metals from discarded automobiles is a sound conservation practice, considering that nearly 90 percent by weight of the average automobile is metal (Figure 12.27).

Recycling appears to be one way that a possible crisis caused by a convergence between rapidly rising population and limited resource base might be delayed or partly alleviated. However, recycling of the wide variety of materials found in urban waste is not an easy task, and before it can become a widespread practice, improved technology and more economic incentives are needed. Figure 12.28 shows a possible flow chart with necessary equipment to recycle urban refuse. The results of experiments conducted at a pilot plant by the U.S. Bureau of Mines using these techniques have been encouraging, and while refinements are necessary, urban refuse has been successfully separated into concentrates of light gauge iron, massive metals, glass, paper, plastics, and organic waste and other combustible wastes (15).

SUMMARY AND CONCLUSIONS

An exponential increase of a population whose needs are filled by a finite resource base may eventually explode as a food, energy, or other resource crisis. There are simply too many people for worldwide affluence similar to that of the United States and other industrial nations based on present technology. Therefore, affluence in many developed areas of the world may have to be reduced to a more conservation-oriented, less materialistic existence (but not necessarily with a lower quality of life) while underdeveloped nations increase the availability of goods and services to their people.

An important concept in analyzing resources and reserves is that *resources are not reserves!* They are "birds in the bush" or "fish in the pond," and unless discovered and captured they cannot be used to solve present shortages.

The geology of mineral resources is complex and intimately related to various aspects of the geologic cycle. In a broad-brush approach, the occurrence of many, but certainly not all, metallic mineral deposits may be correlated with dynamic earth processes found at junctions of lithospheric plates. On a more practical scale, mineral resources for construction and industrial uses are concentrated in the geologic environment by first, igneous or magmatic processes such as crystal settling or hydrothermal activity; second, metamorphic processes such as contact and regional metamorphism; third, sedimentary processes, including oceanic and lake processes, running water, wind, and moving ice; and fourth, weathering processes, including soil-forming activity and insitu concentrations of insoluble minerals.

Recycling of resources is technically feasible, and for conservation purposes, a high-priority goal of urban areas should be conversion of waste to energy or useful resources. However, before it is practical at a large scale, increased economic incentives and better procedures will be necessary.

References

1. Park, C. F., Jr. 1968. *Affluence in jeopardy*. San Francisco: Freeman, Cooper & Company.
2. U.S. Geological Survey. 1975. *Mineral resource perspectives 1975*. U.S. Geological Survey Professional Paper 940.
3. McKelvey, V. E. 1973. Mineral resource estimates and public policy. In *United States mineral resources*, D. A. Brobst and W. P. Pratt, eds., pp. 9–19. U.S. Geological Survey Professional Paper 820.
4. Brobst, D. A.; Pratt, W. P.; and McKelvey, V. E. 1973. *Summary of United States mineral resources*. U.S. Geological Survey Circular 682.
5. Foster, R. J. 1973. *General geology*. Columbus, Ohio: Charles E. Merrill.
6. Bateman, A. M. 1950. *Economic ore deposits*. 2nd ed. New York: John Wiley & Sons.
7. Park, C. F., Jr., and MacDiarmid, R. A. 1970. *Ore deposits*. 2nd ed. San Francisco: W. H. Freeman.
8. Lawrence, R. A. 1973. Construction stone. In *United States mineral resources*, D. A. Brobst and W. P. Pratt eds., pp. 157–62. U.S. Geological Survey Professional Paper 820.
9. Yeend, W. 1973. Sand and gravel. In *United States mineral resources*, D. A. Brobst and W. P. Pratt, eds., pp. 561–65. U.S. Geological Survey Professional Paper 820.
10. Smith, G. I.; Jones, C. L.; Culbertson, W. C.; Erickson, G. E.; and Dyni, J. R. 1973. Evaporites and brines. In *United States mineral resources*, D. A. Brobst and W. P. Pratt, eds., pp. 197–216. U.S. Geological Survey Professional Paper 820.
11. Cornwall, H. R. 1973. Nickel. In *United States mineral resources*, D. A. Brobst and W. P. Pratt, eds., pp. 437–42. U.S. Geological Survey Professional Paper 820.
12. Van, N.; Dorr, J.; Crittenden, M. D.; and Worl, R. G. 1973. Manganese. In *United States mineral resources*, D. A. Brobst and W. P. Pratt, eds., pp. 385–99. U.S. Geological Survey Professional Paper 820.
13. Secretary of the Interior. 1975. *Mining and mineral policy, 1975*. Washington, D.C.: U.S. Government Printing Office.
14. Fagan, J. J. 1974. *The earth environment*. Englewood Cliffs, New Jersey: Prentice-Hall.
15. Sullivan, P. M.; Stanczyk, M. H.; and Spendbue, M. J. 1973. *Resource recovery from raw urban refuse*. U.S. Bureau of Mines, R.I. 7760.
16. Davis, F. F. 1972. Urban ore. *California Geology*, May 1972: 99–112.

13

ENERGY AND ENVIRONMENT

Energy and People

Our discussion of mineral resources established that resources are not infinite, and it is impossible to support an exponential increase in population on a finite resource base. The same is true for energy derived from mineral resources.

Citizens in the United States are experiencing the effects of energy shortages for the first time, including great increases in prices of energy and products produced from petroleum. Before the energy shortage, energy seemed unlimited, and people in the United States, while totaling only a small percentage of the world's population, consumed a disproportionate share of the total electric energy produced in the world. The appetite of the American people for a higher standard of living and their accompanying energy consumption are shown in Figure 13.1.

It is frightening to compare the projected population increase of the United States with the projected energy needs (Figure 13.2). With energy demand expected to increase about three times faster than the population increases, one has to wonder where the energy will come from. For many years the vast majority of

ENERGY AND ENVIRONMENT

FIGURE 13.1 *Relation between energy consumption and standard of living for various countries. (Source: Pennsylvania Department of Education,* The Environmental Impact of Electrical Power Generation: Nuclear and Fossil, *1973.)*

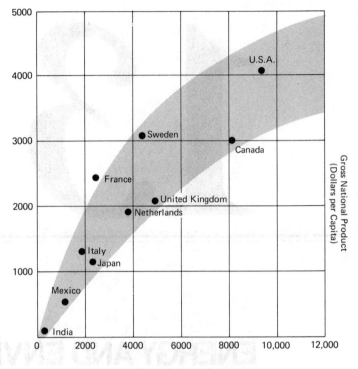

FIGURE 13.2 *Comparison of United States energy consumption and population projected to the year 2000. (Data from the Commission on Population Growth and American Future and the U.S. Department of the Interior.)*

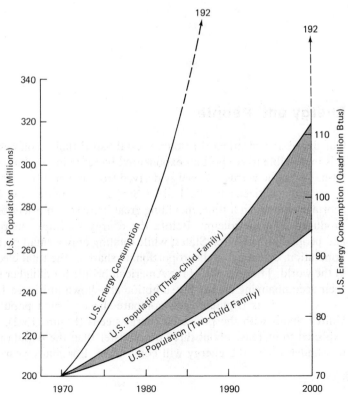

energy has been produced from coal, natural gas, and petroleum (Figure 13.3), with small amounts of hydropower and, more recently, nuclear power. Projections of future energy consumption assume that most of the increased demand will be met by the nuclear power industry (Figure 13.4), which by the year 2000 will supply nearly 25 percent of the total energy. Is this likely? In 1974 there were 50 nuclear generating plants in the United States supplying about 2 percent of the total energy consumed (1, 2). To meet the projected demands, approximately 33 times more power from nuclear reactors will have to be produced by the year 2000 (Table 13.1). Assuming that future reactors will be able to produce twice as much energy as the 50 reactors operating in 1974, over 800 operating reactors will be needed. As of 1974, about 70 new reactors were in some phase of construction, and 100 more were

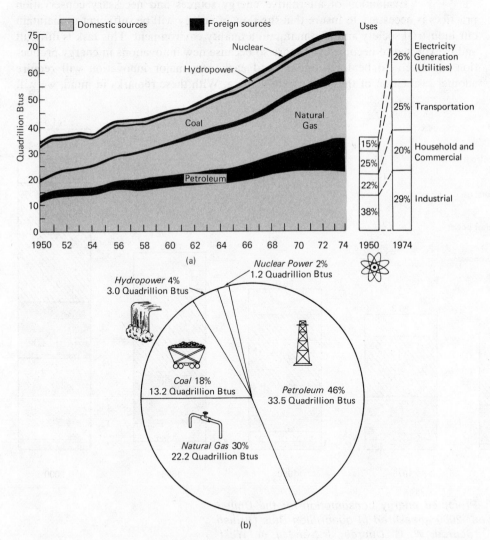

FIGURE 13.3 *(a) United States demand for major energy sources from 1950 to 1974 and (b) consumption of energy by sources for 1974. (Source: U.S. Bureau of Mines,* Mining and Mineral Policy, *1975.)*

planned. Considering the years of lead time necessary for planning and construction, only about 220 reactors will be in operation by 1988 (2). Another 600 or so reactors will have to be completed in the 12 years after 1988 to meet the projected energy demands. Even by the most optimistic of projections, this seems improbable and serves to illustrate the pitfalls of trying to forecast demand and supply of energy resources. That is, there is a considerable gap between projected consumption of energy and probability that the power industry can produce it. Hopefully, the expected demand for power is too high, or the rate of construction of power plants and the amount of power they produce can be greatly increased by the year 2000. More likely, people in industrialized countries are going to have to realize that the quality of life is not directly related to ever-expanding energy needs.

　　　　Evaluation of alternative energy sources and necessary conservation practices is necessary to insure that the flow of energy will be sufficient to maintain our industrial society and also maintain a quality environment. This task is difficult and is expected to become even more so because new innovations in energy production are likely to be forthcoming, and each new major innovation will require another evaluation of the total available data. With these remarks in mind, we will

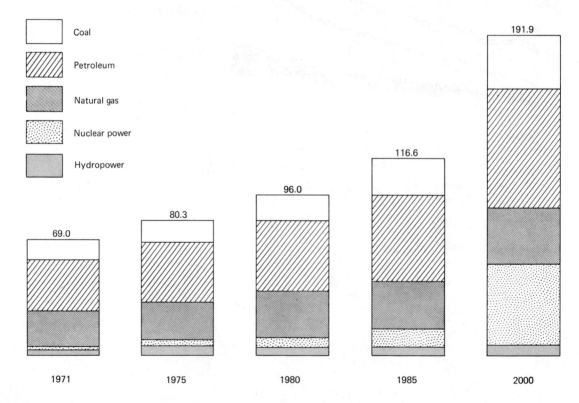

FIGURE 13.4 *Projected energy consumption for the United States to the year 2000, measured in quadrillion Btus (British thermal units). [Source: W. G. Dupree, Jr., and J. A. West,* United States Energy through the Year 2000 *(U.S. Department of the Interior, 1972).]*

TABLE 13.1 *Nuclear energy in 1974 and projected energy demand by the year 2000.*

Year	Total Energy Consumed	Percent of Total Energy Consumed by Nuclear Reactors	Total Energy Consumed by Nuclear Reactors	Reactors in Operation	Reactors Under Construction or on Order	New Reactors Needed to Supply Demand for Nuclear Power
1974 actual	73[a]	2	1.5[a]	52	221[c]	—
2000 projected	192[a]	25	48[a]	?800[b]	—	600 Improbable?

SOURCES: Data from the Department of the Interior and the Office of Industry Relations.
[a]Measured in quadrillion Btus
[b]Assuming each reactor by the year 2000 will produce twice the power of a 1974 reactor
[c]To be completed and in operation by 1988

cautiously explore some selected geologic and environmental aspects of well-known energy resources such as coal and petroleum, nuclear sources, and other possibly important sources, including oil shale, tar sands, and geothermal resources. In addition, we will also briefly discuss the expected increase in demand for water as a result of energy production and several other alternative energy sources such as hydropower (river and tidal) and solar power.

Coal

Geology of Coal   Like other fossil fuels, coal is made up of organic materials that have escaped oxidation in the carbon cycle. Coal is essentially altered residue of plants that have flourished in ancient freshwater or brackish-water swamps, typically found in estuaries, coastal lagoons, and low-lying coastal plains or deltas (3).

Processes that form coal (Figure 13.5) involve the development of a swamp rich in plants that are partially decomposed in an oxygen-deficient environment and that accumulate slowly to form a thick layer of peat. Such swamps and accumulations of peat may then be inundated by a prolonged slow rise of sea level (relative rise as the land may be sinking) and be covered by sediments such as sand, silt, clay, and carbonate-rich material. As more and more sediment is deposited, water and organic gases (volatiles) are squeezed out, and the percentage of carbon increases in the compressed peat. As this process continues, the peat is eventually transformed to coal. Because there are often several layers of coal in the same area, it is believed that sea level rose and fell alternately, allowing the development of coal swamps and then drowning them.

Classification and Distribution of Coal   Coal is commonly classified according to rank and sulfur content. The rank of coal is generally based on the percentage of carbon, which increases from lignite to bituminous to anthracite, as shown in Figure 13.6. This figure also shows that heat content is maximum in bituminous coal, which

FIGURE 13.5 *Idealized diagram showing the processes by which buried plant debris (peat) is transformed into coal.*

(a) *Coal swamp forms.*

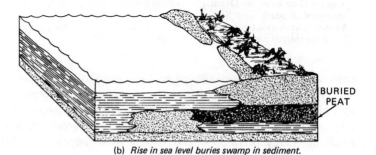

(b) *Rise in sea level buries swamp in sediment.*

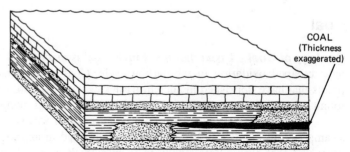

(c) *Compression of peat forms coal.*

has relatively few volatiles (oxygen, hydrogen, and nitrogen) and low moisture content. Heat content is minimum in lignite, which has a high moisture content. The distribution of the common coals (bituminous, subbituminous, and lignite) in the conterminous United States and Alaska is shown in Figure 13.7.

The sulfur content of coal is generally classified as low (zero to 1 percent), medium (1.1 to 3 percent), or high (greater than 3 percent). Most coal in the United States is of the low-sulfur variety (Table 13.2). However, by far the greatest part of the low-sulfur coals is a relatively low-grade, subbituminous variety found west of the Mississippi River (Figure 13.8). The location of coal reserves has environmental significance because with all other factors equal, low-sulfur coal is more desirable as a fuel for power plants because it causes less air pollution. Therefore, to avoid air pollution, thermal power plants on the highly populated East Coast

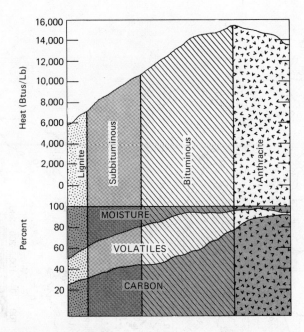

FIGURE 13.6 *Generalized classification of different types of coal based upon their relative content (in percentage) of moisture, volatiles, and carbon. The heat values of the different types of coal are also shown. (Source: D. A. Brobst and W. P. Pratt, eds., U.S. Geological Survey Professional Paper 820, 1973.)*

will have to continue to treat some of the local coal to lower its sulfur content. This treatment is expensive but may be more economical than shipping low-sulfur coal long distances.

Impact of Mining   A discussion of the environmental effects of surface mining has been deferred until now because mining of bituminous coal in the United States accounts for 40 percent of all land utilized by mining (Figure 13.9). From 1930 to 1971, approximately 3.65 million acres of land in the United States was used for mining purposes, and of this, nearly 60 percent was for mine excavations (Figure 13.10). Of the land utilized for mining in this 42-year period, only approximately 40 percent was reclaimed (Figure 13.11). It is encouraging to note that the percentage of land reclaimed in 1971 was 80 percent, or twice that of the previous 42 years. Figure 13.12 shows the magnitude of the problem and the geographical distribution of unreclaimed mining land in the conterminous United States.

Most mining of coal in the United States is still done underground, but strip mining (Figure 13.13), which started in the late nineteenth century, has steadily

| Rank | Sulfur Content (Percent) | | |
	Low 0–1	Medium 1.1–3.0	High 3+
Anthracite	97.1	2.9	—
Bituminous coal	29.8	26.8	43.4
Subbituminous coal	99.6	.4	—
Lignite	90.7	9.3	—
All ranks	65.0	15.0	20.0

TABLE 13.2 *Distribution of United States coal resources according to their rank and sulfur content.*

Source: U.S. Bureau of Mines Circular 8312, 1966.

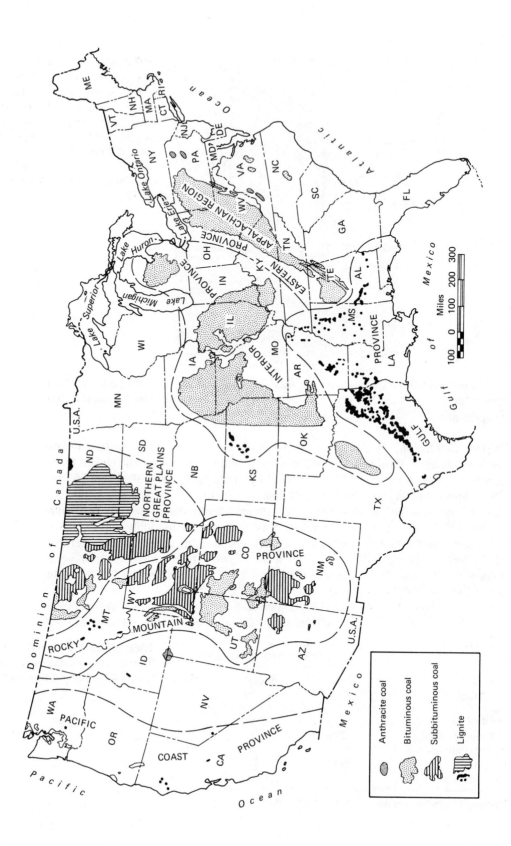

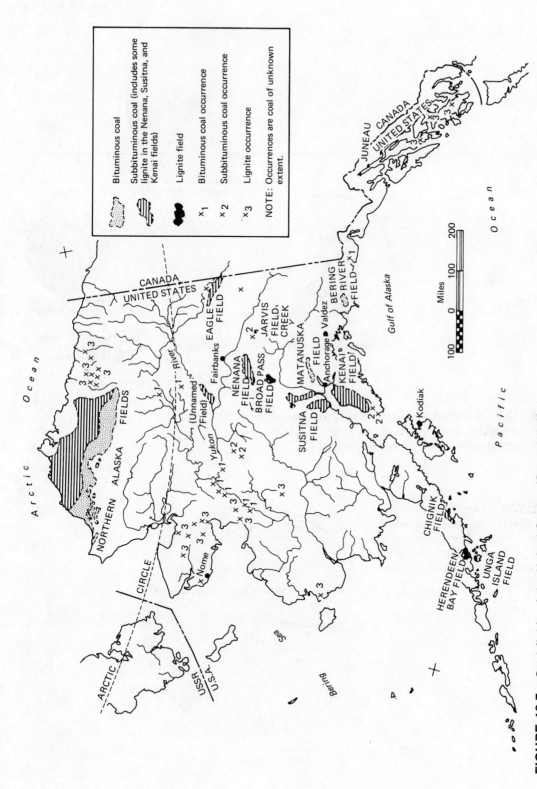

FIGURE 13.7 *Coal fields of the United States. (Source: U.S. Bureau of Mines Information Circular 8531, 1971. Adapted from U.S.G.S. Coal Map of Alaska, 1960.)*

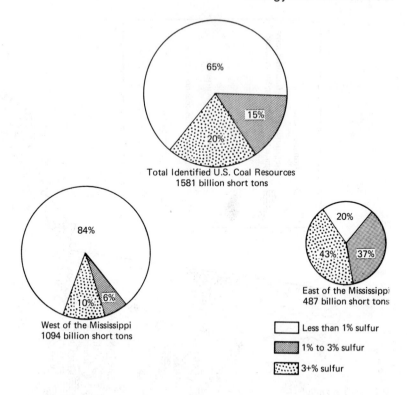

FIGURE 13.8 *Geographical distribution of identified U.S. coal resources by sulfur content. (Source: Federal Energy Administration,* Monthly Energy Review, *July 1975.)*

FIGURE 13.9 *Land utilized in the United States by selected mine activities from 1930 to 1971. (Source: J. Paone et al., U.S. Bureau of Mines Information Circular 8642, 1974.)*

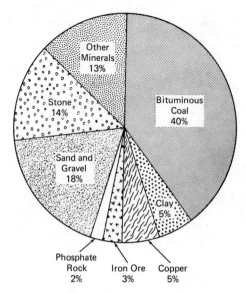

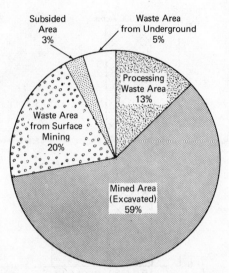

Subsided Area 3%

Waste Area from Underground 5%

Processing Waste Area 13%

Waste Area from Surface Mining 20%

Mined Area (Excavated) 59%

FIGURE 13.10 *Land utilized in the United States by mining broken down into various mining functions from 1930 to 1971. (Source: J. Paone et al., U.S. Bureau of Mines Information Circular 8642, 1974.)*

increased while the production from underground mines has stabilized. The trend toward strip mining has developed because the method is technologically and economically more advantageous in many cases than underground mining. The increased demand for coal will probably lead to more and larger strip mines to extract the estimated 45 billion tons of coal reserves that are now accessible to surface mining techniques. In addition, approximately another 100 billion tons of coal within 150 feet of the surface is potentially available for stripping if need demands its extraction.

The impact of large strip mines varies from region to region depending upon topography, climate, and, most importantly, reclamation practices. In humid areas with abundant rainfall, mine drainage of acid water is a serious problem (Figure 13.14). Surface water infiltrates into the spoil banks (material left after the coal or other minerals are removed) where it reacts with sulfide minerals such as pyrite (FeS_2) to produce sulfuric acid, which then runs into and pollutes streams and groundwater resources. Acid water also drains from underground mines and road cuts and areas where coal and pyrite is abundant, but the problem is magnified when large areas of disturbed material remain exposed to surface waters. Acid drainage can be minimized by proper utilization of water diversion practices that collect surface runoff and groundwater before they enter the mined area and divert them around the potentially polluting materials (spoil banks; see Figure 13.15). This practice reduces erosion, pollution, and water-treatment cost (4).

In arid and semiarid regions, water problems associated with mining are not as pronounced as in wetter regions, but the land may be more sensitive to mining activities such as exploration and road building. In some areas in arid environments, the land is so sensitive that even tire tracks across the land survive for years. Furthermore, soils are often thin, water is scarce, and reclamation work is difficult. In general, there is less regulation of surface mining in the western and southwestern United States than in the eastern.

Common methods of strip mining include *area mining,* which is practiced on relatively flat areas (Figure 13.16), and *contour mining,* which is used in

FIGURE 13.11 *Relationship of land utilized by mining and then reclaimed in the United States from 1930 to 1971 compared to mining and reclamation activity in 1971. Photograph of a strip mine reclaimed for agricultural land use. (Photo by Jerry D. Vineyard, courtesy of the Missouri Geological Survey. Source of graph: J. Paone et al., U.S. Bureau of Mines Information Circular 8642, 1974.)*

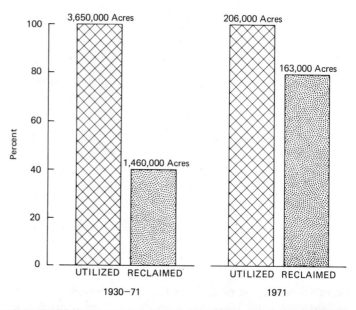

hilly terrain (Figure 13.17). The width of the cut in contour mining is dependent on the ease of excavation and topography. More and more overburden must be removed as the width increases, which increases the cost. Therefore, topography limits the total width of the cut. The objective is to confine the cut to a given elevation (contour) and work around the hill while maintaining that elevation. Both methods of strip mining severely disturb the landscape by removing indigenous vegetation, overburden, and desired minerals, and unless the mined land is reclaimed, unsightly conical piles or ridges of waste remain as a source material that facilitates groundwater and surface-water pollution. Area mining is used only on relatively flat ground where the velocity and erosion potential of runoff are low. Therefore, the potential to pollute streams by siltation is less than for contour mining. However, potential groundwater pollution is greater for area mining because there is more area and more time for precipitation to infiltrate and slowly migrate through the spoil piles (4).

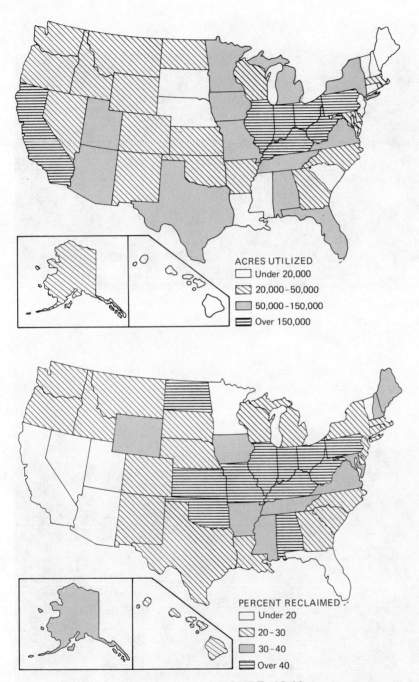

FIGURE 13.12 *Geographic distribution of land utilized by mining and reclaimed from 1930 to 1971. (Source: J. Paone et al., U.S. Bureau of Mines Information Circular 8642, 1974.)*

All methods of strip mining have the potential to pollute or destroy scenic, water, biologic, or other land resources. However, good reclamation prac-

FIGURE 13.13 *Coal strip mine, Henry County, Missouri. (Photo by J. D. Vineyard, courtesy of Missouri Geological Survey.)*

tices can minimize the damage (Figure 13.18). One potentially good method is to segregate the overburden such that topsoil orginally removed can be replaced

FIGURE 13.14 *The stream in the foreground is flowing through waste piles of a Missouri coal mine. The water reacts with sulfide minerals and forms sulfuric acid. This is a serious problem in coal mining areas. (Photo by J. D. Vineyard, courtesy of Missouri Geological Survey.)*

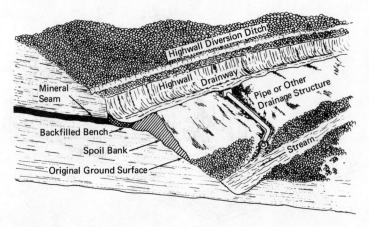

FIGURE 13.15 *Idealized diagrams showing how diversion ditches and drains can be used to minimize water pollution caused by mining activities. (Source: U.S. Environmental Protection Agency, EPA-430/9-73-011, 1973.)*

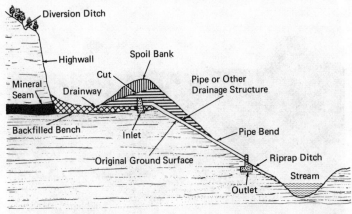

(Figure 13.19). This method has been used to a considerable extent in the coal fields of the eastern United States, and experience has shown that it is a successful way to control water pollution provided it is combined with regrading and revegetation.

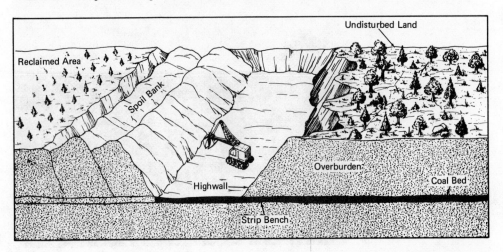

FIGURE 13.16   *Idealized diagram showing area strip mining.*

FIGURE 13.17 *Idealized diagram showing contour strip mining.*

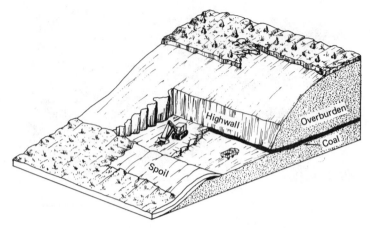

FIGURE 13.18 *A small open-pit mine before (a) and after (b) reclamation. (Photos by J. D. Vineyard, courtesy of Missouri Geological Survey.)*

(a)

(b)

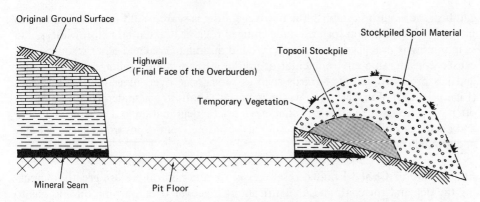

Original Ground Surface

Highwall
(Final Face of the Overburden)

Temporary Vegetation

Topsoil Stockpile

Stockpiled Spoil Material

Mineral Seam

Pit Floor

FIGURE 13.19 *Idealized diagram showing overburden segregation. The basic idea is to segregate the topsoil from the spoil material so that the topsoil may be later utilized during reclamation of the land. (Source: U.S. Environmental Protection Agency, EPA-430/9-73-011, 1973.)*

However, it is more expensive, and in a competitive market, it will probably have to be required by law in order to prevent inequalities from arising because one company segregates the overburden and another does not (4).

      Low-sulfur coal deposits in the western United States, particularly in Montana, Wyoming, Colorado, Utah, New Mexico, and Arizona, will probably be mined by the area method of strip mining. Reclamation will be difficult even if the overburden is segregated because soils are thin and revegetation is extremely diffi-

FIGURE 13.20 *Escaping gases from holes drilled into a burning abandoned coal mine in western Pennsylvania. (Photo courtesy United States Bureau of Mines.)*

cult because water to establish the new vegetation is scarce. Additional research into methods of reclamation for arid and semiarid regions is essential before widespread strip mining is practiced (4, 5). Underground mining of coal and other resources has caused considerable environmental degradation from mine drainage of acid water and has produced serious hazards such as subsidence over mines or fires in mines (Figure 13.20). Furthermore, waste material (spoil) from underground mines is often piled up on the surface, producing aesthetic degradation and causing sediment and chemical pollution of water resources resulting from exposure to surface water and groundwater.

Future Use of Coal Limited resources of oil and natural gas are greatly increasing the demand for coal, and a significant changeover from burning oil and gas to burning coal in thermoelectric power plants and industrial heating generating units is forthcoming. About 60 plants have been, or shortly will be, converted at a savings of about 250,000 barrels of oil per day. Total savings on the order of one million barrels per day could be realized if all thermoelectric power plants burn coal. There are sufficient coal reserves to meet the increased demand from such conversion and to supply coal for new plants for many hundreds of years (6).

The real crunch on oil and gas is still a few years away, but when it does come, it will put tremendous pressure on the coal industry to open more and larger mines in both the eastern and the western coal beds of the United States. This may have tremendous environmental impacts for several reasons. First, more and more land will be strip mined and thus will require careful restoration. Second, unlike oil and gas, burned coal leaves ash (5 to 20 percent of the original amount of the coal) that must be collected and disposed of. Some ash can be utilized for landfill or other purposes, but most, 85 percent, is useless at present. Third, handling of tremendous quantities of coal through all stages—mining, processing, shipping, combustion, and final disposal of ash—will have potentially adverse environmental effects, such as aesthetic degradation, noise, dust pollution, and, most significant, release of trace elements into the water, soil, and air likely to cause serious health problems (6). These environmental effects, while significant enough to cause concern, are not necessarily insurmountable, and careful planning could minimize them. At any rate, there may be few other alternatives other than to mine tremendous quantities of coal to feed thermoelectric power plants, and to provide oil and gas by gasification and liquefaction processes in the future.

Oil and Gas

Geology of Oil and Gas Oil and natural gas, methane, are hydrocarbons. Like coal, they are fossil fuels in that they form from organic material that has escaped complete decomposition after burial. Next to water, oil is probably the most abundant fluid in the earth's crust, and yet processes that form it are only partly understood. Most earth scientists accept that oil and gas are derived from organic materials that are buried with marine or lake sediments. Favorable environments where organic debris might escape oxidation include *near-shore areas* characterized by rapid deposition that quickly buries organic material or *deeper-water areas* characterized by a deficiency in oxygen at the bottom which promotes an anaerobic

Nile delta. The construction of the Aswan Dam has caused considerable environmental disruption in the delta region, including coastal erosion and reduction in number of fisheries. (Photo courtesy of NASA)

Cape Hatteras, North Carolina. There is considerable controversy surrounding the land use of this unique and scenic landscape. (Photo courtesy of the National Park Service)

decomposition. Beyond this, the locations where oil and gas form are generally classified as subsiding, depositional basins in which older sediment is continuously buried by younger sediments, thus progressively subjecting the older, more deeply buried material to higher temperatures and pressures (7).

The major source material for oil and gas is a fine-grained, organic-rich sediment that is buried to a depth of at least 1500 feet and is subjected to increased heat and pressure which physically compress the source rock. The elevated temperature and pressure, along with other processes, start the chemical transformation of organic debris into hydrocarbons (oil and gas). As the pressure increases, the porosity of the source rock is reduced, and the higher temperatures thermally energize the hydrocarbons and induce them to begin an upward migration to a lower-pressure environment with increased porosity. The initial movement of the hydro-carbons upward through the source rock is termed *primary migration,* which merges to *secondary migration* as the oil and gas move more freely into and through coarser-grained, more permeable rock such as sandstone or fractured limestone. These porous, permeable rocks into which the oil and gas migrate are called *reservoir rocks.* If the path is clear to the surface, the oil and gas will migrate and escape there; this perhaps explains why most of the oil and gas is found in geologically young rocks. That is, hydrocarbons in older rocks have had a longer period of time in which to reach the surface and leak out (7). However, the lower amounts of hydrocarbon in very old rocks might also be explained by tectonic processes that uplift rocks containing oil and gas, exposing them to erosion.

If in their upward migration, oil and gas are interrupted or stopped because they encounter a relatively impervious barrier, then they may accumulate. The barrier is called a *trap* or *cap rock,* and if a favorable geometry (structure), such as is found in a dome or an anticline, exists, the oil and gas will be trapped in their upward movement at the crest of the dome or anticline below the cap rocks (7). Figure 13.21 shows an anticlinal trap and two other possible traps caused by faulting or by an unconformity (buried erosion surface). These are not the only possible types of traps. Any rock that has a relatively high porosity and permeability and that is connected to a source rock containing hydrocarbons may become a reservoir, provided that the upward migration of the oil and gas is impeded by a cap rock oriented such that the hydrocarbons are entrapped at a central high point (7).

Distribution and Amount of Oil and Gas   The distribution of oil and gas in time (geologic) and space is rather complex, but in general the following three principles apply. First, commercial oil and gas are produced almost exclusively from sedimentary rocks deposited during the last 500 million years of the earth's history (7). Second, although there are many oil fields in the world, approximately 85 percent of the total production plus reserves occurs in less than 5 percent of the producing fields; 65 percent occurs in about 1 percent of the fields; and, as remarkable as it seems, 15 percent of the world's known oil reserves is in two accumulations in the Middle East (7). Third, the geographic distribution of the world's giant oil and gas fields (Figure 13.22) shows that most are located near tectonic belts (plate junctions) that are known to have been active in the last 60 to 70 million years.

It is a very complex and difficult task to assess the petroleum and gas reserves of the United States, much less assess those of the entire world. Recent (1975) estimates of proven oil and gas reserves of the United States suggest that at

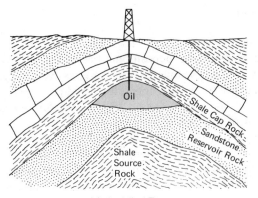

(a) Anticlinal Trap

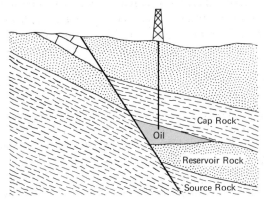

(b) Fault Trap

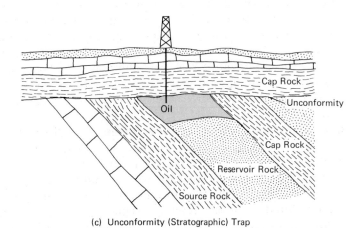

(c) Unconformity (Stratographic) Trap

FIGURE 13.21 *Idealized diagrams showing several types of oil and gas traps.*

362

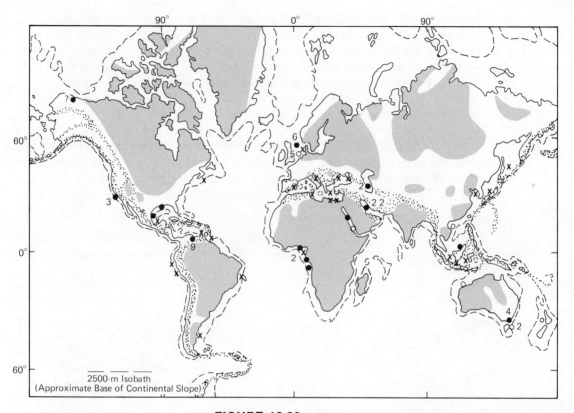

FIGURE 13.22 *Giant oil and gas fields of the world relative to generalized tectonic belts. Active tectonic areas are shown by the stippled pattern, and regions not subjected to tectonic activity for the last 500 million years are shaded. Giant oil fields are denoted by the solid circles; giant gas fields by the open circles; and general areas of discovery either of less than giant size or areas under field development are indicated by an X. (Modified from U.S. Geological Survey Circular 694, as adapted from C. L. Drake,* American Association of Petroleum Geologists Bulletin, *vol. 56, no. 2, 1972, with permission.)*

present production rates, the oil will last 13 years and the gas a little more than 11 years. Using estimates of inferred and undiscovered resources, one can extend these projections to 35 and 32 years, respectively. However, the latter figures are estimates subject to large errors. For example, in 1975 the undiscovered recoverable oil resources were estimated at 105 billion barrels, but from a probability standpoint, there is a 95 percent chance that at least 73 billion barrels will be recovered and only a 5 percent chance that as much as 150 billion barrels will be recovered, giving a possible range of 77 billion barrels (8). The significance of these figures and projections is that they suggest that it will be extremely difficult for the United States to be economically self-sufficient in oil and gas (7). The last time the United States was essentially self-sufficient in oil was about 1950 (Figure 13.23), and we have been importing oil since then despite a peak period of exploratory drilling in 1956. Figure 13.23 also shows the increased interest in offshore development since 1950 as illustrated by drilling activity in ever-increasing water depths.

What is the future of oil and gas? Unless large, new accumulations of petroleum are discovered in the United States, and there is always this possibility, known reserves and projected recoverable resources will last only a few more years before a real shortage is evident. When a significant shortage does occur, we will most likely go to large-scale gasification and liquefaction of our tremendous coal reserves, extract oil and gas from oil shale, and rely more on atomic and solar energy. These changes will strongly affect our petroleum-based society, but there

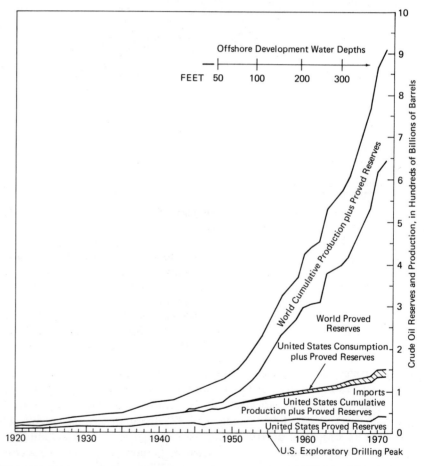

FIGURE 13.23 *Crude oil production and proven reserves of the United States and the world from 1920 until 1971. Also shown is the increased interest in offshore development since 1950 as indicated by drilling activity at ever-increasing water depths. (Source: D. A. Brobst and W. P. Pratt, eds., U.S. Geological Survey Professional Paper 820, 1973.)*

appears to be no insurmountable problem provided meaningful short- and long-range plans to phase out oil and gas and phase in alternative energy sources are implemented. Unfortunately, the phasing in of alternative energy sources will require many years of research, exploration, and development. Therefore, it is crucial that we begin the task now! We are living in an interesting time from an energy standpoint, and it will be exciting to see new innovations, technology, and standards of living. Right now, no one knows the answer or even all the questions.

Impact of Oil and Gas Exploration and Development   The environmental impacts of exploration and development of oil and gas vary from negligible for remote sensing techniques in exploration to significant, unavoidable impacts for projects such as the Trans-Alaska Pipeline. Impacts of exploration for oil and gas may involve building roads, exploratory drilling, and building a supply line (for camps, airfields, etc.) to remote areas. These activities, except in sensitive areas such as some semiarid to arid environments and some permafrost areas, generally cause few adverse effects to the landscape and resources compared to development and consumption activities. Development of oil and gas fields involves *drilling* wells on land or beneath the sea; *transporting* the oil by tankers, pipelines, or other methods to refineries; and *converting* the crude oil into useful products. All along the way, the possibility for environmental disruption from accidental oil spills, shipwreck of tankers, air pollution at refineries, and other impacts are well known and well documented. Tragic oil spills have desecrated the coastlines of Europe and America, spoiling beaches, estuaries, and harbors; killing marine life and birds; and causing economic problems for coastlines dependent upon tourist trade. These unfortunate aspects of developing oil are serious, and additional research and legislation are needed to minimize their occurrence.

   The best-known serious impact associated with oil and gas utilization is air pollution, which is produced in urban areas when fossil fuels are burned to produce energy for electricity, heat, and automobiles. The adverse effects of smog on vegetation and people's health are well documented and need not be restated here.

Oil Shales and Tar Sands

Geology of Oil Shale   Oil shale is a fine-grained sedimentary rock containing organic matter. On heating (destructive distillation), oil shale yields significant amounts of hydrocarbons that are otherwise insoluble in ordinary petroleum solvents (9).

   Like that of other fossil fuels, the origin of oil shale involves the deposition and only partial decomposition of organic debris. Favorable environments for oil shale are lakes, stagnant streams or lagoons in the vicinity of organic-rich swamps, and marine basins (10).

   The best-known oil shales in the United States are those in the Green River Formation, which is about 50 million years old and underlies approximately 17,000 square miles of Colorado, Utah, and Wyoming (Figure 13.24). The Green River Formation consists of oil shale interbedded with variable amounts of sandstone, siltstone, claystone, and compacted volcanic ash (tuff). Variable amounts of

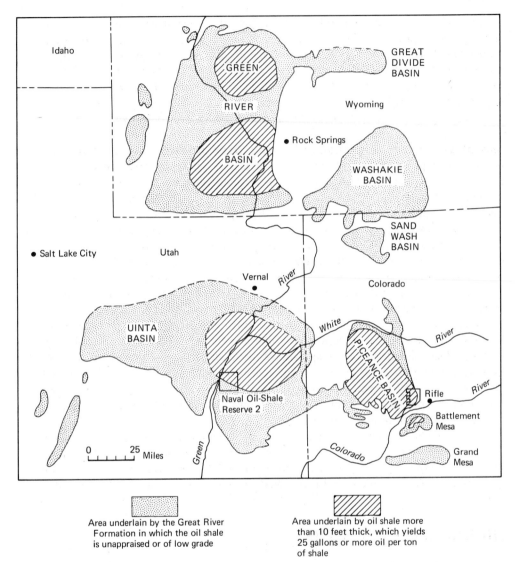

FIGURE 13.24 *Distribution of oil shale in the Green River Formation of Colorado, Utah, and Wyoming. (Source: D. C. Duncan and V. E. Swanson, U.S. Geological Survey Circular 523, 1965.)*

halite (rock salt), trona (hydrous sodium carbonate), nahcolite (sodium bicarbonate), dawsonite (sodium, aluminum, hydroxl, carbonate), and other materials or minerals are also found as nodules, lenses, thin beds, or disseminated crystals (9). Nahcolite is a potentially valuable source of sodium carbonate, and dawsonite is a possible source of aluminum. Their recovery with the oil may help reduce the cost of mining.

The geology of the Green River Formation and the depositional environment in the large lakes obviously varied significantly as the nature and pattern of deposition varied. The Green River Formation varies in thickness from a few feet to several thousand feet, and some beds are rich in organic materials while others are not. Generally, the thickest accumulation of organic material is in the central parts of the large, shallow lakes where microscopic algae and other microorganisms flourished, died, and were buried with a mixture of sediment and other land-derived organic material. Toward the margin of the basins, less organic material accumulated, and the oil shale there is of a lower grade (Figure 13.24) and is interbedded with land-derived, inorganic sediments (sand, silt, etc.) transported into the lake by streams, wind, and other surface processes. Few of these inorganic sediments reached the center of the lake, facilitating the depositions there of rich accumulations of organic material from which the oil shale formed (9).

The large, shallow, ancient lakes of Wyoming, Utah, and Colorado, in which the organic material accumulated, varied from fresh to quite alkaline water. The lake basins subsided slowly and irregularly, causing the center of deposition of the oil shale to shift slowly during the millions of years that the deposition continued. For millions of years after the sediment was buried, little tectonic activity disturbed the sediments. More recently, uplift and local tilting of the rocks have exposed some of the oil shale to erosion (9).

Geology of Tar Sands Tar sands are rocks that are impregnated with tar oil, asphalt, or other petroleum materials and from which recovery of petroleum products by usual methods such as oil wells is not commercially possible. The term *tar sand* is somewhat confusing because it includes several rock types such as shale and limestone as well as unconsolidated or consolidated sandstone. The one thing all these have in common is that they contain a variety of semiliquid, semisolid, and solid petroleum products, some of which ooze from the rock outcrops and others of which are difficult to remove with boiling water (11).

Oil in tar sands is nearly the same as the heavier oil pumped from wells. The only real difference is that tar-sand oil is much more viscous and therefore more difficult to recover. A possible conclusion concerning the geology of tar sands is that they form essentially the same way that the more fluid oil forms, but much more of the volatiles and accompanying liquids in the reservoir rocks have escaped, leaving the more viscous materials behind.

Distribution of Oil Shale and Tar Sands Many countries have deposits of oil shale (Figure 13.25), but there is a tremendous range in thickness of the rock, the quality of oil contained, and the areal extent (12).

Total identified shale-oil resources of the world land areas are estimated to contain about 3 trillion barrels of oil. However, total evaluation of the grade and the feasibility of economic recovery with today's technology and economic situation is not completed. Shale-oil resources in the United States amount to about 2 trillion barrels of oil, or two-thirds of the total identified in the world (Table 13.3), and of this, 90 percent, or 1.8 trillion barrels, is located in the Green River Oil Shales. However, even these rich deposits are not equally distributed: The Unita Basin contains approximately two-thirds (1.2 trillion barrels) of the identified oil in the Green River Oil Shales (9).

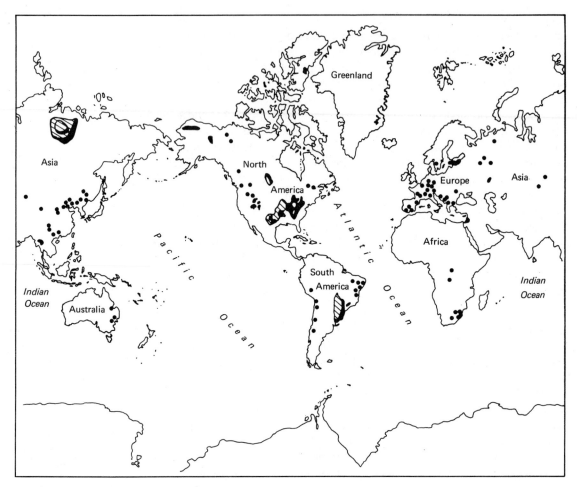

FIGURE 13.25   *Oil-shale deposits of the world. Areas denoted by the diagonal lines are those where some major deposits may be extended. (Source: D. C. Duncan and V. E. Swanson, U.S. Geological Survey Circular 523, 1965.)*

Unfortunately, identified resources of oil shale are not all recoverable with today's economics. However, this could change rapidly, and probably around 80 billion barrels of oil can now be extracted at a profit (10).

Large accumulations of tar sands are known. For example, the Athabasca Tar Sands of Alberta, Canada, cover an area of approximately 30,000 square miles and contain an estimated reserve of 300 billion barrels of oil that might be recovered (13). In addition, smaller tar-sand resources are known in Utah (1.8 billion barrels), California (100 million barrels), Texas, Wyoming, and other areas (11).

Impact of Exploration and Development of Shale Oil and Tar Sands   Recovery of petroleum from surface or near-surface oil shale and tar sands involves use of well-established exploration techniques. Numerous deposits are known—what must be developed are reliable techniques of developing the oil-shale resources that will cause a minimum of environmental disruption. The really significant problem associated with mining of oil shale is insufficient water supplies, a topic that will be discussed in a later section of the text on energy and water demand.

It is hoped that eventually oil from underground oil shales and tar sands may be recovered in place with little surface disturbance. However, at present, the most likely mining procedure is surface strip mining, and this method definitely has environmental problems.

The Athabasca Tar Sands in Alberta are now yielding about 50,000 barrels of synthetic crude oil per day from one large open-pit mine operated by the Great Canadian Oil Sands Limited. The average thickness of the tar sands being mined is about 140 feet, and the amount of overburden removed varies considerably but is generally less than 150 feet. Approximately 140,000 tons of tar sand and 130,000 tons of overburden are removed in each day's activity (13).

Methods of mining the Canadian Tar Sands are atypical for surface mining for four reasons. First, the indigenous vegetation is a water-saturated, organic matte (muskeg swamp) of decayed and decaying vegetation that can be removed easily only in the winter when it is frozen. Second, the muskeg must be drained before excavation, and this takes about two years. Third, once removed, the

TABLE 13.3 *Shale oil resources of the United States in billions of barrels.*

Deposit	Identified[a]		Hypothetical[b]		Speculative[c]	
	25–100 gal/ ton	10–25 gal/ ton	25–100 gal/ ton	10–25 gal/ ton	25–100 gal/ ton	10–25 gal/ ton
Green River Formation, Colorado, Utah, and Wyoming	418[d]	1400	50	600	—	—
Chattanooga Shale and equivalent formations, central and eastern United States	—	200	—	800	—	—
Marine shale, Alaska	Small	Small	250	200	—	—
Other shale deposits	—	Small	Ne	Ne	600	23,000
Total	418[d]	1600	300	1600	600	23,000

SOURCE: D. A. Brobst and W. P. Pratt, eds., U.S. Geological Survey Professional Paper 820, 1973.
[a]Identified resources: Specific, identified mineral deposits that may or may not be evaluated as to extent and grade, and whose contained minerals may or may not be profitably recoverable with existing technology and economic conditions.
[b]Hypothetical resources: Undiscovered mineral deposits, whether of recoverable or subeconomic grade, that are geologically predictable as existing in known districts.
[c]Speculative resources: Undiscovered mineral deposits, whether of recoverable or subeconomic grade, that may exist in unknown districts or in unrecognized or unconventional form.
[d]The 25 to 100 gal/ton category is considered virtually equivalent to the category "average of 30 or more gallons per ton."

overburden and other unwanted material with the tar and sand occupy considerably more volume than before they were removed, creating a disposal problem and a final land surface that is more than 70 feet above the original surface. Fourth, restoration of the land after mining is a serious problem in this fragile environment that is so difficult to work with (13).

Fortunately, the actual processes of recovering oil from tar sands is relatively easy and involves washing the viscous oil out of the sand with hot water. The oil then flows to the surface and is removed (11). However, only about 10 percent of the total oil from the tar sands can be economically recovered by open-pit mining, and therefore recovery of the oil in place (without removal by surface or subsurface mining) should become a necessary goal for obtaining additional oil (13). Hopefully, in-place recovery techniques will involve a minimum of surface disturbance.

Depending on the technique, development of oil-shale resources may involve significant disruption of the environment. Two methods of recovery are being considered: traditional strip mining or subsurface mining, and in-place recovery. In addition, combinations of subsurface mining and in-place recovery are also under experimentation. From an environmental standpoint, the in-place method is preferred.

The first method that will be used to recover oil from the shale is probably surface mining and processing of the shale by crushing it and then heating it to release the oil. The waste material will then have to be disposed of. Unfortunately, the volume of waste will exceed the original volume of shale mined by 20 to 30 percent. Therefore, the mine from which the shale was removed will not be able to accommodate the waste, and it will have to be piled up or otherwise disposed of. The impact of the waste disposal is expected to be considerable. For example, if surface mining is used to produce 100,000 barrels of shale oil per day for 20 years, the operation will produce 20 billion cubic feet of waste. If 50 percent of this is disposed of on the surface, it could fill an area 5 to 10 miles long, 2000 feet wide, to a depth of 200 feet. Therefore, we will have to determine ways to contour and vegetate shale-oil waste to minimize the visual and pollutional impacts (12).

The concept of mining oil shale in place requires several general stages or processes: *drilling* a planned number of wells into the oil shale; *fracturing* the shale to increase permeability [this may be done by explosions or by injecting water under pressure (hydrofracturing)]; *heating* the oil shale in place to release the oil by actually igniting the shale or by circulating hot gases; and *recovering* the oil from other wells (14).

In-place mining of oil shale would obviously cause less surface disruption than open-pit mines. However, additional research into possible consequences of in-place mining is needed. In any event, both mining methods will produce an oil that may need treatment before transportation (12).

The human environment will also change significantly if oil shale is mined in Colorado, Wyoming, or Utah. It will bring more roads, pipelines, airports, etc.; economic activity; more facilities of all kinds; and rapid urbanization as the population increases, all of which will affect the physical environment.

Nuclear Energy

The first controlled nuclear fissions were demonstrated in 1942 and led the way to the development of the primary uses of uranium, that is, in explosives and as a heat source to provide steam for generation of electricity. It now appears that the energy from uranium (one pound of uranium oxide produces a heat equivalent of approximately 16,000 pounds of coal) is soon likely to become one of the principal sources of energy in the United States (15).

Fission is the splitting of uranium (U^{235}) by neutron bombardment (Figure 13.26). The reaction produces three more neutrons released from uranium, fission fragments, and heat. The released neutrons each strike other U^{235} atoms, releasing more neutrons, fission products, and heat. As the process continues, a chain reaction develops as more and more uranium is split, releasing ever more neutrons.

Three types of uranium occur in nature: U^{238}, which accounts for approximately 99.3 percent of all natural uranium; U^{235}, which makes up about 0.7 percent; and U^{234}, which makes up about 0.005 percent. Uranium-235 is the only naturally occurring fissionable material, and it therefore is essential to the production of nuclear energy. However, uranium is processed to increase the amount of U^{235} from 0.7 percent to 3.5 to 5.0 percent before it is used in a reactor. The processed fuel is called *enriched uranium*. Uranium-238 is not naturally fissionable, but it is "fertile material" because upon bombardment by neutrons, it is converted to plutonium-239, which is fissionable (15).

Most reactors now being used consume more fissionable material than they produce. In order to conserve U^{235}, work is progressing on a reactor that actually produces more fissionable material (nuclear fuel) than it uses. These *breeder reactors,* utilizing a fuel core of fissionable material (plutonium-239) surrounded by a blanket of fertile material (U^{238}), will produce or breed additional plutonium-239 (a man-made fissionable material) from the U^{238}. The transformation from U^{238} to P^{239} occurs in the breeder reactor at the same time that the plutonium nuclei in the core are undergoing fission, providing heat that produces steam to drive turbines and generators which produce electricity. Development of the breeder reactor will greatly extend the limited supply of natural U^{235}.

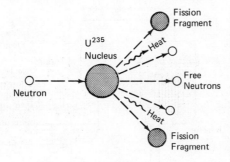

FIGURE 13.26 *Idealized diagram showing fission of U^{235}. A neutron strikes the U^{235} nucleus, producing fission fragments and free neutrons and releasing heat. The released neutrons may then each strike another U^{235} atom, releasing more neutrons, fission fragments, and energy. As the process continues, a chain reaction develops.*

It is the heat from the fission reaction that is used to produce steam and drive electrical generators at power plants. The function of the nuclear reactor is to maintain controlled fission and provide the necessary heat to produce the desired quantity of steam. Therefore, the reactor has exactly the same function as the boiler has in producing heat from burning coal or oil (Figure 13.27).

The demand for electric power is growing at a rapid rate, and an increase in the amount of power generated by nuclear energy is expected to grow even faster, until by the year 2000, when it is expected, heat from nuclear reactions will produce about 60 percent of the electrical energy in the United States (Figure 13.28). Over 50 reactors are now in operation (Figure 13.29), but many more will be needed if the projected energy from uranium is to be realized. Probably due to increased costs, environmental considerations, and other factors, the projected figures of Figure 13.28 are too high and will have to be revised. Therefore, the full impact of what began in 1942 is still to be determined.

Geology and Distribution of Uranium   The natural concentration of uranium in the earth's crust is about 2 ppm. Uranium originates in magma and is concentrated to about 4 ppm in granitic rock, where it is found in a variety of minerals. Some

FIGURE 13.27   *Idealized diagram comparing a fossil fuel power plant and nuclear power plant. Notice that the nuclear reactor has exactly the same function as the boiler in the fossil fuel power plant. (Reprinted, by permission, from* Nuclear Power and the Environment, *American Nuclear Society, 1973.)*

FOSSIL FUEL POWER PLANT

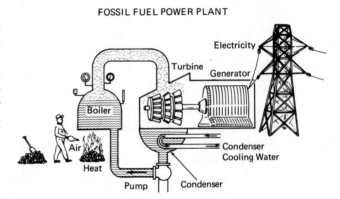

NUCLEAR POWER PLANT
Boiling Water Reactor (BWR)

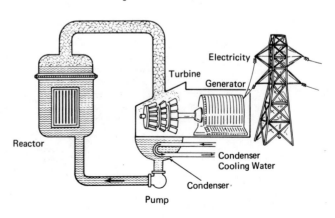

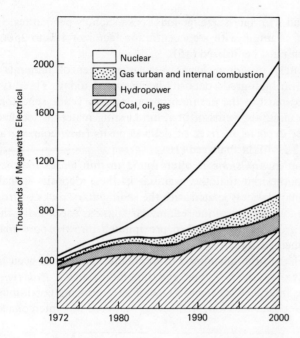

FIGURE 13.28 *Projected electrical power generating capacity in the United States to the year 2000. (Source: Office of Industry Relations,* The Nuclear Industry, 1974.)

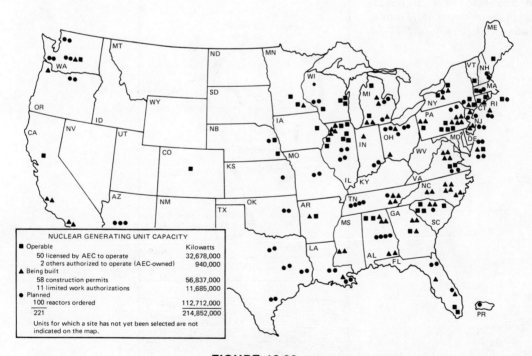

FIGURE 13.29 *Nuclear power reactors in the United States as of 1974. (Source: Office of Industry Relations,* The Nuclear Industry, 1974.)

uranium is also found with late-stage igneous rocks such as pegmatites. In order to be mined at a profit, uranium with a concentration factor of 400 to 2500 times the natural concentration must be utilized (15).

Fortunately, uranium forms a large number of minerals, many of which can be found in high-grade deposits being mined today. Three types of deposits have produced most of the uranium in the last few years: sandstone impregnated with uranium minerals; veins of uranium-bearing materials localized in rock fractures; and placer deposits in river or delta deposits (now coarse-grained sedimentary rock) over 2.2 billion years old (16).

Uranium in *sandstone* is often found in thin lenses interbedded with mudstone. It is hypothesized that the uranium in these deposits was derived by leaching from volcanic glass associated with the sedimentary rocks or from granitic rocks exposed along the margins of the sedimentary basins. Supposedly, the uranium was transported by groundwater and then precipitated into the pore spaces of the sandstone under reducing (oxygen-deficient) conditions (15).

Most of the uranium mined in the United States has been from sandstone deposits of two types: roll type and tabular deposits. *Roll-type* uranium deposits form at the interface between oxidizing and reducing conditions. Characteristically, the ore deposits are elongated bodies scattered like interconnected beads

FIGURE 13.30 *The geologic cross section and plan view of typical roll-type uranium deposits. (Source: U.S. Geological Survey, INF-74-14, 1974.)*

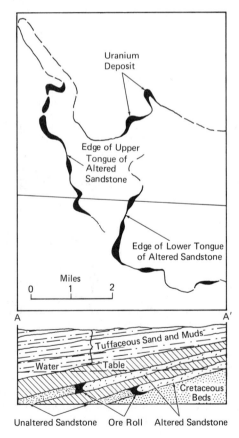

along miles of interface between altered (oxidized) sandstone and unaltered sandstone (Figure 13.30). *Tabular* uranium deposits are discrete masses completely enclosed by altered (reduced) sandstone which is itself enveloped in oxidized sandstone, as shown in Figure 13.31 (15, 16).

Uranium deposits in *veins* generally fill fissures (rock fractures) in many types of rocks of varying geologic age. The veins usually vary from several inches to a few feet in width and several hundred feet in length. Veins may coalesce to form vein systems that may extend for several thousand feet (15).

Uranium in ancient river or delta sediment was deposited before the occurrence of abundant free oxygen in the atmosphere. Therefore, it is believed, the uranium mineral was deposited as stream-rounded particles along with gold, pyrite, and other typical placer material (16).

The distribution of uranium deposits being mined in the United States is shown in Figure 13.32. Although sandstone ores have been the main source of uranium in this country, vein deposits are a significant source in Australia, Canada, France, and Africa, and ancient placer deposits are being mined in Africa and Canada (16).

The amount of uranium from the known deposits is sufficient to last until the 1980s. Beyond that, additional exploration for deposits and research of techniques to locate new reserves will be required (15). However, if by the 1980s the breeder reactor program is successful and shown to be environmentally safe, then the supply of fissional material will be greatly extended.

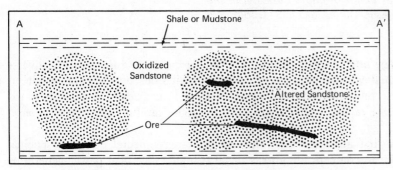

CROSS-SECTIONAL VIEW

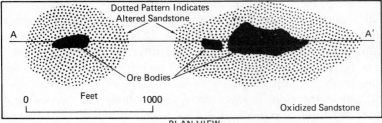

PLAN VIEW

FIGURE 13.31 *Geologic cross section and plan view of tabular uranium deposits. (Source: U.S. Geological Survey, INF-74-14, 1974.)*

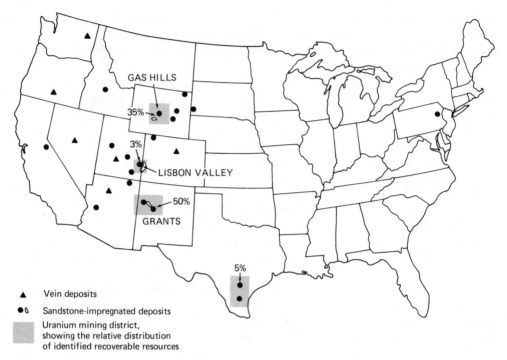

FIGURE 13.32 *Distribution of major uranium deposits in the United States. (Modified from A. P. Butler, Jr., U.S. Geological Survey Circular 547, 1967.)*

Nuclear Energy and the Environment   Nuclear energy and the possibly adverse effects associated with it have been subjects of vigorous debate. The debate is healthy because the number of reactors producing heat to drive generators is expected to increase rapidly in the future, and we should carefully examine the consequences.

A first approach to evaluating the environmental effects of nuclear power might be to compare the effects of all common electrical power generating methods by power source: coal, oil, gas, and uranium (Table 13.4). All methods affect the environment in various ways during their normal operation. Disregarding the magnitude of a particular effect (as, for example, thermal pollution, which is greater for nuclear power because the generator capacity is generally larger and no heat is directly dissipated to the atmosphere), the real difference of nuclear power is radioactivity. It is the possible radiation hazard from nuclear energy production and waste disposal that really concerns people.

Throughout the entire nuclear cycle, from mining and processing of uranium to controlled fission, to reprocessing of spent nuclear fuel, to final disposal of radioactive waste, various amounts of radiation enter and, to a lesser or greater extent, affect the environment.

The chance of a really serious nuclear accident is estimated to be very remote. Nevertheless, the chances of problems increase with every reactor put into operation. Furthermore, there are other potential hazards associated with transporting nuclear materials and supplying other nations with reactors. Terrorist activity and the possibility of irresponsible persons in governments add a risk that is present in no other forms of energy production.

Nuclear energy may indeed be the answer to our energy problems, and perhaps someday it will provide unlimited cheap energy. However, with nuclear power must come an increased responsibility to insure that nuclear power is used for, not against, people and that future generations inherit a quality environment free from worry concerning hazardous nuclear waste.

TABLE 13.4 *Comparison of environmental effects of coal, oil, gas, and uranium electrical power generation.*

Energy Source	Effect on Land	Effects on Water	Effects on Air	Biological Effects	Supply
Coal	Disturbed land; large amounts of solid waste	Acid mine drainage; increased water temperature	Sulfur oxides; nitrogen oxides; particulates; some radio-active gases	Respiratory problems from air pollutants	Large reserves
Oil	Wastes in the form of brine; pipeline con-struction	Oil spills; increased water temperature	Nitrogen ox-ides; carbon monoxide; hydrocarbons	Respiratory problems from air pollutants	Limited domestic reserves
Gas	Pipeline con-struction	Increased water temp-erature	Some oxides of nitrogen	Few known effects	Extremely limited domestic reserves
Uranium	Disposal of radioactive waste	Increased water temperature; some radio-active liquids	Some radio-active gases	None detectable in normal operation	Large reserves if breeders are developed

SOURCE: Pennsylvania Department of Education. *The Environmental Impact of Electrical Power Generation: Nuclear and Fossil,* 1973.

Geothermal Energy

The useful conversion of natural heat from the interior of the earth *(geothermal energy)* to heat buildings and generate electricity is a very exciting application of geologic knowledge and engineering technology. The idea of harnessing the earth's internal heat is not new. As early as 1904, geothermal power was developed in Italy utilizing dry steam, and natural internal heat is now being used to generate electricity in the USSR, Japan, New Zealand, Iceland, Mexico, and California (Figure 13.33). However, the existing geothermal facility probably constitutes only a small portion of the total energy that might eventually be tapped from the earth's reservoir of internal heat.

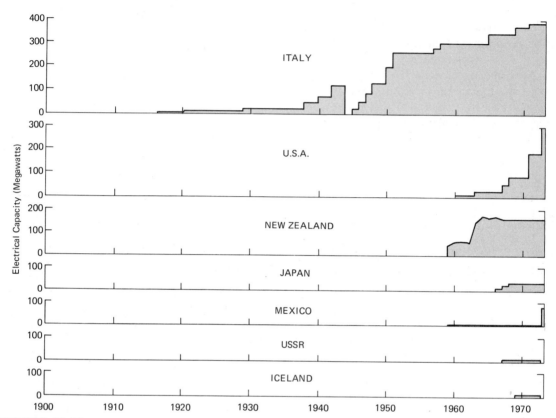

FIGURE 13.33 *Growth of geothermal generating capacity by countries from 1900 to 1972. (Source: D. A. Brobst and W. P. Pratt, eds., U.S. Geological Survey Professional Paper 820, 1973.)*

Geology of Geothermal Energy   Natural heat production within the earth is only partly understood. We do know that some areas have a higher flow of heat from below than others, and that for the most part these locations are associated with the tectonic cycle. Oceanic ridge systems (divergent plate boundaries) and convergent plate boundaries, where mountains are being uplifted and volcanic island arcs are forming, are areas where this natural heat flow from the earth is anomalously high.

      The relative distribution of heat flow from within the earth for the United States is shown in Figure 13.34. The data for this generalized classification of hot, normal, and cold crustal areas are not sufficient to accurately define the limits of the regions (17), and therefore the map has limited value for locating specific sites where geothermal energy resources could be developed.

      It is interesting to note that the region of high heat flow is concentrated in the western United States where tectonic activity and volcanic activity have been recent. The wide width of the belt is somewhat of an anomaly, and for years geologists have been vigorously debating the origin of the rock structure and tectonic

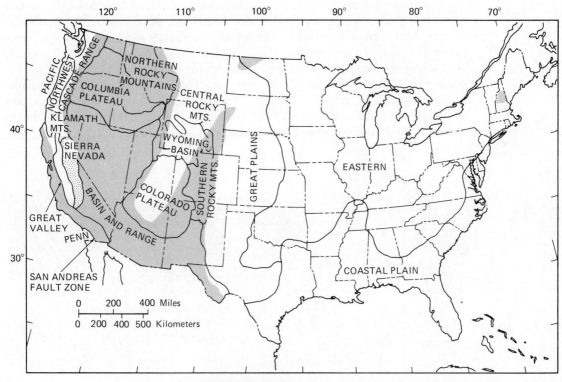

FIGURE 13.34 *Map showing the generalized extent of hot (shaded), normal (white), and cold (stippled) crustal regions of the United States. Also shown are the major physiographic provinces such as the Colorado Plateau. (Source: D. F. White and D. L. Williams, eds., U.S. Geological Survey Circular 726, 1975.)*

activity of the Basin and Range region (Figure 13.34), which is characterized by tensional stress and volcanic activity. Some geologists believe that the region represents a place where the plate on which North America rides (American plate) is overlapping the Pacific plate, including the East Pacific Rise (oceanic ridge). With this interpretation, the Basin and Range is the surficial expression of oceanic ridge activity. Other geologists believe that the East Pacific Rise has not been overridden by the American plate but is separated along the San Andreas fault. Regardless of what the tectonic picture truly is, the entire western region is generally characterized by recent uplift and volcanic activity, and therefore, with limited exceptions, it is a good prospect for geothermal energy exploration.

Within the region of relatively high heat flow, commercial development of geothermal power is most likely where a heat source such as convecting magma is relatively near the surface (3 to 10 kilometers) and in thermal contact with circulating groundwater. Therefore, the likely sites for exploration are threefold: first, areas with naturally occurring hot springs and geysers that reflect near-surface hot spots; second, areas of recent volcanic activity, particularly those characterized by high-silica magma because they are more likely to have stored heat near the surface where it might be accessible to drilling (17); and third, other localized hot spots with little or near-surface expression that are discovered by direct and indirect (geophysical) subsurface exploration. However, the best location might not necessarily be that with the most prominent surface expression of geyser and hot-spring activity because geothermal systems with a high rate of upflow, feeding geysers, and hot springs might not be as well insulated as a system with little or no leakage, as shown in Figure 13.35 (18). Therefore, there is likely to be more stored heat at shallow depths in the well-insulated geothermal reservoir.

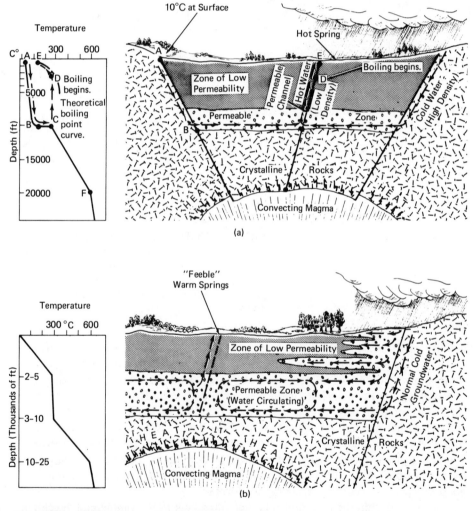

FIGURE 13.35 *Generalized diagrams showing (a) a hot-spring type of geothermal system with a high rate of upflow characterized by vigorous hot-spring or geyser activity at the surface; and (b) an insulated geothermal reservoir with little or no leakage. (Source: U.S. Geological Survey Publication GPO: 1969 O-339-536.)*

Geothermal Energy

381

Based on geologic criteria, several geothermal systems may be defined: hydrothermal convection systems, hot igneous systems, and geopressured systems (19, 20). Each of these systems has a different origin and different potential as an energy source.

Hydrothermal convection systems are characterized by a permeable layer in which a variable amount of hot water circulates. They are of two basic types: vapor-dominated systems and *hot-water* systems. Vapor-dominated hydrothermal convection systems are geothermal reservoirs in which both water and steam are present at depth. However, near the surface where pressure is less, the water flashes to superheated steam, which when tapped can be piped directly into turbines to produce electricity. These systems characteristically have a sufficiently slow recharge of groundwater so the hot rocks can convert the water to steam. That is, the heat supply is great enough to boil off more water than can be replaced by natural recharge (19).

Vapor-dominated systems are not very common. Only three have been identified in the United States: The Geysers, 90 miles north of San Francisco, California; Mt. Lassen National Park, California; and Yellowstone National Park, Wyoming. The parks are not available for energy development. However, steam from The Geysers in California has been producing electric energy for years (Figure 13.36).

FIGURE 13.36 *The Geysers Power Plant north of San Francisco, California, where approximately 502 megawatts of electrical power is produced each year. (Photo courtesy of Pacific Gas and Electric Company.)*

In the United States, hot-water systems are about 20 times more common than vapor-dominated systems (Figure 13.37). In these systems, with surface temperatures greater than 150 degrees Celsius, there is a zone of circulating hot water (no steam) that when tapped moves up with reduced pressure to yield a mixture of steam and water at the surface. The water must be removed from the steam before the steam can be used to drive the turbine (20). One problem with this system is disposal of the water. However, as ideally shown in Figure 13.38, the water could be injected back into the reservoir to be reheated.

Hot igneous systems may involve the presence of molten magma at temperatures of 650 degrees Celsius to about 1200 degrees Celsius depending on the type of magma. Furthermore, even if the igneous mass is not still molten, it may involve a large quantity of hot, dry rocks. These systems contain more stored heat

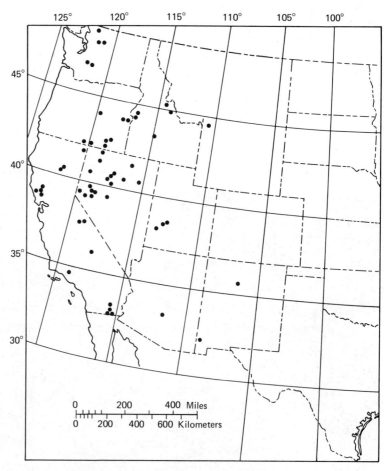

FIGURE 13.37 *Distribution of hydrothermal convection systems in the conterminous United States where subsurface temperatures are thought to be in excess of 150 degrees Celsius. (Source: D. F. White and D. L. Williams, eds., U.S. Geological Survey Circular 726, 1975.)*

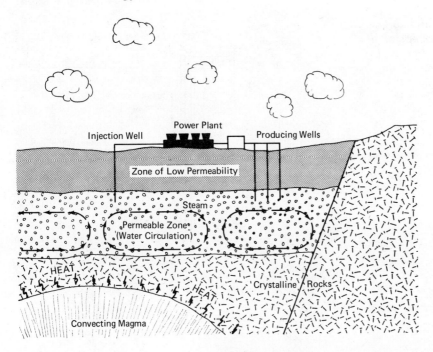

FIGURE 13.38 *Idealized diagram showing a hot-water geothermal system. At the power plant, the steam will be separated from the water and used to generate electrical power. The water will be injected back into the geothermal system by a disposal well. (Diagram courtesy of Pacific Gas and Electric Company.)*

per unit volume than any of the other geothermal systems. However, they lack the circulating hot water of the convection systems, and innovative methods to utilize the heat will have to be developed (19). Some of these geothermal reservoirs may have hot, dry rocks accessible to drilling, in which case the rocks might first be drilled and then fractured by using explosives or hydrofracturing techniques. Then cold water might be injected into the rock at one location and pumped out at elevated temperatures at another to recover the heat. The induced circulation of water would effectively mine the heat of the dry rock system (21). A pilot project by the Los Alamos Scientific Laboratory is currently drilling and testing hot, dry rock systems in the southwestern United States.

Recovery of heat directly from magma, accessible to drilling (less than 10 kilometers deep), is an exciting prospect. However, the necessary technology is yet to be developed, and in fact it may not be even remotely possible (22). Nevertheless, the possibility is being studied and evaluated. A generalized diagram of the concept to tap magma is shown in Figure 13.39. Some of the major problems to be solved involve drilling, heat-extraction technology, and, perhaps most important, learning more about first, how heat is transferred in magma; and second how the introduction of a heat-exchange device would affect the proportion of crystals, liquid, and vapor, and other physical and chemical properties of the magma in the

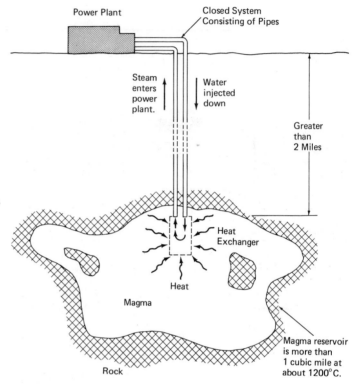

FIGURE 13.39 *Idealized diagram showing the general concept of a magma tap. (Modified after J. L. Colp, Sandia Laboratories Energy Report SAND-74-0253, 1974.)*

vicinity of the device (23). Extraction of heat directly from magma, even if theoretically possible, is a long way off.

Geopressured systems exist where the normal heat flow from the earth is trapped by impermeable clay layers that act as an effective insulator. The favorable environment is characterized by rapid deposition of sediment and regional subsidence. Deeply buried water trapped in the sediments in such systems develops considerable fluid pressure and is heated (20).

Perhaps the best-known regions where geopressurized systems develop are in the Gulf Coast of the United States where temperatures of 150 degrees Celsius to 273 degrees Celsius at depths of 4 to 7 kilometers have been identified. These systems have the potential to produce large quantities of electricity for three reasons. First, they contain hot-water thermal energy that could be extracted. Second, they contain mechanical energy from the high-pressure water that could be used to turn hydraulic turbines to produce electricity. Third, as an added asset, the waters contain considerable amounts of dissolved methane gas, up to 30 cubic feet per barrel of water, that could be extracted (20, 21). Of the three, the energy potential from the heat and methane gas far exceeds that of the high-pressure water (24).

Environmental Impact of Geothermal Energy Development   Although the potentially adverse environmental impact of intensive geothermal energy development is perhaps not as extensive as that of other sources of energy, it is nevertheless considerable. Geothermal energy is developed at a particular site, and environmental problems include on-site noise, emissions of gas, and industrial scars. Fortunately, development of geothermal energy does not require the extensive transportation of raw materials or refining typical of the fossil fuels. Furthermore, geothermal energy does not produce atmospheric particulate pollutants associated with burning fossil fuels; nor does it produce any radioactive waste. However, geothermal development, except for the vapor-dominated systems, does produce considerable thermal pollution from hot waste waters, which may be saline or highly corrosive. The plan is to dispose of these waters by reinjecting them back into the geothermal reservoir, but such disposal has problems because injection of fluids may activate fracture systems in the rocks and cause earthquakes. In addition, original withdrawal of fluids may compact the reservoir, causing surface subsidence. It is also feared that subsidence might occur because as the heat in the system is extracted, the cooling rocks will contract (20).

Future of Geothermal Energy   It appears that geothermal energy has a bright future. The estimated yield from this vast resource, which is both identified and recoverable at this time (discarding cost) from hydrothermal convection systems and geopressured systems, over a 30-year period is about equivalent to that of 140 modern nuclear power plants. In addition, it is expected that many more systems with presently recoverable energy are yet to be discovered (19). Furthermore, geothermal energy in the form of hot water (not steam) utilizing normal heat flow may be a future source of energy to heat homes and other buildings along the Atlantic Coast, where alternative energy sources are scarce.

   At present, geothermal energy supplies only a small fraction of one percent of electrical energy produced in the United States. With the exception of the unusual vapor-dominated systems such as The Geysers in California, the production of electricity from geothermal reservoirs is still rather expensive, generally more so than from other sources of energy. For this reason, commercial development of the energy will not develop rapidly until the economics are equalized. Even given this, the total energy output (electrical) from geothermal sources is not likely in the near future to exceed a few percent (10 percent at most), even in states like California where it has been produced and where expanding facilities are likely (21).

Water Power

Since at least the time of the Roman Empire, the use of running water as a source of power has historically been a successful venture. Water wheels that harness water power and convert it to mechanical energy were turning in Western Europe in the seventeenth century, and during the eighteenth and nineteenth centuries large water wheels provided energy to power grain mills, sawmills, and other machinery in the United States.

   Although hydroelectric power plants in the last few years have provided only about 15 percent of the total electricity produced in the United States, and this percentage is expected to be reduced to about 10 percent by the year 2000, it nevertheless remains an important source of power. Furthermore, the reduction in the percentage of electrical power generated by running water does not mean that there will be a reduction in the total number of hydroelectric plants. In fact, the actual power produced from water is expected to increase about two to three times the present production. However, this increase is small compared to that of nuclear power which is expected to increase by several hundred times.

Water power is clean power. That is, it requires no burning of fuel, does not pollute the atmosphere, produces no radioactive or other waste, and is efficient. However, there is an environmental price to pay. Water falling over high dams may pick up nitrogen gas which enters the blood of fish, expands, and kills them. Nitrogen has killed many migrating game fish in the Pacific Northwest. Furthermore, dams trap sediment that would otherwise reach the sea and replenish the sand on beaches. In addition, for a variety of reasons, there is a growing desire of many people not to turn all of the wild rivers into a series of lakes. Therefore, for a number of reasons, including the facts that many good sites for dams are already utilized and others are extremely limited and that many rivers will be saved by legislation, the growth of hydroelectric power from rivers is likely to be curtailed somewhat in the future.

Another form of water power might be derived from ocean tides in a few places where there is favorable topography, such as the Bay of Fundy region of the northeastern United States and Canada. The tides in the Bay of Fundy have a maximum rise of about 15 meters. A minimum rise of about 8 meters is necessary to even consider developing tidal power.

The idea in developing tidal power is to build dams across the entrance to bays, creating a basin on the landward side such that a difference in water level between the ocean and the basin can be created. Then, as the water in the basin is filling or emptying, it can be used to turn hydraulic turbines which will produce electricity (25).

Solar Energy

The solar power input to the earth is about 180 quadrillion watts per year, which is an enormous amount of power. Using the latitudes and weather conditions of the United States, scientists have computed that approximately 450 thermal megawatts of thermal energy fall on each square mile. Therefore, with present solar collection technology (20 percent efficiency and 50 percent spacing), an area of about 10,000 square miles (about one-tenth the size of Nevada) would produce 450,000 electrical megawatts, which is equivalent to the entire electrical generating capacity of the United States (26). Although it may be technically feasible to cover thousands of square miles with solar collectors and build a system to store the energy and transform it into electrical energy, the cost at present is prohibitive. Other ideas concerning large-scale production of solar energy involve using special satellites to collect the energy which would then be beamed down to earth and converted to electrical power. It appears that large-scale solar energy development is a long way off.

From a pragmatic view, solar energy on a small scale will probably soon be commonly used to help heat and cool buildings. However, support systems using fossil fuels or electricity will have to be used to augment solar power variations caused by changes in daily and seasonal weather. From examining solar power potential, one can conclude that it is indeed a viable alternative to help alleviate energy shortages, and if it is widely used in homes and buildings for heating and cooling, perhaps we will not need so many nuclear power plants.

Nuclear Fusion

Nuclear fusion may be a future alternative source of energy. The source of energy in the sun and other stars is the fusion of light hydrogen atoms to more complex, heavier helium atoms. For many years scientists have been interested in applying the principle of fusion to produce a controlled reaction as an energy source. However, fusion takes place only under very high temperatures, and it is explosive. Therefore, controlled fusion remains at best a hope of the future (27).

Energy and Water Demand

The recent push in the United States to rapidly accelerate energy development has raised an important question: Is there enough water to support the projected development? This complex question has recently been addressed by a U.S. Geological Survey evaluation, and our discussion here will summarize some of the conclusions from that report (28).

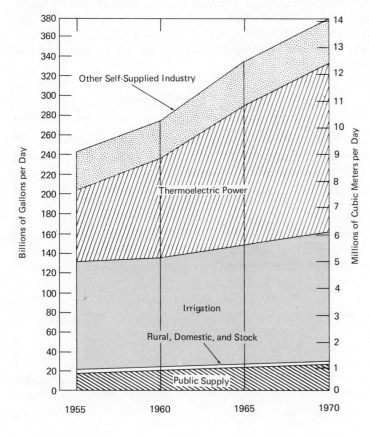

FIGURE 13.40 *Withdrawal of water for major uses from 1955 to 1970 in the United States. (Source: G. H. Davis and L. A. Wood, U.S. Geological Survey Circular 703, 1974.)*

The withdrawal of water for major uses is shown in Figure 13.40. Note that in 1955 the largest withdrawal was for irrigation. But by 1970, water used for thermoelectric power generation (fossil fuel, nuclear, or geothermal heat sources) had increased by almost 150 percent and exceeded irrigation demands for water. Most of the water for the thermoelectric power generation was withdrawn to cool and condense the steam on the outlet side of the turbine. Most of the water used is withdrawn from a river or lake, is circulated once through the condenser, and carries the waste heat to a point of discharge back in the river or lake. Where thermal pollution is unacceptable or there is no large lake or river from which to obtain abundant cooling water, cooling towers, evaporation ponds, and water-sprayed evaporation ponds may be utilized in the cooling process (28).

The above discussion established that a great deal of water is used in the thermoelectric generating industry. However, water is needed in all stages of energy development from mining and land reclamation, to on-site processing, refining, and converting fuels to other energy forms. Figure 13.41 compares water consumption in refining and conversion processes for the presently used commercial energy sources and those likely to be used in the future.

The amount of water required for mining and reclamation varies for different fuels and different climatic conditions. For example, water demand for mining of coal or uranium is generally low, but obtaining the water necessary in arid regions to establish vegetation as part of the reclamation program may be a serious problem. On the other hand, reclaiming of a surface mine in a humid climate would probably not have a problem of lack of water. Another example can be drawn from the oil industry. Little water is used in drilling for oil, but during the recovery phase, a tremendous amount of water may be pumped down into the well to float the oil as part of a secondary recovery technique (28).

FIGURE 13.41 *Consumption of water for various energy-related refining and conversion processes. (Source: G. H. Davis and L. A. Wood, U.S. Geological Survey Circular 703, 1974.)*

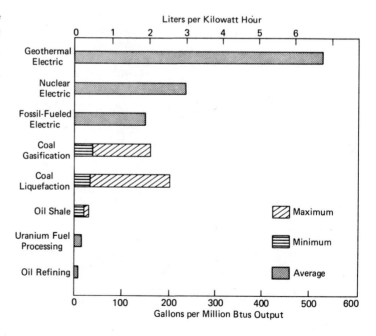

Mining of oil shale and particularly the disposal of spent shale and surface reclamation in Colorado, Utah, and Wyoming will also be associated with a high demand for water. In fact, water requirements will probably limit the production of shale oil to about one million barrels per day. A much larger production would require the purchasing and transfer of water that is now used for agricultural purposes (28).

The conclusion concerning the demand for water and production of energy in the United States is that in the entire nation, there is sufficient water to support accclerated energy development. However, in arid regions, limited water will affect site location and design of energy recovery and conversion facilities (28).

SUMMARY AND CONCLUSIONS

The ever-increasing population and appetite of the American people and the world for energy in all forms is staggering. Projections suggest that by the year 2000 most electrical energy in the United States will be produced from steam utilizing nuclear reactors as heat sources. Hundreds of new nuclear power plants will be needed, and since only about 200 are expected to be in operation by 1988, only 12 years will remain to complete a large number of other new reactors. However, it is time to seriously question the need and desirability of an ever-increasing demand for electrical and other sources of energy in industrialized societies. Quality of life is not necessarily directly related to energy consumption.

The origin of fossil fuels (coal, oil, and gas) is intimately related to the geologic cycle. These fuels are, essentially, stored solar energy in the form of organic material that has escaped total destruction by oxidation. The environmental disruption associated with exploration and development of these resources must be weighed with the benefits gained from the energy. However, this development is not an either-or proposition, and good conservation practices combined with pollution control and reclamation can help minimize the environmental disruption associated with fossil fuels.

Nuclear energy is here to stay, and the only question is how fast new plants will be constructed. The troublesome parts of the nuclear energy cycle concern the possible radiation hazard, disposal of waste, and, in particular, the possibility of tragic misuse of nuclear materials by terrorists.

Use of geothermal energy in the future will become much more widespread in the western United States where natural heat flow from the earth is relatively high. Although the electric energy produced from the internal heat of the earth will probably never exceed 10 percent of the total electric power generated, it nevertheless will be significant. However, geothermal energy also has an environmental price, and the possibility of causing surface subsidence due to withdrawal of fluids and heat, as well as possibly causing earthquakes from injection of hot waste water back into the ground, must be considered.

The use of water power from rivers to generate electricity is expected to increase until about the year 2000. However, due to the expected rapid expansion of nuclear power, the proportion of electric power produced by running water to other sources of power may actually decrease. Limiting factors for future development of water power result from the fact that many of the best sites are already used. Furthermore, more rivers are being protected from development by legislation.

Electricity produced from tidal (water) power is expected to increase in the future. However, the amount generated will remain very small in proportion to the total power generated by all sources because the number of good sites is extremely limited.

Large-scale solar energy developments, although technically feasible, appear a long way off. However, it is expected that many new buildings and homes in the future will be partially heated and cooled by solar energy with supporting fuel-burning heating for times when solar energy is not sufficient. Solar energy, geothermal energy, and other alternative sources are the viable alternatives that might reduce the number of nuclear power plants that would otherwise be needed.

References

1. U.S. Bureau of Mines. 1975. *Mining and minerals policy, 1975.*
2. Office of Industry Relations. 1974. *The nuclear industry: 1974.*
3. Averitt, P. 1973. Coal. In *United States mineral resources,* eds. D. A. Brobst and W. P. Pratt, pp. 133–42. U.S. Geological Survey Professional Paper 820.
4. U.S. Environmental Protection Agency. 1973. *Processes, procedures and methods to control pollution from mining activities.* EPA-430/9-73-001.
5. U.S. Bureau of Mines. 1971. *Strippable reserves of bituminous coal and lignite in the United States.* U.S. Bureau of Mines Information Circular 8531.
6. Committee on Environment and Public Planning. 1974. Environmental impact of conversion from gas or oil to coal for fuel. *The Geologist,* Newsletter of the Geological Society of America. Supplement to vol. 9, no. 4.
7. McCulloh, T. H. 1973. Oil and gas. In *United States mineral resources,* eds. D. A. Brobst and W. P. Pratt, pp. 477–96. U.S. Geological Survey Professional Paper 820.
8. Federal Energy Administration. 1975. FEA reports on proven reserves. *Geotimes,* August 1975, pp. 22–23.
9. Culbertson, W. C., and Pitman, J. K. 1973. Oil shale. In *United States mineral resources,* eds. D. A. Brobst and W. P. Pratt, pp. 497–503. U.S. Geological Survey Professional Paper 820.
10. Duncan, D. C., and Swanson, V. E. 1965. *Organic-rich shale of the United States and world land areas.* U.S. Geological Survey Circular 523.
11. Office of Oil and Gas, U.S. Department of the Interior. 1968. *United States petroleum through 1980.* Washington, D.C.: U.S. Government Printing Office.
12. Committee on Environmental and Public Planning. 1974. Development of oil shale in the Green River Formation. *The Geologist,* Newsletter of The Geological Society of America. Supplement to vol. 9, no. 4.
13. Allen, A. R. 1975. Coping with oil sands. In *Perspectives on energy,* eds. L. C. Ruedisili and M. W. Firebaugh, pp. 386–96. New York: Oxford University Press.

14. Dinneen, G. U., and Cook, G. L. 1975. Oil shale and the energy crisis. In *Perspectives on energy,* eds. L. C. Ruedisili and M. W. Firebaugh, pp. 377–85. New York: Oxford University Press.
15. Finch, W. I. et al. 1973. Nuclear fuels. In *United States mineral resources,* eds. D. A. Brobst and W. P. Pratt, pp. 455–76. U.S. Geological Survey Professional Paper 820.
16. U.S. Geological Survey. 1973. *Nuclear energy resources: a geologic perspective.* U.S.G.S. INF-73-14.
17. Smith, R. L., and Shaw, H. R. 1975. Igneous-related geothermal systems. In *Assessment of geothermal resources of the United States—1975,* eds. D. F. White and D. L. Williams, pp. 58–83. U.S. Geological Survey Circular 726.
18. White, D. F. 1969. *Natural steam for power.* U.S. Geological Survey Publication, GPO-19690-339-S36.
19. White, D. F., and Williams, D. L. 1975. In *Assessment of geothermal resources of the United States—1975,* eds. D. F. White and D. L. Williams, pp. 1–4. U.S. Geological Survey Circular 726.
20. Muffler, L. J. P. 1973. Geothermal resources. In *United States mineral resources,* eds. D. A. Brobst and W. P. Pratt, pp. 251–61. U.S. Geological Survey Professional Paper 820.
21. Worthington, J. D. 1975. *Geothermal development.* Status Report—Energy Resources and Technology, A Report of The Ad Hoc Committee on Energy Resources and Technology, Atomic Industrial Form, Incorporated.
22. Colp, J. L. 1974. *Magma tap—the ultimate geothermal energy program.* Sandia Laboratories Energy Report, SAND74-0253.
23. Colp, J. L. et al. 1975. *Magma workshop: assessment and recommendations.* Sandia Laboratories Energy Report, SAND75-0306.
24. Papadopulus, S. S. et al. 1975. Assessment of onshore geopressured-geothermal resources in the northern Gulf of Mexico basin. In *Assessment of geothermal resources of the United States,* eds. D. F. White and D. L. Williams, pp. 125–46. U.S. Geological Survey Circular 726.
25. Committee on Resources and Man, National Academy of Science. 1969. *Resources and man.* San Francisco: W. H. Freeman.
26. Interagency Task Force on Solar Energy. 1974. *Solar energy.* National Science Foundation. Washington, D.C.: U.S. Government Printing Office.
27. Fagan, J. J. 1974. *The earth environment.* Englewood Cliffs, New Jersey: Prentice-Hall.
28. Davis, G. H., and Wood, L. A. 1974. *Water demands for expanding energy development.* U.S. Geological Survey Circular 703.

Spectacular scenery such as that in Monument Valley must be preserved for future generations to enjoy. (Photo by Emil Muench)

Cape Hatteras, North Carolina, a lagoon-type estuary. Plumes of turbid water are discharged through inlets in the barrier beaches.

Coastal landscape long valued for its scenic quality near Morro Bay, California. Coastal land-use planning is now viewed as an important activity in California.

Part Five

LAND USE AND DECISION MAKING

Perhaps the most controversial environmental issue today is the determination of the most appropriate uses of our land and resources. The controversy is basically a human issue that involves responsibility to future generations as well as compensation for landowners today.

Before we decide how to plan the use of the land and its resources, it is necessary to develop reliable methods to evaluate the land and develop a legal framework within which sound environmental decisions can be made. Chapter 14 discusses several aspects of landscape evaluation such as land-use planning, site selection, landscape aesthetics, and environmental impact. Chapter 15 introduces the relatively new and expanding field of environmental law. Although environmental law is obviously not geological, there is a real need for applied earth scientists to know something about our legal system and the laws and legal theory that affect the natural environment.

14

LANDSCAPE EVALUATION

Evaluation of the landscape for purposes such as land-use planning, site selection, construction, and determination of environmental impact is a common practice. Although in various stages of quantification and testing for validity, methodology and techniques to accomplish the evaluation are a significant part of applied environmental work.

The role of the geologist in landscape evaluation is to provide geologic information and analysis before planning, design, and construction. Table 14.1 shows some of the possible interrelationships between and among geologists and other professional disciplines during the planning phase of landscape evaluation. This table clearly indicates that the geologist is an important member of the planning team (1).

The specific information that geologists apply as part of the landscape evaluation process varies from area to area and project to project, but it generally may include the following: first, the physical and chemical properties of earth

materials, including seismic stability, potential for building materials, and accept-
ability for waste disposal; second, slope stability; third, presence of active or possibly active faults (at the centers of historical earthquakes) and fracture systems;
fourth, depth to the water table and groundwater flow characteristics; fifth, depth to
bed rock; and sixth, extent of any floodplains present. Table 14.2 lists the importance of some of the geologic conditions for typical land uses (1).

TABLE 14.1 *Professional interrelationships during planning.*

Legend:
- ▨ Primary interrelationship
- ▦ Secondary interrelationship
- ☐ No interrelationship

Selected Professional Disciplines

Major Study Groups	Typical Study Topics	Geologists	Geographers	Civil Engineers	Sanitary Engineers	Architects	Landscape Architects	Planners	Recreation Planners	Conservationists	Lawyers	Public Administrators	Sociologists
Regional economy	Economic base resource potential												
Regional population	Population studies, socio-economic studies												
Transportation	Transport facilities, public transport, parking facilities												
Natural environment and public utilities	Natural resources, hazards protection, land reclamation, public utilities												
Community facilities	Schools, libraries police and fire, parks, recreation												
Land use	General plans, neighborhood plans, commercial developments, industrial developments												
Housing and public bldgs.	Private dwellings, public buildings												
Aesthetics	History, cultural values, community improvement, legislative controls												
Administration and legislation	Legislation administration												
Finance	Capital improvements, federal-aid programs												
Other planning studies	Defining goals, urban renewal programs, waste disposal, public health, civil defense												

SOURCE: Turner and Coffman, "Geology for Planning: A Review of Environmental Geology," *Quarterly of the Colorado School of Mines,* vol. 68, no. 3, 1973.

	Typical Land Uses						
X Primary importance **O** Secondary importance Geologic Conditions	Light Structure Construction	Heavy Structure Construction	Waste Disposal	Building Material Resources	Ease of Excavation	Road Performance	Agriculture
Physical properties of soils and rocks	X	X	X	X	X	X	X
Slope stability	X	X	O	O	O	X	O
Thickness of surficial materials	X	X	O	X	O	O	O
Depth to groundwater	X	X	X	X	O	O	X
Supply of surface water	O	O	O	X	O	O	X
Danger of flooding	X	X	X	X	X	X	X

TABLE 14.2 *Geologic conditions that affect typical land uses.*

SOURCE: Turner and Coffman, "Geology for Planning: A Review of Environmental Geology," *Quarterly of the Colorado School of Mines,* vol. 68, no. 3, 1973.

Land-Use Planning

Traditional land utilization in the conterminous United States from 1900 to 1969 is shown in Figure 14.1. Although the increase in conversion of rural land to non-agricultural uses appears to be very slow, it currently amounts to about 2 million acres per year (Figure 14.2). The intensive conversion of rural land to urban development, transportation networks and facilities, and reservoirs is nearly matched by the extensive conversion of rural land to wildlife refuges, parks, and wilderness and recreation areas.

The contribution of the earth scientists to land-use planning is primarily in emphasizing that all land is not the same and that particular physical and chemical characteristics of the land may be more important to society than geographic location. Earth scientists recognize that there is a limit to our supply of land, and, therefore, we should strive to plan so that suitable land is available for specific uses for this generation and those following (2).

The need for land near urban areas in recent years has led to the concepts of *multiple* and *sequential* land use rather than permanent, exclusive use. Multiple land use, as, for example, active subsurface mining below urban land, is probably less common than sequential land use which involves changing use with time. The concept of sequential use of the land is consistent with our earlier discussed fundamental principle that the effects of land use are cumulative, and, therefore, we have a responsibility to future generations. The basic idea is that after a particular activity (say, mining or a sanitary landfill operation) is completed, the land is reclaimed for another purpose.

FIGURE 14.1 *Land utilization from 1900 to 1969 for the conterminous United States. (Source: Council on Environmental Quality.)*

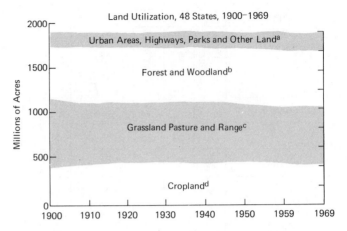

a Urban and other built areas, highways, railroads, airports, parks, and other land.
b Excludes forested areas reserved for parks and other special uses.
c Includes grassland pasture and range, private and public.
d Cropland planted, cropland in summer fallow, soil-improvement crops, and land being prepared for crops and idle.
Cropland acreages before 1954 are for the year preceding the date of the inventory.

There are several known examples of sequential land use. Sanitary landfill sites in California, North Carolina, and other locations are being planned so that when the site is completed, the land will be used for a golf course. The city of Denver used abandoned sand and gravel pits for sanitary landfill sites that today are the sites of a parking lot and the Denver Coliseum (Figure 14.3) (3). Enormous

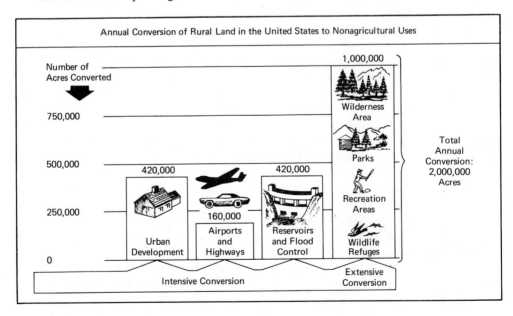

FIGURE 14.2 *Annual conversion of rural land in the United States to nonagricultural uses. (Source: Council on Environmental Quality.)*

(a)

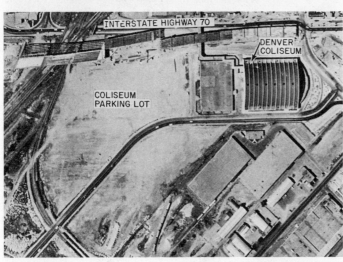

(b)

FIGURE 14.3 *Examples of sequential land use in Denver, Colorado. Gravel pits (a) were used as a sanitary landfill site as long ago as 1948. Today (b) the Denver Coliseum and parking lots cover the landfill site. (Photos courtesy of the Missouri Geological Survey.)*

underground limestone mines in Kansas City, Springfield, and Neosho, Missouri, have been profitably converted to warehousing and cold-storage sites, offices, and manufacturing plants (3). Other possibilities also exist, as, for example, the use of the abandoned surface mines for parking below shopping centers, chemical storage, and petroleum storage (Figure 14.4) (3).

The Land-Use Plan There are four basic elements of a land-use plan: first, a statement of land-use issues, goals, and objectives; second, a summary of data collection and analysis; third, a land classification map; and fourth, a report that describes and indicates appropriate development of areas of special environmental concern (4).

      The *statement* of major land-use issues to assist planning future development for at least a ten-year period is significant and should be prepared in cooperation with citizens and public agencies. The statement would include information on issues such as impact of economic and population trends; housing and

FIGURE 14.4 *Examples of sequential land use of underground mines in Missouri, for freezer storage (a) and warehousing (b and c). (Photos by J. D. Vineyard, courtesy of Missouri Geological Survey.)*

(a)

(b)

(c)

services; conservation of natural resources; protection of important natural environments; and protection of cultural, historical, and scenic resources (4).

   Data collection and *analysis* include a variety of information, including the following: analyses of present population, economy, and land use, and current plans, policies, and regulations; constraints, such as land potential, capacity of community facilities (water and sewer service, schools, etc.); and estimated demand from changes in population and economy, future land needs, and demand for facilities. Table 14.3 lists in detail required and optional data-analysis items for land-use plans submitted under the North Carolina Coastal Management Act of 1974.

The *land classification map* is the heart of the land-use plan. It serves as a statement of land-use policy and has five aims: first, to achieve and encourage coordination and consistency between local and state land-use policies; second, to provide a guide for public investment in land; third, to provide a useful framework for budgeting and planning of construction of facilities, such as schools, roads, and sewer and water systems; fourth, to coordinate regulatory policies and decisions; and fifth, to help provide guidelines for development of equitable land tax. An example of a currently used land classification system is shown in Table 14.4

The *report* accompanying the land classification map gives special attention to areas of environmental concern—those areas in which uncontrolled or incompatible development might produce irreparable damage. Examples of coastal areas include coastal marshland, estuaries, renewable resources (watersheds or aquifers), and fragile, historic, or natural resource areas (Outer Banks, Barrier Islands, etc.). The report should be precise as to the permissible land uses of these areas (4).

The above discussion demonstrates that the preparation of a land-use plan is complex and requires a team approach. The input of the applied earth scientist is especially valuable in evaluating natural resources and suitability of land for a particular purpose and in defining areas of environmental concern. To this end, it is important to make maximum use of soils information and to develop environmental geology maps to assist planners.

Soil Surveys and Land-Use Planning   Properly utilized information from detailed soil maps can be extremely helpful in land-use planning. Soils can be rated according to their limitations for land uses, such as housing, light industry, septic-tank systems, roads, recreation, agriculture, and forestry. Soil characteristics that help in determining possible limitations for a particular land use include slope, water content, permeability, depth to rock, susceptibility to erosion, shrink-swell potential, bearing strength (ability to support a load, such as a building), and corrosion potential.

The limitations can be mapped from detailed soil maps and the accompanying description of the individual soil types. An example of a soil evaluation to determine limitations for buildings in recreation areas will emphasize the value of the soil survey (5). Figure 14.5 is a detailed soil map and short description of soils for 1000 acres in Aroostock County, Maine. Table 14.5 lists soil limitations for buildings in recreational areas. Information from Figure 14.5 and Table 14.5 is combined to produce the map in Figure 14.6 showing areas where buildings could be located. The limitation classes are standard and defined as follows: *none to slight* —soils are relatively free of limitations that affect the planned use, or soils have limitations that can readily be overcome; *moderate*—soils with moderate limitations resulting from effects of soil properties. The limitations can normally be overcome with correct planning, careful design, and good construction; and *severe*—soils with severe limitations that make the proposed use doubtful. Careful planning and more

TABLE 14.3 *Required and optional data analysis items for the land-use plan.*

Element	Required	Optional
1. Present conditions a. Population and economy	Brief analysis, utilizing existing information.	More detailed analyses relating to human resources (population composition, migration rates, educational attainment, etc.) and economic development factors (labor force characteristics, market structure, employment mix, etc.).
b. Existing land use	Mapped at generalized categories.	Mapped with more detailed categories, including more detailed analyses, building inventory, etc.
c. Current plans, policies, and regulations		
1) Plans and policies	1) Listing and summary.	1) Detailed impact analysis of plans and policies upon land-development patterns.
2) Local regulations	2) Listing and description of their enforcement mechanism.	2) Detailed assessment of adequacy and degree of enforcement.
3) Federal and state regulations	3) Listing and summary (to be provided by N. C. Dept. of Natural and Economic Resources).	
2. Constraints a. Land potential		

1) Physical limitations	1) Analysis of following factors (maps if information available):
	—Hazard areas
	—Areas with soil limitations
	—Sources of water supply
	—Steep slopes
		1) Detailed analysis and mapping of required items.
		
		Analysis and mapping of additional factors:
		    —Water-quality limited areas
		    —Air-quality limited areas
		    —Others as appropriate

2) Fragile areas	2) Analysis of following factors (maps if information available): —Wetlands —Frontal dunes —Beaches —Prime wildlife habitats —Scenic and prominent high points —Unique natural areas —Other surface waters —Fragile areas	2) Detailed analysis and mapping of required items. Analysis and mapping of additional factors.

Element	Required	Optional
3) Areas with resource potential	3) Analysis of following factors (maps if information available): —Areas well suited for woodland management —Productive and unique agricultural lands —Mineral sites —Publicly owned forests, parks, fish and game lands, and other outdoor recreational lands —Privately owned wildlife sanctuaries	3) Detailed analysis and mapping of required items Analysis and mapping of additional factors: —Areas with potential for commercial wildlife management —Outdoor recreation sites —Scenic and tourist resources
b. Capacity of community facilities	—Identification of existing water and sewer service areas —Design capacity of water-treatment plant, sewage-treatment plant, schools, and primary roads —Percent utilization of the above	Detailed community facilities studies or plans (housing, transportation, recreation, water and sewer, police, fire, etc.).
3. *Estimated demand* a. Population and economy 1) Population	1) 10-yr estimates based upon Dept. of Administration figures as appropriate.	1) Detailed estimate and analysis, adapted to local conditions using Department of Administration model.
2) Economy	2) Identification of major trends and factors in the economy.	2) Detailed economic studies.
b. Future land needs	Gross 10-yr. estimate allocated to appropriate land classes.	Detailed estimates by specific land-use category (commercial, residential, industrial, etc.).
c. Community facilities demand	Consideration of basic facilities needed to service estimated growth.	Estimates of demands and costs for some or all community facilities and services.

SOURCE: North Carolina Coastal Resources Commission, 1975.

extensive design and construction measures are required. This may include major soil reclamation work (5).

Environmental Geology Mapping   The primary goal of environmental geology mapping is assistance to the planner (1), and, therefore, geologic information and engineering properties of earth materials must be presented in a way that is readily

TABLE 14.4 *Example of a land classification system.*

a. *Developed*—Lands where existing population density is moderate to high and where there are a variety of land uses which have the necessary public services.

b. *Transition*—Lands where local government plans to accommodate moderate- to high-density development and basic public services during the following ten-year period will be provided to accommodate that growth.

c. *Community*—Lands where low-density development is grouped in existing settlements or will occur in such settlements during the following ten-year period, and which will not require extensive public services now or in the future.

d. *Rural*—Lands whose highest use is for agriculture, forestry, mining, water supply, etc., based on their natural resource potential. Also included are lands for future needs not currently recognized.

e. *Conservation*—Fragile, hazard, and other lands necessary to maintain a healthy natural environment and necessary to provide for the public health, safety, or welfare.

SOURCE: North Carolina Coastal Resources Commission, 1975.

interpreted. That is, the environmental geology map must contain important information necessary for planning, and it must eliminate extraneous data (1).

From a pragmatic view, the best environmental geology maps are those that combine geologic and hydrologic information in terms of engineering properties. The landscape may then be mapped in terms of a specific land use as is currently done for soils. The idea is to produce a series of maps, one for each possible land use. Therefore, the "environmental geology map" is essentially a combination of geologic and hydrologic data expressed in less technical terms to facilitate a general understanding by a large audience.

The Illinois State Geological Survey has been a leader in developing methods for mapping for environmental geology. An example of such mapping in DeKalb County, Illinois, will illustrate the procedure. The county was glaciated during the Pleistocene Ice Age, and surficial deposits include water-transported sands and gravels as well as glacial till (mixture of clay and larger particles).

The preparation of interpretive maps of suitability for a particular land use generally includes a color code: green for "go" (favorable conditions); yellow for "caution"; and red for "stop" (unfavorable conditions or problem area). Three shades for each color represent different kinds of limitations. Figure 14.7 is a suitability analysis for solid-waste disposal (sanitary landfill) sites in DeKalb County. In this case, the letters *G, Y,* and *R* are used in place of the colors, and the numbers 1, 2, and 3 are used in place of the shades of each color. In theory, G-1 (Green-1) indicates areas with the least limitation, and R-3 indicates problem areas with the most unfavorable conditions (6).

The evaluation of a site for solid-waste disposal requires information on topography, type and thickness of unconsolidated material, present and potential sources of surface water and groundwater, and type of bed rock. The best sites have

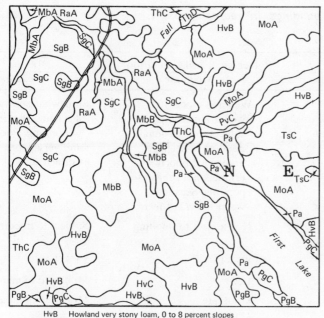

FIGURE 14.5 *Detailed soil map of a 1000-acre tract of land, Aroostock County, Maine.* [*Reproduced from* Soil Surveys and Land Use Planning *(1966) by permission of the Soil Science Society of America.*]

HvB	Howland very stony loam, 0 to 8 percent slopes
HvC	Howland very stony loam, 8 to 15 percent slopes
MbA	Madawaska fine sandy loam, 0 to 2 percent slopes
MbB	Madawaska fine sandy loam, 2 to 8 percent slopes
MoA	Monarda and Burnham silt loams, 0 to 2 percent slopes
Pa	Peat and Muck
PgB	Plaisted gravelly loam, 2 to 8 percent slopes
PgC	Plaisted gravelly loam, 8 to 15 percent slopes
PvC	Plaisted and Howland very stony loam, 8 to 15 percent slopes
RaA	Red Hook and Atherton silt loams
SgB	Stetson gravelly loam, 2 to 8 percent slopes
SgC	Stetson gravelly loam, 8 to 15 percent slopes
ThC	Thorndike shaly silt loam, 8 to 15 percent slopes
ThD	Thorndike shaly silt loam, 15 to 25 percent slopes
TsC	Thorndike and Howland soils, 8 to 15 percent slopes

at least 30 feet of impermeable material between the base of the fill material and the top of any existing or potential groundwater aquifer. The areas labeled *G* on Figure 14.7 are areas where it is believed that at least 50 feet of relatively impermeable till overlies any aquifer. However, several test borings at a particular proposed site would be necessary to verify this. Less favorable areas, Y-1 and Y-2, primarily in the southern part of the map, are characterized by layers of sand and gravel interbedded with glacial till. There are acceptable sites within these areas, but they require careful, preliminary investigation. Areas designated R-1, R-2, or R-3 are the least favorable sites for solid-waste disposal. R-1 areas have surficial deposits of silt, sand, or sand and gravel underlain by till less than 20 feet thick. These areas are found along the streams and rivers, and surface-water pollution from landfill leachate would be a problem. R-2 areas are those in which the thickness of the glacial deposits over bed rock is thought to be less than 50 feet. The bed rock does supply groundwater to individual residences, and, therefore, at least 30 feet of relatively impermeable material above the rock is necessary. Test borings will be required before siting of a landfill in these areas. R-3 areas have severe limitations in that the upper 20 feet of surficial material is silt, sand, or sand and gravel which are likely to cause a pollution problem if landfill leachate is present. However, even

TABLE 14.5 *Soil limitations for buildings in recreational areas.*

Soil Items Affecting Use	Degree of Soil Limitation [a]		
	None to Slight	Moderate	Severe [b]
Wetness	Well to moderately well drained soils not subject to ponding or seepage. Over 4 feet to seasonal water table.	Well and moderately well drained soils subject to occasional ponding or seepage. Somewhat poorly drained, not subject to ponding. Seasonal water table 2–4 feet.[c]	Somewhat poorly drained soils subject to ponding. Poorly and very poorly drained soils.
Flooding	Not subject to flooding	Not subject to flooding	Subject to flooding
Slope	0%–8%	8%–15%	15% +
Rockiness [d]	None	Few	Moderate to many
Stoniness [d]	None to few	Moderate	Moderate to many
Depth to hard bed rock	5 feet [b]	3–5 feet [c]	Less than 3 feet

SOURCE: Reproduced from *Soil Surveys and Land Use Planning* (1966) by permission of the Soil Science Society of America.

[a] Soil limitations for septic-tank filter fields, hillside slippage, frost heave, piping, loose sand, and low bearing capacity when wet are not included in this rating, but must be considered. Soil ratings for these items have been developed.

[b] Soils rated as having severe soil limitations for individual cottage sites may be best from an aesthetic or use standpoint, but they do require more preparation or maintenance for such use.

[c] These items are limitations only where basements and underground utilities are planned.

[d] *Rockiness* refers to the abundance of stones or rock outcrops greater than 10 inches in diameter. *Stoniness* refers to the abundance of stones 3 to 10 inches in diameter.

within the R-3 areas, there probably are areas with thick deposits of silt that would make a satisfactory landfill site (6).

The preparation of environmental geology maps as done in Illinois is valuable to the planner, but a larger-scale, more detailed map showing very accurate geologic boundaries is preferable. However, it may not be practical because map scale is based on the accuracy and amount of geologic data available, and, furthermore, natural geologic boundaries often are gradual rather than sharp (6).

Another interesting approach to environmental mapping that is likely to become more significant in the future is the concept of an *environmental resource unit* (ERU). This method uses a multidisciplinary team approach to define and analyze the total natural environment—geologic, hydrologic, and biologic. A portion of the environment with a similar set of physical and biological characteristics is called an environmental resource unit. Every ERU is supposedly a natural division characterized by specific patterns or assemblages of structural components (rocks, soils, vegetation, etc.) and natural processes (erosion, runoff, soil genesis, productivity of wildlife, etc.) (1).

A study site of 4 square miles near Morrison, Colorado, illustrates the concept of defining and working with environmental resource units. Figure 14.8 shows the topography and the area, and Figure 14.9 shows the four ERUs. The first is the *mountain forest* ERU—ridge and ravine topography supporting a pine and Douglas fir forest. Running-water and mass-wasting processes are important in modifying the land, and the dominant ecological control is the moisture regime which varies with elevation and exposure. The second is the *floodplain forest* ERU —recent alluvial (stream-deposited) material deposited by channel and overbank stream flow. This ERU supports a cottonwood-willow forest that needs abundant water. The third is the *hogback wood* and *grassland* ERU—a series of tilted sedimentary rock units with steep scarp slopes, gentle dip slopes (Figure 14.8), and gentle debris slopes or level bottoms. Mass-wasting processes are controlled by slope and exposure. The scarp slopes and bottomlands support grass, and the dip slopes support junipers. The fourth is the *Pleistocene grasslands* ERU—a complex of older alluvial and mass-wasting material deposited by processes no longer operating. Natural vegetation is grass, but in the study area, much of this ERU is used for agricultural purposes (1).

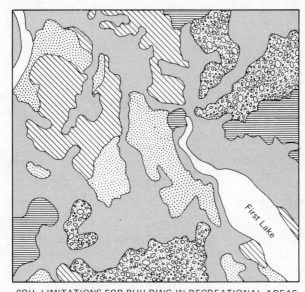

FIGURE 14.6 *Soil limitations for buildings in recreational areas for the 1000-acre tract of land shown in Figure 14.5. [Reproduced from Soil Surveys and Land Use Planning (1966) by permission of the Soil Science Society of America.]*

SOIL LIMITATIONS FOR BUILDING IN RECREATIONAL AREAS

None to Slight
Deep, well-drained, gently sloping (2%–8%) soils

Moderate
Sloping (8%–15%) soils

Severe
Very stony soils

Wet soils

Shallow sloping (8%–25%) soils

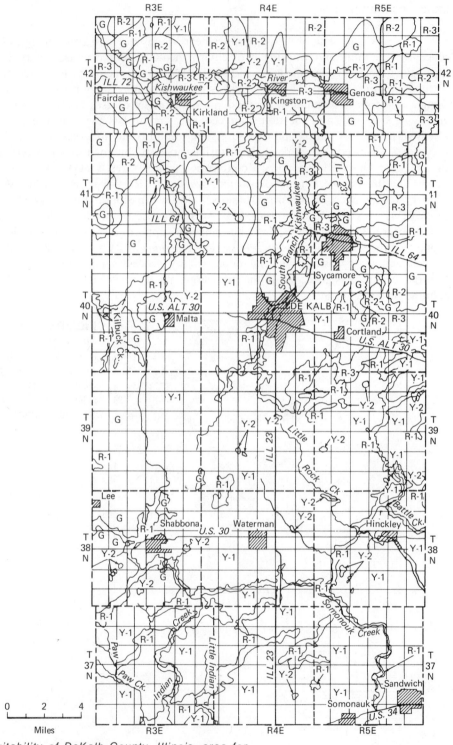

FIGURE 14.7 *Suitability of DeKalb County, Illinois, area for solid-waste disposal. (Source: Gross, Illinois State Geological Survey, Environmental Geology Notes No. 33, 1970.)*

408

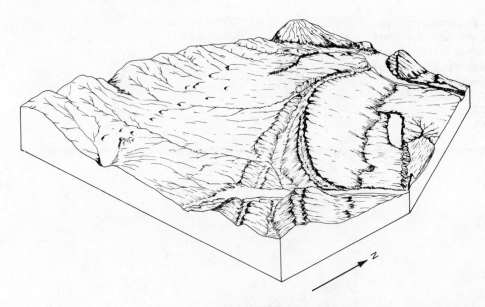

FIGURE 14.8 *Physiographic diagram of the Morrison Test Site. (Source: Turner and Coffman, "Geology for Planning: A Review of Environmental Geology,"* Quarterly of the Colorado School of Mines, *vol. 68, no. 3, 1973.)*

The environmental resource units are used to establish patterns of structure and process, and these are valuable in establishing land capabilities and suitabilities. The ERU concept is most valuable in defining areas in a systems approach that involves the entire environment. These units may then be analyzed separately or collectively for a specific land use.

The ultimate value of the ERU will be realized when multidisciplinary teams including geologists, soil scientists, biologists, landscape architects, engineers, and planners use remote sensing along with field survey techniques to identify different units and then analyze the data for planning purposes by utilizing computer-drawn maps for rapid data analysis and display.

Site Selection and Evaluation

Site selection is the process of evaluating a physical environment that will support activities of man. It is a task shared by professionals in engineering, landscape architecture, planning, earth science, social science, and economics, and, therefore, it involves a multidimensional approach to landscape evaluation.

In recent years, a philosophy of site evaluation known as *physiographic determinism* has emerged (7). The basic thrust of this philosophy is "design with nature." That is, rather than laying down an arbitrary design or plan for an area, it might be advantageous to find a plan that nature has already provided (7). Using this method, planners take advantage of physical conditions at a site in such a way as to maximize the amenities of the landscape while minimizing social and economic

FIGURE 14.9 *Environmental resource units for the Morrison Test
Site. (Source: Turner and Coffman,
"Geology for Planning: A Review of
Environmental Geology," Quarterly
of the Colorado School of Mines,
vol. 68, no. 3, 1973.)*

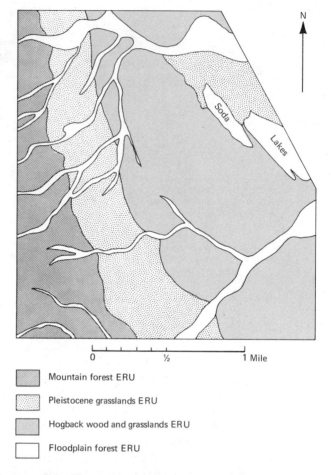

0 ½ 1 Mile

Mountain forest ERU

Pleistocene grasslands ERU

Hogback wood and grasslands ERU

Floodplain forest ERU

expenditures whenever possible. For example, if the soils of an area vary considerably in their ability to corrode, then the best corridor for an underground pipeline might not be the shortest route. Rather, the route might be designed so that the greatest possible length lies in the soils with the least corrosive potential. Although such a route is slightly longer and requires a higher initial cost, it might, by taking advantage of nature, produce tremendous savings in maintenance and replacement costs (8).

Although the philosophy of man and nature working together in harmony is obviously advantageous, it is often overlooked in the siting process. People still purchase land for various activities without considering if the land use they have in mind is compatible with the site they have chosen. There are well-known examples of poor siting resulting in increased expense, limited production, or even abandonment of partially completed construction. Construction of a West Coast nuclear power facility was terminated when fractures in the rock (active faults) and possible serious foundation problems arose (Figure 14.10). The productivity of a large chicken farm in the southeast was greatly curtailed because the property was purchased before it was determined if there were sufficient groundwater to meet the projected needs. A housing developer in northern Indiana pur-

FIGURE 14.10 *Nuclear power development was abandoned due to an earthquake hazard at this site at Bodega Bay, California, after expenditure of millions of dollars and years of time in site preparation. This waste of time and money could have been avoided had data on the fault been available earlier. (Source: U.S. Geological Survey Circular 701, 1974.)*

chased land and built country homes in one of the few isolated areas where bed rock (shale) is at the surface. Thus, septic-tank systems had to be abandoned and a surface sewage treatment facility had to be built. Furthermore, the rock made it much more expensive to excavate for basements and foundations for the homes.

The goal of site selection for a particular land use is to insure that the site development is compatible with both the possibilities and the limitations of the natural environment. The earth scientist's role may be significant in providing crucial geologic information to help accomplish this goal. The type of information provided by geologists includes the following: soils and rock types; rock structure, especially fractures; drainage characteristics; groundwater characteristics; landform information; and estimates of possible hazardous earth events and processes, such as floods, landslides, earthquakes, and volcanic activity. In addition, the engineering geologist takes samples, makes tests, and predicts the engineering properties of the earth materials.

Processes of Site Selection The process of site selection begins with a thorough understanding of what the general purposes and specific requirements of the designed activity are and for whom the site is being selected. The next step is an analysis of possible alternative sites in terms of physical limitations and cost. The final step is making a recommendation as to which site or sites are most advantageous for the desired land use (9).

The geologic aspect requires the following steps: reconnaissance of possible sites to note possible problems and possibilities; collection of existing data, such as geologic reports, topographic maps, aerial photographs, soil maps, water surveys, engineering reports, climate records, etc.; collection of necessary new data, including soil, rock, and water analyses; and determination of the magnitude and importance of geologic limitations and possibilities of the site or sites.

The *magnitude* of geologic limitations and possibilities in site selection and evaluation refer to the degree, extensiveness, or scale, while the *importance* is a measure of the weight a particular limitation or possibility should be assigned. For example, if a site evaluation for the foundation of a large dam or nuclear facility reveals that the bed rock has several small fractures, the *magnitude* of the limitation is small. However, the *importance* may be great if the faults are active, whereas if they are inactive, the importance is less. The presence of an active fault is obviously significant because it might cause the site to be abandoned, whereas inactive fractures can be treated by specific engineering procedures.

Methods of Site Selection and Evaluation

Although specific methods of determining the location and suitability of an area obviously vary with the intended purpose for the particular land use, there are a few methods that are commonly used to evaluate the desirability of a particular site for relatively large-scale land uses, such as construction of reservoirs, highways, and canals and development of parks and other recreational areas. Two such methods are *cost-benefits analysis* and *physiographic determinism,* determining the best site by maximizing physical and social benefits while minimizing social costs. The first approach is strictly economical, while the second balances economics with less tangible aspects, such as aesthetics.

Cost-Benefits Analysis   Cost-benefits analysis is a way to assess the long-range desirability of a particular project, such as construction of a highway or development of a recreational site. The basic idea is to develop the benefits-to-cost ratio. That is, the total benefits in dollars over a period of time are compared to the total cost of the project, and the most desirable projects are those for which the benefits-to-cost ratio is greater than one (10).

Cost-benefits analysis assumes that all relevant costs and benefits can be determined. There are three difficult aspects: which costs and which benefits are to be evaluated; how they are to be evaluated; and how the intangible costs and benefits, such as aesthetic degradation or improvement, are to be evaluated (10).

Cost-benefits analysis is commonly done for three general types of projects: *water resource* projects, such as irrigation systems, flood control, hydro-electric power systems, and multipurpose schemes, such as reservoirs used for flood control, power generation, and recreation; *transportation* projects, such as highways, roads, railways, and river and canal work to facilitate navigation; and *land-use* projects, such as urban renewal and recreation (10).

An example of some of the costs and benefits for a hypothetical flood-control project will further illustrate the method. The initial tangible costs of a flood-control project are relatively straightforward. However, evaluation of repair and maintenance costs and intangible costs over the life of the project are much more difficult. The main benefit from flood control is averted loss of property. *Loss,* here, refers primarily to different types of assets, such as property, furnishings, and crops. The general principle is to evaluate the annual flood loss and damage based on the most probable flood levels for several recurrence intervals, say, the 10-year and 25-year floods. This estimate of annual loss and damage is then used as the maximum yearly amount that the people would be willing to pay for flood control. Other benefits to be considered include reduction in deaths by drowning and elimination of recurring costs of evacuating flood victims, providing emergency sandbag work, and reducing the incidence of sanitary failure (10).

Physiographic Determinism   An interesting graphic method of site selection based on physiographic determinism, or designing with nature, is structured such that the appropriate site literally selects itself (11). The method utilizes physical, social, and aesthetic data, and the objective is to maximize social benefit while minimizing social cost.

The actual methodology in selecting a site or a route for a highway, regional development plan, or other land use involves selecting a group of factors for physical and social values. For each factor, three or more grades of the value are mapped on transparencies in tones from clear to dark grey. The darker tones indicate areas of conflict where physical conditions are not appropriate or social values are in conflict with the desired land use. When the transparencies (one for each factor) are superposed upon one another, the darkest areas represent those places with the greatest physiographic obstructions and social costs, and the lighter areas represent places where the physical limitations (construction costs and social costs) are least. By this method, the best sites emerge as the lightest areas on the set of superposed transparencies. No matter how dark the entire area may look, there will always be relatively lighter areas (11).

The method of determining the lowest social cost of land use in site selection discussed above is innovative in addressing social values as well as traditional physical-economic values. However, it is not above criticism. It is very subjective in that the person doing the evaluation selects the factors and thus essentially builds the formula used in the evaluation. For example, in evaluating the siting of a highway route, the evaluator may decide that property values are important and that the darker tone should be assigned to high property values. This decision essentially prescribes that the highway should be routed away from people, and wealthy people most of all, a point some planners would vigorously argue against (7). Another criticism is that all factors, whether physical or social values, are weighed the same. That is, a geologic factor such as suitability of soil for foundation material or susceptibility to erosion is weighted equally with social values, such as scenic value or recreational property value. Regardless of these shortcomings, the method of physiographic determinism in evaluating both physical-economic factors and social values is a step in the right direction (7).

Site Evaluation for Engineering Purposes

The geologist has a definite role in the planning stages of site development for engineering projects, such as construction of dams, highways, airports, tunnels, and large buildings.

Dams Construction of either masonry (concrete) dams or earth dams (Figure 14.11) requires careful and complete geologic mapping and testing early in the planning process. Such testing should include the following: evaluation of the valley and the dam site to determine the present and expected suitability of the slopes; identification of possible problems, such as fracture zones in rocks or adverse soil conditions where the dam structure is in contact with rock or soil (foundation) that might affect the suitability of the dam; prediction of the rates of sediment accumulation; and assessment of the availability of building material for the construction of the dam. Furthermore, continuous geologic mapping during construction may be important because the geologic history of a river valley often involves several periods of alternating erosion and scour. These processes have produced irregular or unexpected deposits that might cause problems in the construction, operation, and maintenance of the dam and reservoir. However, it is emphasized that major irregularities or problems that might affect the dam or reservoir should be recognized by direct and indirect geologic investigation before the construction phase of the project.

Foundations for dams vary with rock types encountered. Generally, igneous rocks (such as granite) and some pyroclastic rocks are satisfactory foundations for a dam site. Leakage, if any, will be along fractures, and fractures can be filled with a mixture of cement, sediment, and water (grouted). Metamorphic rocks also are good foundation material, and the best sites are those in which the foliation of the rocks is parallel to the axis of the dam. Sedimentary rocks such as limestone

FIGURE 14.11 *Garrison Dam and Lake Sakakawea on the Missouri River in North Dakota. The earth fill dam is 202 feet high and 11,300 feet long. The width at the base is 3400 feet. The construction required 66,500,000 cubic yards of earth fill and 1,500,000 cubic yards of concrete for the spillway, power plant, etc. The power plant generates 400,000 kilowatts. (Photo courtesy of the U.S. Army Corps of Engineers, Omaha District.)*

and particularly compaction shale may be troublesome. Limestone may have large underground cavities, and compaction shales are noted for problems stemming from their tendency to deform and settle when loaded (12, 13). Particularly important is the stability of the sedimentary rocks following wetting and drying or freezing and thawing. Failure of the St. Francis Dam in California was in part due to the deterioration of the sedimentary rocks in the foundation upon wetting.

Highways   Site evaluation for highways requires considerable geologic input, including mapping the geology along the proposed route and isolating any problems of slope stability, topography, flooding, or weak earth materials that would cause foundation problems. It is also helpful to locate and estimate the amount of possible construction materials in close proximity to the proposed route (13).

Airports   Important geologic aspects of site selection for airports are threefold: first, topography as it relates to necessary grading, drainage, and surfacing; second, soils—the best are coarse-grained with high bearing-capacities and good drainage; and third, availability of construction materials. Since good surface drainage is very important, areas subject to flooding should be avoided (12).

Tunnels   Construction of tunnels requires a very detailed knowledge of the geology, and evaluation of geologic information is basic to the selection of the route and determination of proper construction methods. Furthermore, the geology is a major factor that determines the economic feasibility of the project. Therefore, a geologic profile along a center line of the proposed tunnel route is essential before tunneling. This profile is developed from mapping the surface geology and utilizing subsurface data whenever possible (12, 13). Furthermore, as with dam investigations, it is important to continue the geologic mapping as construction proceeds to insure that unexpected hazards, not recognized earlier, do not produce problems.

      Two distinct types of tunnel conditions are defined by whether the tunnel is in solid rock or in cohesionless, or plastic, earth (soft ground) that may tend to flow and fill the tunnel (13). In any one tunnel, both conditions may be present.

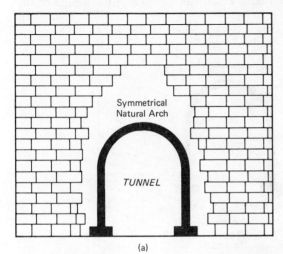
(a)

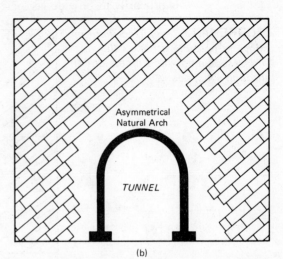
(b)

FIGURE 14.12   *Idealized diagram showing the overbreak which develops in horizontally layered rocks (a) and inclined rocks (b).*

Problems in rock tunnels are determined primarily by the nature and abundance of fracturing or other partings in the rock. The fracturing greatly affects how much overbreak (excess rock that falls into the tunnel) is present. In general, the closer the fractures and other partings are, the greater is the overbreak (12). Tunnels in granite or horizontally layered rocks generally develop a symmetrical, naturally arch-shaped overbreak, whereas inclined rocks produce an irregular (asymmetric) roof (Figure 14.12). The above discussion suggests that when possible, it is advisable to select tunnel routes through rock that has a minimum of fractures.

Rock structure, particularly folds and faults, is also significant in location and orientation of tunnels. When tunnels are built through synclines, water problems are likely because the natural inclination of the rocks facilitates drainage into the tunnel (Figure 14.13). In addition, rock fractures in synclines tend to form inverted keystones, and, thus, the roof of the tunnel may require substantial support. On the other hand, tunnels through anticlines are less likely to have water problems, and the fractures tend to form normal keystones that are more nearly self-supporting (12).

The occurrence of faults in tunnels usually causes water problems because water often migrates along open fractures associated with fault systems. Therefore, whenever possible, tunnels should be driven at right angles to faults to minimize the length of tunnel in contact with the fault (12). Tunneling in soft ground is generally done at shallower depths, thus facilitating the collection of more detailed subsurface information before construction than is possible in deep rock tunnels (12, 13).

Most problems in earth tunneling are compounded when groundwater saturates the material through which the tunnel is being driven. Sandy material when saturated may slowly flow and fill a tunnel or suddenly develop a hazardous condition known as a "blow or boil" in which the material quickly fills the tunnel. A thorough evaluation of soil and hydrogeology is necessary when tunneling in soft ground (12, 13).

Large Buildings   The role of the geologist in evaluating a site for a large building is primarily to provide geologic information that can be used in the design of the

FIGURE 14.13   *Idealized diagram showing possible groundwater hazards associated with tunnels driven through synclines.*

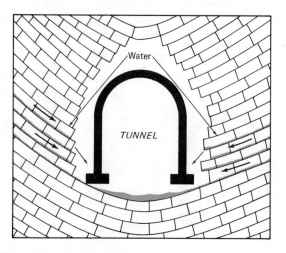

building foundation. The site investigation includes drilling for soil or rock samples (Figure 14.14) to be evaluated for engineering properties to assist engineers in recognizing unanticipated geologic problems that might arise during construction of the foundation. The fact that field investigation in most cases is very simple does not mean that detailed investigation can be routinely shortened (12, 13).

Geologic investigation of foundation sites for nuclear facilities is very important and may involve detailed geologic mapping of rock structure. Even small fracture systems must be carefully evaluated to insure there are no active or recently active faults. In addition, the physical characteristics of old fractures, including their length, width, and degree of rock alteration (weathering) to clay and other material along the fractured planes, must be measured and evaluated for possible engineering treatment before construction. Figure 14.15 shows the McGuire Nuclear Power Plant during construction (a). The square boards bolted to the rock (b) along with dewatering points stabilize the slope during excavation.

FIGURE 14.14 *Drilling for soil samples as part of site investigation. (Photo courtesy of Williams and Works.)*

Landscape Aesthetics: Scenic Resources

Scenery in the United States has been recognized as a natural resource since 1864, when the first state park, Yosemite Valley in California, was established. The early recognition and subsequent action primarily related scenic resources to outdoor recreation, with the focus on preservation and management of discreet parcels of unique scenic landscapes. Public awareness of and concern for scenic values over a broader range of resource issues is a recent development. Society now fully recognizes scenery as a valuable resource and, furthermore, a recognition that the "everyday" nonurban landscape also has scenic value that enhances the total resource base of the region is emerging. The basic assumption is that there are varying

418

FIGURE 14.15 *Construction of the McGuire Nuclear Power Plant near Charlotte, North Carolina (a). Stability of the slope was maintained during excavation by use of rock bolts and dewatering points, as shown by the arrows in the photograph (b).*

(a)

(b)

scenic values just as there are varying economic values relating to other more tangible resources (14).

Quantitative evaluation of tangible natural resources such as water, forests, or minerals is standard procedure before economic development or management of a particular land area. Water resources for power or other uses may be evaluated by the flow in the rivers and storage in lakes; forest resources may be evaluated by the number, type, and size of the trees, which will yield so many million board-feet in lumber; and mineral resources may be evaluated by estimating the number of tons of economically valuable mineral material (ore) at a particular location. We can make a statement of the quality and quantity of each tangible resource compared to some known low quality or quantity. Ideally, we would like to make similar statements about the more intangible resources, such as scenery. That is, we would like to compare scenery to specific standards (15). Unfortunately, this is a difficult task for which few standards are available.

Scenic resources of a region or an area can be defined as the visual portion of an aesthetic experience. That is, the scenic base consists of those aspects of the landscape that produce visual amenities. Areas with recognized visual amenities are generally considered unique. Therefore, they are of particular public concern and possibly subject to protection.

Landscape evaluation of scenic resources as part of land-use planning or assessing environmental impact generally rests on a rather subjective methodology. Figure 14.16 shows generally how a quantitative evaluation might proceed from a recognition of scenic resources to relative or absolute indices to define their scenic values.

The general philosophical framework used to evaluate scenic resources includes the following concepts. First, scenic resources are visual amenities that can be evaluated in terms of an aesthetic judgment. Second, scenic resources, like soil and other resources, vary in quality from place to place. Third, topographic relief, presence of water, and diversity of form and color are three of the significant positive characteristics of scenic quality, whereas man-induced change is the most negative characteristic. Fourth, landscape that is unique is of more significance to

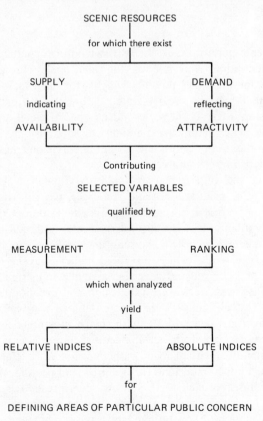

FIGURE 14.16 *Idealized diagram showing how a quantitative evaluation of scenic resources might proceed from recognition of scenic resources to development of indices to hierarchically rank various landscapes. (Modified after Coomber and Biswas, 1972, Ecological Systems Branch, Canada.)*

society than common landscapes. Fifth, although the determination of what is aesthetically pleasing varies from person to person, the judgment of what is really displeasing is more universal.

Virtually all schemes to evaluate scenic resources involve measuring or observing specifically selected factors. Such observation is admittedly subjective but is not necessarily a bad practice as long as relevant factors based on sound judgment are chosen. Furthermore, a completely objective method of analyzing scenic resources based on our present knowledge of how people perceive the landscape is probably not possible and certainly not practical. Examination of presently used methods and procedures to evaluate scenic resources suggests that three general categories of factors or variables are most appropriate. The first is *landform characterization,* assuming that as the terrain becomes steeper and has greater altitudinal differences (relief) and is more diverse, the scenic quality increases. The second is *natural characterization,* assuming that scenic quality increases as both the percentage of an area in natural surfaces (water, forest, marshland, or swampland) and the diversity of natural surface types increase, the scenic quality also increases. The third is *man-made characterization,* assuming that as the amount of area covered by man-made features increases, scenic quality decreases. Agricultural land and nonforested rural land can be considered neutral. Using this very general model, a scenic quality index can be developed.

$$SQI = LC + NC - MC$$

where *SQI* is the scenic quality index, *LC* is the landform characterization, *NC* is the natural characterization, and *MC* is the man-made characterization. The determination of the numbers and possible weights for each individual factor is a subjective decision that must be clearly stated before the evaluation. Several examples of factors using a five-point scale and no special weighting of factors are shown in Table 14.6. The factors using numerical subdivisions, as, for example, relief, will vary from region to region, and proper subdivisions must be determined by the nature of the terrain to be analyzed.

Realizing the importance of quantitative and at least procedural-objective methodology of evaluating scenic resources, we will discuss two rather different approaches. The first is a regional study which attempted to evaluate the scenic resources of Scotland (15). Because of the scale, a "broad-brush" approach was necessary. The method included mapping and assigning numerical values to landforms such that high-relief forms were rated higher than the low-relief forms, as shown in Figure 14.17. Land use was also assigned numerical values in maps such that wild or wilderness areas were rated high and urban/industrial areas low (see Figure 14.18). The two maps were then combined to give a final rating of scenic resources, shown in Figure 14.19. Although it uses a general approach, the map clearly shows rather abrupt change in scenic quality. This is not unusual, as changes in relief in landforms tend to be rather abrupt in Scotland, for example, emerging from forests to open moorland (hills, mires, and peat bogs).

The second approach to evaluating scenic resources is based on the premise that unique landscapes are more significant to society than common landscapes (16, 17). Explanation of how unique landscape is mathematically determined

TABLE 14.6 *Examples of possible factors which could be used to evaluate scenic resources of a landscape.*

1. *Landform characterization*

   A) Convex landforms

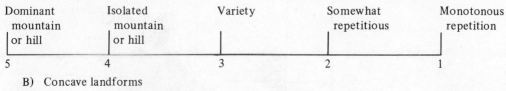

Dominant mountain or hill	Isolated mountain or hill	Variety	Somewhat repetitious	Monotonous repetition
5	4	3	2	1

   B) Concave landforms

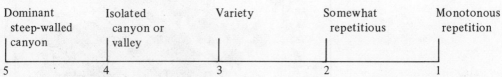

Dominant steep-walled canyon	Isolated canyon or valley	Variety	Somewhat repetitious	Monotonous repetition
5	4	3	2	1

   C) Relief (feet)

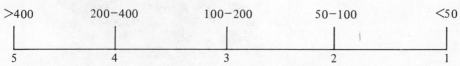

>400	200–400	100–200	50–100	<50
5	4	3	2	1

2. *Natural characterization*

   D) Percent of land covered with indigenous vegetation and natural rock or soil

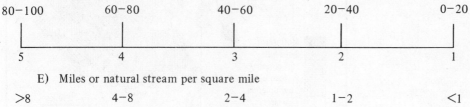

80–100	60–80	40–60	20–40	0–20
5	4	3	2	1

   E) Miles or natural stream per square mile

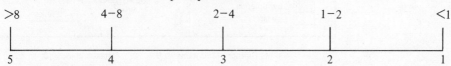

>8	4–8	2–4	1–2	<1
5	4	3	2	1

3. *Man-made characterization*

   F) Number of buildings per square mile

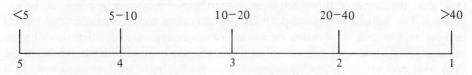

<5	5–10	10–20	20–40	>40
5	4	3	2	1

   G) Miles of road per square mile

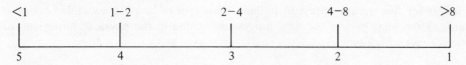

<1	1–2	2–4	4–8	>8
5	4	3	2	1

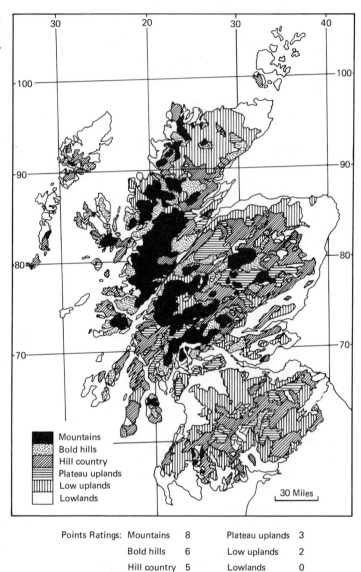

Points Ratings:	Mountains	8	Plateau uplands	3
	Bold hills	6	Low uplands	2
	Hill country	5	Lowlands	0

is beyond the scope of this discussion. Nevertheless, the major aspects can be explained. The method is most appropriate for corridors such as river valleys, coasts, and highway routes. It involves measuring or observing variables that describe the physical, biological, and human use of the landscape being evaluated. The data are then analyzed to determine uniqueness indices that indicate how different, in a relative sense, one landscape is from others (17). Until this point, the only subjective part is the selection of the variables to be used. Figure 14.20 shows uniqueness indices for five stream valleys in Indiana. This type of graph is valuable because it also shows what part of the total uniqueness is due to the physical, biological, or human-use variables.

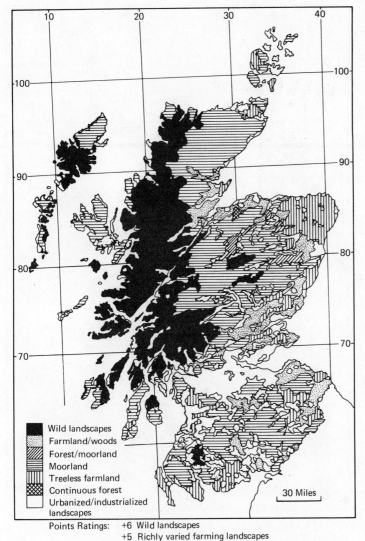

FIGURE 14.18 *Point rating for land use in Scotland. (Source: Linton,* Scottish Geographical Magazine, *vol. 84, 1968.)*

Legend:

- Wild landscapes
- Farmland/woods
- Forest/moorland
- Moorland
- Treeless farmland
- Continuous forest
- Urbanized/industrialized landscapes

30 Miles

Points Ratings:
+6 Wild landscapes
+5 Richly varied farming landscapes
+4 Varied forest and moorland with some farms
+3 Moorland
+1 Treeless farmland
−2 Continuous forests
−5 Urbanized and industrialized landscapes

If only "relative uniqueness" were determined, then the method would have little application. For example, the most unique landscape might be the only polluted and altered river valley, or it might be the most scenic. Therefore, the second step of the evaluation requires determining why a particular landscape is unique (17). With scenic resources, we are most interested in determining if the most unique landscapes are also the most scenic. To accomplish this, we use our premise that the judgment of what is pleasing may vary, but what is really ugly to

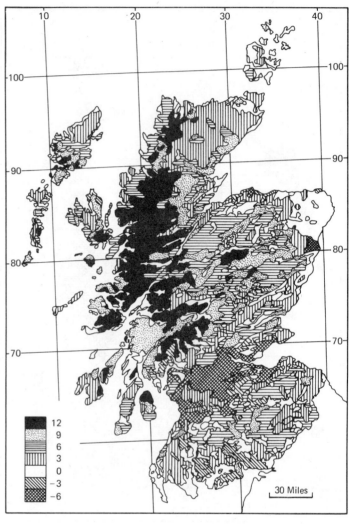

Resource Ratings: More than 12 Points (Black)

+9 to +12	0 to +3
+6 to +9	0 to −3
+3 to +6	−3 to −6

one person is usually ugly to many people. In the case of streams, the Wild and Scenic River Act, Public Law 90–542, of 1968 defines a scenic river as "those rivers or sections of rivers that are free from impoundments, with shorelines or watersheds still largely primitive and shorelines largely undeveloped, but accessible in places by roads." This definition implies that the stream is also clear-running, unpolluted, and unlittered. Using this information, we can compare the data used to determine the uniqueness of the idealized scenic river and lower a river valley's original uniqueness in proportion to how much of the original uniqueness is antithetical to our "ideal" scenic river (17). By this philosophy, even a very unique stream, if it is heavily developed or polluted, will rank very low in terms of a scenic river index. Figure 14.21 shows the scenic river indices for the same five streams in Indiana shown earlier. The upper line for each stream valley is the uniqueness

BAR GRAPH OF UNIQUENESS

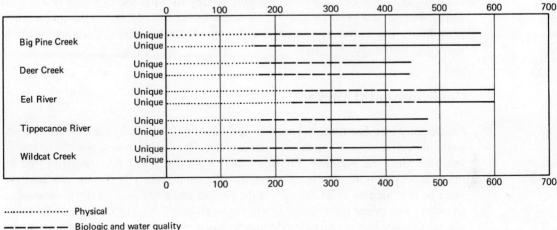

.... Physical

— — — — Biologic and water quality

————— Human use and interest

FIGURE 14.20 *Bar graph of uniqueness indices for five Indiana streams. (Source: Melhorn, Keller, and McBane, Purdue University Water Resources Research Center, Technical Report No. 37, 1975.)*

BAR GRAPH OF SCENIC RIVERS

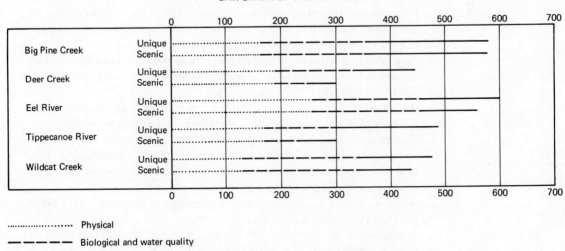

.... Physical

— — — — Biological and water quality

————— Human use and interest

FIGURE 14.21 *Bar graph of scenic river indices for the same five Indiana streams shown in Figure 14.20. (Source: Melhorn, Keller, and McBane, Purdue University Water Resources Research Center, Technical Report No. 37, 1975.)*

index, and the lower is the scenic river index. This graph is valuable because it summarizes succinctly the uniqueness and scenic river indices.

Quantifying scenic resources is valuable because the results can be visually displayed and different landscapes hierarchically ranked. Quantification is basically another tool to help the decision making for land-use planning or environmental-impact analyses. In addition, it tends to separate facts from emotion while providing a way to balance the intangibles with the more readily evaluated tangible aspects of landscape evaluation.

Environmental Impact

Philosophy and Methodology   The probable effects of human use of the land are generally referred to as *environmental impact*. This term became popularly known in 1969 when the National Environmental Policy Act (NEPA) required that all major federal actions that could possibly affect the quality of the human environment be preceded by an evaluation of the project and its impact on the environment. In order to carry out both the letter and the spirit of NEPA, the Council on Environmental Quality has prepared guidelines to assist in the preparing of environmental impact statements. The major components of the statement are as follows (18):

1. A description of the project or proposed action.
2. Probable impact of the proposed action on the environment, including impact on ecological systems such as wildlife, fish, and marine life. Both primary and secondary significant consequences should be included in the analysis. For example, any implications of the action for population distribution or concentration should be estimated, and an assessment should be made of the effect of any possible change in population patterns on the resource base of the area in question.
3. Any probable adverse environmental effects which cannot be avoided, for example, water pollution, threat to health, urban congestion, damage to wildlife, etc.
4. Alternatives to the proposed project or proposed action that might avoid or minimize some or all of the adverse environmental effects.
5. The relationships between local short-term uses of the environment and maintenance and enhancement of long-term productivity. This basically requires an assessment of cumulative, long-range effects to insure that future generations inherit a quality environment.
6. Any irreversible or irretrievable commitments of resources involved in the project or proposed action.
7. A discussion of problems and objections raised by federal, state, or local agencies; private organizations; or individuals during the review process.

There is no accepted methodology for assessing environmental impact of a particular project or action. Furthermore, because of the wide variety of topics such as hunting of migratory birds or construction of a large reservoir, there is no one methodology of impact assessment appropriate for all situations. What is important is that those responsible for preparing the statement strive to minimize personal bias and maximize objectivity. The analysis must be scientifically, technically, and legally defendable and prepared to highly objective scientific inquiry (19).

Environmental-impact analysis for major projects or actions generally benefits from the combined effort of a task force or a team of scientists, each of which is assigned specific resource topics or disciplines. It is very important to remember that the specific function of the task force is to *evaluate;* they are not to decide the issue. The work of a task force is to provide information necessary for those with authority to make just decisions. At this time in the development of our environmental awareness, knowledge of the nature and extent of information necessary to make decisions is still in the formative stage. However, as more and more work is done and decisions are made, our concept of the critical elements of environmental information will be better understood, and eventually some will be acknowledged as requirements in the same way as cost and profit data are now accepted practice in economic analysis (19).

The spirit of NEPA developed in response to the recognized need to consider environmental consequences of a particular action before its implementation. The objective is to recognize potential conflicts and problem areas as early as possible. Hopefully, such consideration will minimize regrettable and expensive environmental deterioration. Unfortunately, however, there is no guarantee that recommendations in even a well prepared, valuable impact assessment will be followed. Furthermore, because environmental-impact statements are required for such a wide variety of essential projects in urban areas, where serious degradation is unlikely, considerable time and money are being wasted on unnecessary environmental analyses. There is a real need for a careful screening of projects to insure that an impact analysis is really needed. Regardless of these limitations, two examples, the closing of a soda-ash operation in Saltville, Virginia, and the Aswan Dam disaster in Egypt, will emphasize the extremes in scale of environmental degradation that might hopefully be avoided in the future if careful environmental impact analysis is completed before initiation of projects.

Saltville, Virginia   Saltville is a small, historic Appalachian mountain town on the north fork of the Holston River. Before 1972, the community was strictly a company town. The Olin Chemical Works of the Olin Corporation built and maintained almost everything in the town. The plant, which produced soda ash (sodium carbonate), prospered for about 75 years before it finally closed down because it could not economically meet new water-quality standards of the Virginia Water Control Board. The plant disposed of chloride wastes into the north fork of the Holston River (Figure 14.22), and reportedly up to 5000 ppm of chlorine was measured in the river water. The salts from which the sodium was derived came from solution mining, a procedure in which subsurface salts are dissolved and pumped up from deep solution wells. The wells in Saltville were up to 3000 feet deep, and the combination of available sodium and carbonate rock facilitated the production of the sodium carbonate. The plant started production when the dilution and dispersion philosophy of waste disposal was accepted, and, tragically for the town, it ended

FIGURE 14.22 *Abandoned chemical works plant in Saltville, Virginia, on the banks of the Holston River.*

production when this philosophy was no longer acceptable. Seventy-five years ago, people seldom worried about long-range environmental effects. Hopefully, requirements of environmental-impact analysis will help alleviate future problems that might not be considered as problems today.

High Dam at Aswan   The situation with the High Dam at Aswan and Lake Nasser in Egypt is a dramatic example of what can happen when environmental impact is ignored or inadequately evaluated. The High Dam project is causing many serious problems, including health problems that could one day be catastrophic, as the following case study reported by Claire Sterling (20) will demonstrate.

The High Dam at Aswan, completed in 1964, is one of the biggest and most expensive dams in the world, and, from a hydrologic view, it is tragic that after harnessing one of the world's mightiest rivers, Egyptians have less water than they had before the dam was constructed. The main idea of the dam was to store a sufficient amount of the Nile's yearly floodwaters to irrigate existing lands, reclaim additional land from the desert, and protect against drought and famine. The lake was to be filled by 1970, but by 1971, it was not yet even half full, and water resource experts feel that it is unlikely that the lake level will rise much more in the next 100 or so years (20).

The unexpected water loss has resulted because of two factors: first, a 50 percent error in computing the evaporation loss; and second, tremendous water losses to underground water systems. The error in computation of evaporation loss, which takes an unanticipated 5 billion cubic meters of water per year, resulted from overlooking the evaporation loss induced by high-velocity winds traveling over the tremendous expanse of water in this very hot, dry region of the earth. A chain of smaller and less expensive dams would have avoided this tragic waste of water resources. However, even this loss is small compared to the water lost in the geologic environment (20).

It has long been known that for hundreds of kilometers up river from Aswan, the Nile cut across an immense sandstone aquifer that fed the Nile an incalculable amount of water. When the first and much smaller Aswan Dam was built in 1902, the flow of groundwater was reversed; the water pressure caused by the reservoir forced the water to move elsewhere through numerous fissures and fractures in the sandstone. From 1902 to 1964, when the High Dam was completed, the Aswan Reservoir stored about 5 billion cubic meters of water per year, but it resulted in the loss of 12 billion cubic meters of water per year through reversed groundwater flow. The amount of water escaping from Lake Nasser is unknown, but because the new reservoir is designed to store 30 times more water than the old Aswan Reservoir, and because seepage tends to vary directly with lake depth, the amount of water being lost must be tremendous (20).

It might be argued that in time the clay settling out from the lake will plug fractures in the rock and the lake will fill. However, if the fractures are very large and numerous, the water could essentially escape forever, and Egypt might well end up with less water than it had before the dam was constructed (20).

Unfortunately, the direct water problems with the High Dam and Lake Nasser are only the tip of the iceberg when the total impact of the project is analyzed. These include five factors. First, the High Dam lacks sluices to transport sediment through the reservoir, and therefore the reservoir traps the Nile's sediment, 134 million tons per year, which historically has produced and replenished the fertile soils along the banks of the Nile. No practical, man-made substitute is available to counteract this. Second, marine life in the eastern Mediterranean is deprived of the nutrients in the Nile's sediment, resulting in a 1/3 reduction of the plankton, which is the food base for sardine, mackerel, lobster, and shrimp. As a result, the fishing industry in the region of the Nile delta has suffered a tremendous setback. Supposedly, an emerging fisheries industry in Lake Nasser was to compensate for the setback, but this wish has not yet developed. Third, the lake and associated canals are becoming infested with snails that carry the dreaded disease, schistosomiasis (snail fever). This has always been a problem in Egypt, but the swift currents of the Nile floodwaters flushed out the snails each year. The tremendous expanse of waters in Lake Nasser and the irrigation canals are now providing a home for these snails. Fourth, there is an increased threat of a killer malaria carried by a particular mosquito often found only 80 kilometers from the southern shores of Lake Nasser. Malaria has historically migrated down the Nile to Egypt. The last epidemic in 1942 killed 100,000 Egyptians. Authorities fear that the larger reservoir and the irrigation canal system might be invaded on a more permanent basis by the disease-carrying insects. Fifth, salinity of soils is increasing at rather alarming rates in middle and upper Egypt. Soil salinity has been a long-standing problem on the Nile delta but was alleviated upstream by the natural flushing of the salt by the floodwaters of the Nile River. Millions of dollars will have to be spent to counteract the rising soil salinity that threatens the productivity of millions of acres of land.

On the positive side, the High Dam and lake have converted 700,000 acres from natural floodwater irrigation to canal irrigation, and they allowed double cropping—but at a tremendous cost. In the future, tremendous amounts of money will have to be spent on fertilizers to replenish soil nutrients to counteract the rise in soil salinity and to control water - and insect-borne disease.*

The Trans-Alaska Pipeline—An Example of Environmental-Impact Analysis

The controversy surrounding the Trans-Alaska Pipeline which went into the construction phase in 1975 provides a good example of environmental-impact analysis. Hopefully, experience gained from this project will set precedents and establish procedures for future impact analyses.

In 1968, vast subsurface reservoirs of oil and gas were discovered near Prudhoe Bay in Alaska. Since the Arctic Ocean is frozen much of the year, it is impractical to ship the oil by tankers, and thus a 789-mile (1270-kilometer) pipeline from Prudhoe Bay to Port Valdez where tankers can dock and load oil the entire year was suggested (19, 21).

The general route of the Trans-Alaska Pipeline is shown in Figure 14.23. Over 80 percent of the total length is across federal lands. Because of the

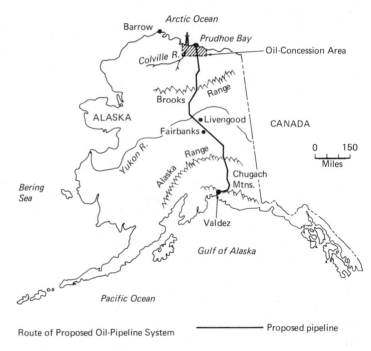

Route of Proposed Oil-Pipeline System   ——————— Proposed pipeline

FIGURE 14.23 *Approximate corridor for the Trans-Alaska Pipeline. (Source: U.S. Geological Survey Circular 695, 1974.)*

sensitive nature of the Arctic environment and the certainty of an unknown amount of irreversible environmental degradation, a comprehensive impact analysis was required to evaluate both the negative and the positive aspects of the project. A

---

*Claire Sterling, "The Aswan Disaster," National Parks & Conservation Magazine 45 (1971): 10–13.

summary of the evaluation published by the U.S. Geological Survey emphasizes the critical aspects of the natural and social economical impacts of the project (19).

The corridor for the Trans-Alaska Pipeline (Figure 14.23) traverses rough topography, large rivers, areas with extensive permafrost, and areas with a high earthquake hazard.

The pipeline crosses three *major mountain ranges:* the Brooks Range, which is the Alaskan continuation of the Rocky Mountains; the Alaska Range; and the very active (tectonically) Chugach Mountains. The pipeline also crosses the Denali fault on the north side of the Alaska Range. This is an active fault zone that has experienced recent displacement of the ground.

The corridor also transverses a number of *large rivers,* and it is feared that scour or lateral erosion at meander bends might damage the pipeline. A study was conducted to evaluate the river crossings (22). The purpose was to predict areas that might experience excessive bank erosion or scour during the 30-year expected life of the pipeline. Figure 14.24 shows the Gulkana River crossing in the Copper River Basin between the Alaska Range and the Chugach Mountains. The map shows that the lateral bank erosion of the meander bend upstream from the pipeline crossing has been rapid (over 100 feet) in the last 20 years. This meander bend is expected to be cut off during the life of the pipeline. The cutoff would shorten the course of the Copper River by about 1/2 mile, increasing the channel slope and causing several feet of scour at the pipeline crossing (22). Even smaller rivers (such as Hess Creek, approximately 65 miles northwest of Fairbanks) that are actively eroding laterally at a rapid rate will need bank protection to protect the pipeline crossing (Figure 14.25.)

The *permafrost* is also a potential hazard in that once frozen ground melts, it is extremely unstable. This aspect of the project has been thoroughly engineered, and hopefully problems with permafrost will be minimized. Nevertheless, the many miles of a large hot-oil pipeline crossing hundreds of miles of permafrost provides sufficient cause for concern.

Analysis of possible physical, biological, and social economical impacts associated with the pipeline establishes that there are three areas of concern (19): the construction, operation, and maintenance of the pipeline system, including access roads and highways; development of the oil fields; and operation of the marine tanker system at Port Valdez. Furthermore, some of the effects would result in predictable, unavoidable disruptions, while other effects are more speculative and present a threat, for example, the impact of an oil spill caused by a break in the pipeline. These threatened effects are not easily evaluated and predicted (19).

A list of primary and secondary effects of the trans-Alaska hot-oil pipeline, gas pipeline, and tanker transport system considered by the task force to evaluate environmental impact is shown in Table 14.7. In addition to these effects, it was necessary to evaluate possible linkages between the effects that might have an impact on the environment. For example, there are obvious possible linkages between oil spills or unavoidable release of oil in the tanker operation into the marine environment and the effect on marine resources. This, however, is difficult to evaluate because of the variable or unknown aspects of the hazard, the dynamic hydrologic environment, and seasonal and other changes in fish and marine resources. In general, the task force concluded that the impact linkage data were hard

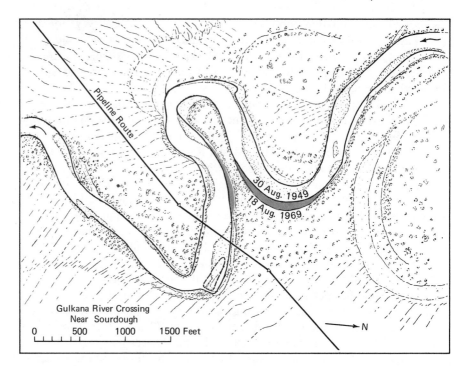

FIGURE 14.24 *Gulkana River pipeline crossing near Sour-dough, Alaska. Recent lateral erosion on the bends of the river suggest that a meander cutoff may occur in the next 20 years just upstream from the pipeline crossing. (Source: Brice, U.S. Geological Survey Publication, Alaska District, 1971.)*

to obtain and evaluate and so were considered as the largest problem area in quantitatively determining possible impacts (19).

Based on the environmental analysis, both unavoidable and threatened impacts of the Trans-Alaska Pipeline were determined (19). The main unavoidable effects were threefold: first, disturbances of terrain, fish and wildlife habitats, and human use of the land during construction, operation, and maintenance of the entire project, including the pipeline itself and access roads, highways, and other support facilities; second, discharge of effluents and oil into Port Valdez from the tanker transport system; and third, increased human pressures on services, utilities, and many other areas, including cultural changes of the native population.

The major threatened effects are accidental loss of oil from the oil field, pipeline, or tanker system. Accidental loss of oil from the pipeline could be caused by slope failure (landslides), differential settlement in permafrost areas, stream bed or bank scour at river crossings, or destructive sea waves damaging the pipeline, causing a leak or rupture. Oil loss from tankers may be caused by ship-wreck or accidental loss during transfer operations at Port Valdez. The potential loss of oil from the pipeline and tanker systems involves a great number of variables, and therefore predictions concerning loss of oil are difficult—but some loss is inevitable. Estimates place maximum oil loss at 1.6 to 6 barrels per day at Valdez

FIGURE 14.25   *Tortuous meander bends of Hess Creek down-stream from the pipeline crossing. (Photo courtesy of James Brice.)*

for tanker operations, and 384 barrels per day from tanker accidents. The latter, of course, could be either a series of small spills or several large spills at unknown times, locations, and intervals (19).

Before the final impact statement was completed early in 1972 until construction started in 1975, the pipeline generated (and continues to generate) controversy. This controversy results because of the conflicts in balancing the need for resource development and the known or predicted environmental degradation. Although alternative routes and transport systems, including trans-Canada routes and railroads, other Alaska routes, and marine routes, were extensively evaluated (Figure 14.26), the earlier proposed route to Port Valdez was ultimately approved, perhaps because this route led to the most rapid resource development while maintaining national security. However, it is emphasized that no one route is superior in all respects to the others (19). Comparison of the alternative routes suggests the following. First, all the trans-Alaska routes have less unavoidable adverse impact on the abiotic (nonliving) systems than the trans-Alaska-Canada routes. Second, the trans-Alaska route to the Bearing Sea would probably have the least unavoidable impact on terrestrial-biologic and social-economic systems. The trans-Alaska-Canada coastal route would be next in minimizing unavoidable impact on these systems. Third, the trans-Alaska-Canada routes would have the least unavoidable impact on marine environments because no direct marine transport of oil is involved (19).

Based on the above information and considerable other evaluations and analyses (19), it was concluded that to avoid the marine environment and earth-

TABLE 14.7 *Primary and secondary impacts associated with the Trans-Alaska Pipeline, arctic gas pipelines, and proposed tanker system.*

A. *Primary effects associated with arctic pipelines:*

1. Disturbance of ground
2. Disturbance of water (including treated effluent discharge into water)
3. Disturbance of air (including waste discharged to air and noise)
4. Disturbance of vegetation
5. Solid waste accumulation
6. Commitment of physical space to pipeline system and construction activities
7. Increased employment
8. Increased utilization of invested capital
9. Disturbance of fish and wildlife
10. Barrier effects on fish and wildlife
11. Scenery modification (including erosional effects)
12. Wilderness intrusion
13. Heat transmitted to or from the ground
14. Heat transmitted to or from water
15. Heat transmitted to or from air
16. Heat to or from vegetation
17. Moisture to air
18. Moisture to vegetation
19. Extraction of oil and gas
20. Bypassed sewage to water
21. Man-caused fires
22. Accidents that would amplify unavoidable impact effects
23. Small oil losses to the ground, water, and vegetation
24. Oil spills affecting marine waters
25. Oil spills affecting freshwater lakes and drainages
26. Oil spills affecting ground and vegetation
27. Oil spills affecting any combination of the foregoing

B. *Secondary effects associated with arctic pipelines:*

1. Thermokarst development
2. Physical habitat loss for wildlife
3. Restriction of wildlife movements
4. Effects on sports, subsistence, and commercial fisheries
5. Effects on recreational resources
6. Changes in population, economy, and demands on public services in various communities, including native communities, and in native populations and economies
7. Development of ice fog and its effect on transportation
8. Effects on mineral resource exploration

C. *Primary effects associated with tanker system:*

1. Treated ballast water into Port Valdez
2. Vessel frequency in Port Valdez, Prince William Sound, open ocean, Puget Sound, San Francisco Bay, Southern California waters, and other ports
3. Oil spills in any of those places

D. *Secondary effects associated with tanker system:*

1. Effects on sports and commercial fisheries
2. Effects on recreational resources
3. Effects on population in Valdez and other communities

SOURCE: U.S. Geological Survey Circular 695, 1974.

quake zones, and to place both an oil and a gas pipeline in one corridor, the trans-Alaska-Canada route to Edmonton, Canada, was the route that would cause the least environmental impact. This is past history since the route is going to Port Valdez, but it serves to emphasize first, the alternatives considered and the obligation of scientists to state their opinion based on sound scientific information, even though this opinion may be either unpopular or likely to be overridden in the final balancing of alternatives; and second, the significant power of diverse political maneuvering at all levels in deciding between alternatives.

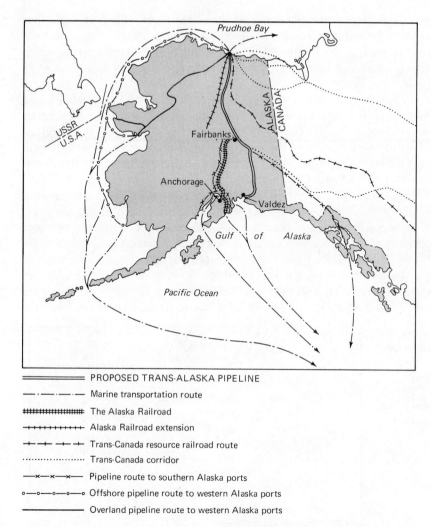

FIGURE 14.26 *Alternative routes for transporting oil from the North Slope of Alaska. (Source: Brew, U.S. Geological Survey Circular 695, 1974.)*

Cape Hatteras National Seashore—An Example of Environmental-Impact Analysis The Outer Banks of North Carolina has for generations been inhabited by people living and working in essentially a marine-dominated environment. Until recently, a way of life that has persisted for generations depended on livestock raising, fishing, hunting, boat building, and other marine pursuits (23).

The landscape of the Outer Banks characteristically can change due to dramatic physical processes in a very short time. Storms, particularly hurricanes, can initiate rapid changes in the relationships between land and water, particularly in the location of ephemeral beaches, inlets, and salt marshes. For these reasons, the people of the Outer Banks have historically taken the philosophy of living with and adjusting to a changing landscape. However, in recent times, this philosophy has changed because of economic pressure to develop coastal property. A new philosophy of coastal protection has developed in an attempt to stabilize the coastal environment. Stabilization or a constant-appearing landscape is a prerequisite to commercial development.

Congress approved in 1937 and amended in 1940 an act establishing the Cape Hatteras National Seashore as the first national seashore. The park consists of 28,500 acres along a 75-mile portion of the Outer Banks (Figure 14.27) and includes portions of three islands of the more than 150 miles of the Barrier Island system. The Barrier Islands essentially bound and protect the largely undeveloped coastal plains lowland of North Carolina (23).

Eight unincorporated villages are located within the Cape Hatteras National Seashore and are spaced along nearly the entire length of the seashore. Legislation authorizing the park provided for the continued existence of these villages, including a maximum beach strip of 500 feet in front of each facing the Atlantic Ocean. This legislation has been interpreted by many as an obligation of the federal government to maintain and stabilize access to and beach frontage of these villages. Unfortunately, stabilization has been very difficult, and in some villages there is less than 200 feet of beach front due to recent coastal erosion. This

FIGURE 14.27 *Map showing the Outer Banks of North Carolina and the Cape Hatteras National Seashore. [Source: Godfrey and Godfrey, in D. R. Coates, ed.,* Publications in Geomorphology *(New York: State University of New York,1971.)]*

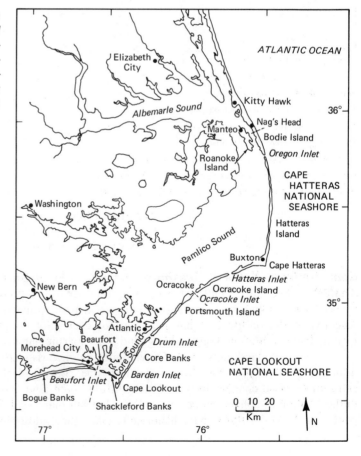

philosophy of protection is contrary to that during the early development of the islands, which was mainly on the sound (inland) side of the islands. However, with the opening of the road link and the construction of the first dune systems in the 1930s, the basic configuration of the villages began to change as communities began to spread toward the ocean away from the more protected sites along the sound. This migration has persisted and has grown at an ever-increasing pace (23).

At one time it was assumed that much of the Outer Banks was heavily forested and that logging and overgrazing had destroyed these forests. Evidence for this is controversial, and in fact extensive forests may have been developed on only a few islands with an orientation more protected from storms. The wooded areas were located in an area protected by a natural dune line. Therefore, the logical conclusion was to construct a man-made barrier dune system to help the area return to a natural state by inhibiting beach erosion. Of course, the dune line would also protect the roads and communities. As a result, there is now an artificial dune system along the entire Cape Hatteras National Seashore at an average dune height of 5 meters and an average distance of 100 meters from the ocean (23). This man-made erosion-control measure, along with programs to artificially keep sand on the beach, costs about $1 million per year. This expensive program, along with an essentially geological argument as to whether man-made dune lines may jeopardize the future of the Barrier Islands, has led to a controversy on how to best manage the park.

The geologic problem and controversy of how barrier islands are developed and maintained by dynamic earth processes is at the philosophical heart of the U.S. Park Services dilemma in developing a long-range program to maintain the natural environment for the use and enjoyment of present and future generations. Although there are several ideas and known ways that coastal processes can develop barrier islands, there is heated debate on the processes that are needed to maintain them.

It is believed that the present Barrier Islands developed about 5000 years ago in response to complex wind and water processes associated with a slowly rising sea level. One belief is that since many of the capes (protruding barrier islands) apparently coincide with mouths of rivers, development of these capes has been facilitated by sediment deposited as deltas during times of relatively low sea levels during the Pleistocene (last several million years). As the Pleistocene sea levels retreated because water was held in the continental glaciers, deltas developed seaward of the present coastline. Then, when the ice melted a few thousand years ago and sea levels began to rise, these deltas became the nucleus of the Barrier Islands and capes. Submergence was accompanied by coastal erosion until it produced the present island system which is still eroding at a rate of 4 to 6 feet per year in response to continuing rise in sea level. Figure 14.28 shows this general model for development of barrier islands and capes (23, 24).

Although the true role of a rising sea level in forming the barrier islands is debated, most investigators do agree that the rise is closely associated with tremendous changes in the coastline and that this is a real and immediate hazard to future coastal development, which demands a static shoreline.

Debate on the nature and extent of geologic processes that maintain the "natural changing" Barrier Islands centers on the importance of whether periodic overwash of the frontal dune system by storm waves is essential to maintaining the

TABLE 14.8 *Possible alternative management plans for the Cape Hatteras National Seashore.*

Alternative 1

Management objectives:	To manage Cape Hatteras National Seashore as a primitive wilderness so not to impede natural processes.
Proposed action:	Let nature take its course, and oppose any attempts to control natural forces or stabilize the shoreline.
Legislative changes:	Classify Cape Hatteras as a recreation area, and officially designate the Seashore a wilderness area by administrative decree.
Major impacts:	A natural environment would be restored; the existing highway could not be maintained; property within the villages would be lost; historic structures would be destroyed unless relocated; and the economic base and economic development of the village enclaves would decline.
Costs of alternative:	No direct federal costs except possible relocation of Cape Hatteras Lighthouse and other historic structures. State costs of maintaining the road would be excessive.

Alternative 2

Management objectives:	To manage Cape Hatteras National Seashore in accordance with the policies established for natural areas of the National Park Service, and to preserve the area in a natural state.
Proposed action:	Develop an ongoing resource management program, revegetate major overwash areas following destructive storms, and investigate alternative transportation modes.
Legislative changes:	Classify Cape Hatteras as a recreation area, and officially designate the Seashore a natural area by administrative decree.
Major impacts:	The Seashore would be preserved in a natural state, and residents of the villages would have to live with and adjust to natural events. Some property would be lost, and economic development of the villages would be retarded. Historic structures would be relocated or abandoned. The road would be damaged or washed out in places but could be maintained over the next several years.
Costs of alternative:	Estimated at $125,000 annually, or $3,125,000 over a 25-year period.

Alternative 3

Management objectives:	Continue present shore erosion-control practices.
Proposed action:	Management practices of the past—dune building, dune repair, and beach nourishment would continue.
Legislative changes:	No changes would be required.
Major impacts:	The Seashore environment would become increasingly artificial as efforts were concentrated on protecting private development. The village enclaves would continue to grow and the highway would have to be widened to serve these areas.
Costs of alternative:	Estimated at $1 million annually, or $25 million over a 25-year period.

TABLE 14.8, continued

	Alternative 4
Management objectives:	To acquire threatened private property within the village enclaves to preserve a public beach fronting the villages and avoid having to protect such property.
Proposed action:	Identify threatened private property, determine fair market value, and purchase the property.
Legislative changes:	Authorize legislation for Cape Hatteras to allow for the acquisition of additional land.
Major impacts:	The economic base of the village enclaves would be reduced, and, as the shoreline continues to recede, future acquisition would be required at much greater cost. This alternative would establish the precedent of compensating unwise development.
Costs of alternative:	Very high—over $25 million for Kinnakeet Township alone (1973 prices). The land involved is the costliest on the Seashore. Loss of tax revenue and revenue derived from visitor expenditures is not included.

	Alternative 5
Management objectives:	To protect the development taking place within the village enclaves and protect the highway running through the Seashore.
Proposed action:	Stabilize the shoreline fronting threatened private property and protect exposed sections of highway by relocating or elevating the road or by stabilizing shoreline structures or building dunes.
Legislative changes:	Amend the authorizing legislation for Cape Hatteras, deleting the section pertaining to managing the area as a primitive wilderness.
Major impacts:	The natural setting of the Seashore would be destroyed and the resource base would be seriously degraded. The village enclaves would develop into urban complexes and dominate the landscape. This alternative would acknowledge a Federal responsibility to subsidize unwise economic development.
Costs of alternative:	Overall initial cost of construction would be $40 to $56 million with annual maintenance of $3.2 to $6.4 million.

SOURCE: National Park Service.

islands. If the overwash is essential, then building a dune line is contrary to natural processes, and the final result over a period of years may be a deterioration of the island as a natural system. This would be contrary to the stated Park Services' objectives to preserve the islands in a natural state.

The argument that all barrier islands are maintained in natural landward migration by periodic overwash of frontal dunes is not convincing because most of the energy expended by storm waves is concentrated in the surf zone rather than splashing over low dunes. On the other hand, overwash of dunes is on some barrier islands a real process that does tend to facilitate a more "natural appearing"

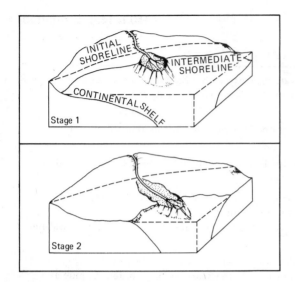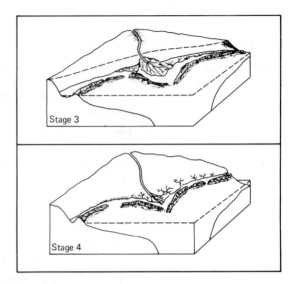

FIGURE 14.28 *Idealized diagram illustrating how some capes and barrier islands may form in four stages. Stage 1 is characterized by deposition of sediment at a river's mouth due to a lowering of sea level. Stage 2 involves continued lowering of sea level and subsequent deposition of sediment. Stage 3 is characterized by a rise in sea level and subsequent submergence of deltaic ridges. Stage 4 shows the present cape and barrier island morphology which has resulted from slow submergence and landward migration of earlier formed barrier island systems. (Source: Hoyt and Henry,* Geological Society of America Bulletin, *vol. 82, no. 1.)*

coastline. However, it is argued that frequent overwash is atypical and takes place only on islands that are already in an unnatural state due to overgrazing and subsequent lowering of dunes due to loss of protective vegetation. Regardless of the cultural influence and historic role of overwash, it is apparent that some overwash is a natural process and as such is probably significant in the geologic history of barrier-island migration. However, this does not mean that it is necessarily always bad to attempt to selectively control rapid coastal erosion by maintaining a protective frontal dune system. Proper placement of dunes and subsequent good conservation practice in their maintenance remains a viable alternative to short-range selected coastal erosion problems, particularly near settlements and critical communication lines.

Faced with the problem of selecting a management policy for the Cape Hatteras National Seashore, the Park Service, in 1974, presented five possible alternatives (Table 14.8) ranging from essentially no control of natural coastline processes (Alternative 1) to attempting complete protection (Alternative 5). Each alternative was analyzed to determine the entire spectrum of possible impacts, and although the final decision has not been resolved, few people want complete protection or complete control of the seashore (23). The position of the Park Service is to attempt to strike a compromise in which the Barrier Islands for the most part may be preserved in a natural state, while the need to maintain a transportation link with the mainland is recognized. Thus, the communities will have to live with and adjust to the dynamic high-risk environment they choose to live in, much as the people of the Outer Banks historically have. This is contrary to prevailing trends in coastal development which assume a more stable environment and acknowledge that natural processes play a significant role in preserving the natural environment (23).

The major impact of the proposed management program if enacted will be threefold. First, the Cape Hatteras National Seashore will be preserved in a natural state. Second, the residents of the villages will have to live with and adjust to the effects of natural events to a greater extent. Third, economic development of the villages will be retarded to a lesser or greater extent. The road will be subject to more periodic damage but can probably be maintained.

SUMMARY AND CONCLUSIONS

Landscape evaluation, including land-use planning, site selection, evaluation of landscape intangibles, and environmental-impact analysis, is one of the most controversial environmental issues of our times. Before these issues can be resolved satisfactorily, it is necessary to develop a sound methodology to insure that land and water resources are evaluated, used, and preserved in a way consistent with the emerging land ethic.

The role of the geologist in landscape evaluation is to provide geologic information and analysis before planning, design, or construction of projects such as reservoirs, large buildings, housing developments, tunnels, pipelines, and parks. In this respect, it is the obligation of the earth scientist to emphasize that all land is not the same and that particular physical and chemical characteristics of the landscape may be more significant to society than the geographic location.

Land-use planning in the future will have to increasingly include concepts of sequential or multiple land-use rather than exclusive use. This results because there is a definite limit to the supply of land for specific purposes, and therefore we must strive to plan so that suitable land for future generations is available for needed purposes. The basic element of land-use planning is developing a statement defining a plan that includes objectives, issues, and goals; analysis and summary of pertinent data; a land classification map; and a report discussing appropriate developments.

The use of soils and engineering geology is significant in determining the possible limitations to land development. This information is best presented in a series of maps that summarize geologic hazards and soil and engineering limitations for specific land uses.

Site selection and evaluation is the process of evaluating the physical environment with the objective to support activities of man. A philosophy of site evaluation based on "physiographic determinism" or "designed with nature" has emerged as a worthy philosophical framework to partially balance the traditional economic aspects of site evaluation issues. From a geologic view, this philosophy essentially requires the determination of the magnitude and importance of geologic limitations of a particular site for a certain land use.

Site evaluation for engineering purposes such as construction of dams, highways, airports, tunnels, and large buildings all require careful geologic evaluation before planning and design of the project. The role of a geologist is to work with engineers and indicate possible adverse or advantageous geologic conditions affecting the project.

Evaluation of scenic resources and other environmental intangibles is becoming more important in landscape evaluation. The significance is in balancing intangibles with the more easily evaluated economic aspects of environmental evaluation. The objective is to be able to quantify and hierarchically rank alternatives, as is done in the evaluation of the more tangible elements of the landscape. All methods that attempt to evaluate environmental intangibles such as scenery depend on the analysis of landscape factors or variables such as topographic relief, presence of water, degree of naturalness, and diversity. Thus, all the methods are somewhat subjective. However, subjectivity in this case is not necessarily a bad attribute. In fact, a completely objective technique is probably impossible to obtain given our present knowledge of landscape perception by individuals. What is important is that the evaluation be based on sound judgment.

Evaluation of environmental impact is now required by law for all federal actions that could possibly affect the quality of the human environment. The result of the evaluation is an environmental-impact statement which describes the proposed action; discusses all probable impacts, including unavoidable adverse impacts and possible alternatives; discusses irreversible or irretrievable commitments of resources; and discusses possible problems or objections to the action raised during public or institutional debate.

There is no one method for determining environmental impact for the wide spectrum of possible actions and projects that might affect the environment, and furthermore no one method is probably possible. The objective of the analysis process before design and construction is to minimize the possibility of causing extensive environmental degradation. In the past, rather serious problems of pollution, loss of resources, or creation of a hazard have accompanied certain projects. This has led to unfortunate closings of industries and has forced people to adjust to possible hazards and economic loss.

Examples of environmental-impact analysis for the Trans-Alaska Pipeline and the management of the Cape Hatteras National Seashore are significant examples of the way possible impacts and alternatives are evaluated. Furthermore, both of these examples stress the importance of considering the geological aspects of the projects.

References

1. Turner, A. K., and Coffman, D. M. 1973. Geology for planning: a review of environmental geology. *Quarterly of the Colorado School of Mines* 68.
2. Bartelli, L. J.; Klingebiel, A. A.; Baird, J. V.; and Heddleson, M. R., eds. 1966. *Soil surveys and land use planning.* Soil Science Society of America and American Society of Agronomy.

3. Hayes, W. C., and Vineyard, J. D. 1969. *Environmental geology in town and country*. Missouri Geological Survey and Water Resources, Educational Series No. 2.

4. North Carolina Coastal Resources Commission. 1975. *State guidelines for local planning in the coastal area under the Coastal Area Management Act of 1974*. Raleigh, North Carolina.

5. Montgomery, P. H., and Edminster, F. C. 1966. Use of soil surveys in planning for recreation. In *Soil Surveys and Land Use Planning*, eds. L. J. Bartelli et al., pp. 104–12. Soil Science Society of America and American Society of Agronomy.

6. Gross, D. L. 1970. *Geology for planning in DeKalb County, Illinois*. Environmental Geology Notes No. 33, Illinois State Geological Survey.

7. Whyte, W. H. 1968. *The last landscape*. Garden City, New York: Doubleday.

8. Flawn, P. T. 1970. *Environmental geology*. New York: Harper & Row.

9. Lynch, K. 1962. *Site planning*. Cambridge, Massachusetts: M.I.T. Press.

10. Prest, A. R., and Turvey, R. 1965. Cost-benefit analysis: a survey. *The Economic Journal* 75: 683–735.

11. McHarg, I. L. 1971. *Design with nature*. Garden City, New York: Doubleday.

12. Schultz, J. R., and Cleaves, A. B. 1955. *Geology in engineering*. New York: John Wiley & Sons.

13. Krynine, D. P., and Judd, W. R. 1957. *Principles of engineering geology and geotechniques*. New York: McGraw-Hill.

14. Zube, E. H. 1973. Scenery as a natural resource. *Landscape Architecture* 63: 126–32.

15. Linton, D. L. 1968. The assessment of scenery as a natural resource. *Scottish Geographical Magazine* 84: 219–38.

16. Leopold, L. B. 1969. *Quantitative comparison of some aesthetic factors among rivers*. U.S. Geological Survey Circular 620.

17. Melhorn, W. N.; Keller, E. A.; and McBane, R. A. 1975. *Landscape aesthetics numerically defined (land system): application to fluvial environments*. Purdue University Water Resources Research Center Technical Report No. 37.

18. Council on Environmental Quality. 1971. Statements on proposed federal actions affecting the environment. *Federal Register,* vol. 36, no. 19, pp. 1398–1402, and no. 79, pp. 7724–29.

19. Brew, D. A. 1974. *Environmental impact analysis: the example of the proposed Trans-Alaska Pipeline*. U.S. Geological Survey Circular 695.

20. Sterling, C. 1971. The Aswan disaster. *National Parks and Conservation Magazine* 45: 10–13.

21. National Research Council. 1972. *The earth and human affairs*. San Francisco: Canfield Press.

22. Brice, J. 1971. *Measurements of lateral erosion at proposed river crossing sites of the Alaskan Pipeline*. U. S. Geological Survey, Water Resources Division, Alaska District.

23. National Park Service. 1974, Cape Hatteras Shoreline Erosion Policy Statement, Department of Interior.

24. Hoyt, J. H., and Henry, V. J., Jr. 1971. Origin of capes and shoals along the southeastern coast of the United States. *Geological Society of America Bulletin* 82: 59–66.

15

ENVIRONMENTAL LAW

History and Development of Environmental Law

A land, or environmental, ethic is becoming an important part of our governmental institutions. It is being incorporated into our legal and economic principles, and no longer is the best use of land that which returns the greatest profit. The new land ethic establishes that man is part of a land community including trees, rocks, animals, and scenery, and he is morally bound to assure the community's continued existence. Thus, this ethic affirms our belief that this earth is our only suitable habitat (Figure 15.1), and it recognizes the rights of people to breathe clean air, drink unspoiled water, and generally exist in a quality environment.

According to the National Environmental Policy Act of 1969, the national environmental policy is to encourage productive and enjoyable harmony between people and their environment; to promote efforts to eliminate or at least minimize degradation of the environment; and to continue investigating relations between ecological systems and important natural resources. This policy suggests that each generation has the moral responsibility to provide the next generation with

FIGURE 15.1 *This earth is indeed our only suitable habitat,
and there are few alternatives to maintaining a quality environment. (Photo courtesy of NASA.)*

a healthful, productive, and aesthetically pleasing surrounding. It implies a doctrine of trust which asserts that all public land and, to a lesser extent, some private land is essentially held in trust for the general public.

The Trust Doctrine as a theory is closely related to constitutional issues involving the right of people to a quality environment. Although nowhere in the Constitution does it state this right, some students of law believe that such a right can be inferred from the Ninth Amendment to the Constitution, which recognizes that the listing of specific rights in the Bill of Rights does not deny the existence of other unlisted rights. That is, there are many rights not specifically listed but always held by the people (1).

Fundamental rights entitled to the protection of the Ninth Amendment are defined as those so basic and important to our society that it is inconceivable that they are not protected from unwarranted interference (2). It might, therefore, be argued that this amendment encompasses the right of people to have clean air, clean water, and other resources necessary to insure a quality environment.

It is important to recognize that there is a certain amount of friction between the Trust Doctrine, which establishes that particularly significant land resources are held in a public trust, and the Fifth Amendment, which establishes that land may not be taken from an individual without due process of law and just compensation. Problems result in interpreting law and determining how to measure just compensation. The real issue is defining at what point the private use of land infringes on the rights of other people and future generations.

Although the term *environmental law* has only recently gained common usage and there are few new laws in this field, it is fast becoming an important part of our jurisprudence (1). The many recent articles devoted to environmental law attest to the fact that the subject is becoming increasingly popular among lawyers and environmental law societies, and environmental courses have recently been established at leading law schools (3).

Air, water, and other resources necessary for our survival are often taken for granted. In America, we still tend to suffer from the myth of super-abundance and to think of resources as inexhaustible. However, in large cities with accompanying large populations of people, cars, and industries, resources are deteriorating, and as a result, regulations and laws concerning pollution of air and water are necessary. This is not new; London's air by 1306 contained large amounts of smoke from the burning of coal that evidently polluted the air to the extent that a royal proclamation was issued to curtail the use of coal. Violations of the law were punishable by death (3).

Air pollution was proclaimed a public nuisance as early as 1611 when an English court essentially ruled that a property owner has the right to have the air he breathes and smells free of pollutants. The case involved a plaintiff who asked for injunction and damages because his neighbor, the defendant, raised hogs, whose odor the plaintiff considered a nuisance. The defendant was found to be creating a nuisance even though he pleaded that the raising of hogs was necessary for his sub-sistence and that neighbors should not have such delicate noses that they cannot bear the smell of hogs (4). It is emphasized that the English Nuisance Law of 1536 involved a type of common law still used in the United States, the main principle of which is that if other people suffer equally from a particular pollution, an individual cannot bring suit against the polluter. In the case of the hog farmer, it is obvious that the effect of the nuisance varied with proximity to the hog pens (3).

It is beyond the scope of our discussion to consider in detail the pro-cesses of law and how they relate to the environment. Suffice it to say that law is a technique for the ordered accomplishment of economic, social, and political pur-poses, and the most desired legal technique generally is one that most quickly allows ends to be reached. However, it is emphasized that law serves primarily the major interest that dominates the culture, and in our sophisticated culture, the major con-cerns are wealth and power. These stress the ability of society to use the resource base to produce goods and services, and the legal system provides the vehicle to insure that productivity (5).

The general trend of law and the courts has been to render judgments based upon the greatest good to the greatest number and to use what is commonly referred to as the *Balancing Doctrine,* which asserts that public benefit from and importance of a particular action should be balanced with potential injury to certain individuals. However, the method of balancing is changing, and courts are now considering possible long-range injury caused by certain activities to large numbers of citizens other than the immediate complainants, and this is more likely to insure a true balancing of the equities (5).

Case Histories

The discussion of particular case histories is useful in understanding some legal aspects of environmental problems. The particular selection of cases is not meant to imply a judgment of any particular activity, but rather to indicate the considerable variability and possibilities in environmental law. The cases chosen include areas of air pollution: aesthetic pollution and land use.

Two cases, *Madison* v. *Ducktown, Sulfur, Copper and Iron Company*, and *Hulbert* v. *California, Portland Cement Company*, demonstrate different approaches. The equities were balanced in one court, and in another the balancing concept was rejected (4). The *Madison* case involved man-made fumes which were discharged from smelting and refining companies. The effect of the pollutants in the vicinity of the plants was to kill crops and timber, thereby accelerating soil erosion (Figure 15.2). As a result, people living in the area were prevented from using their farms and homes as they did before construction and operation of the plants. The court chose to balance the equities in favor of the mining industry, declining to grant injunctive relief. The court reasoned that since it was not possible for the industry to reduce the ores in a different way or move to a more remote location, forcing them to stop air pollution would compel the company to stop operating their plants. This would cause 10,000 people to lose their jobs, destroy the tax base of the county, and make the plant properties practically worthless (4). Therefore, although damages were awarded, the plant was allowed to continue operation.

In the *Hulbert* case, the plaintiffs argued for injunction requiring the cement company to stop discharging cement dust which was falling on their properties. The dust contaminated the plaintiffs' homes and also formed an encrustation on the upper sides of flowers and other plants, especially the citrus fruit trees of the plaintiffs. The dust was durable and not removed by natural processes such as strong winds or rain. As a result, the value of the citrus fruit declined, and it was more difficult and expensive to cultivate and harvest the crop. The company argued that

FIGURE 15.2 *Fumes from the smelter at Ducktown, Tennessee, killed nearby vegetation and initiated a rapid, man-made erosion cycle. (Photo by A. Keith, courtesy of U.S. Geological Survey.)*

they were doing all possible to stop the discharge of dust and that damages were sufficient to compensate the plaintiffs for their injuries. Furthermore, they argued that the large payroll of the company and other benefits to the community must be balanced. Nevertheless, the court rejected the Balancing Doctrine and granted the injunction (4).

When the *Ducktown* and *California* cases are compared, it is important to recognize three facts. First, compared to the cement rock (limestone), the citrus fruit has a relatively high unit value and can be grown only in certain favorable areas. Second, the cement rock has a lower unit value than the Ducktown ores and is found in many more locations. Third, reclamation of an area mined for cement rock is possible and would retain land for other productive uses, an alternative that is neither likely nor very feasible for Ducktown.

Water, land, and air pollution is relatively easy to measure and evaluate compared with the intangible aesthetic pollution involving visual, audio, and other senses that affect a person's well-being. However, we do recognize some of the possible harmful effects of aesthetic pollution, and the courts are forming attitudes on these matters. The precedent that aesthetics is a valid subject of legislative concern was set in New York courts in cases involving aesthetic degradation, particularly loss of value of property because the view of a lake or mountain is spoiled. It was recognized that reduction in property value due to loss of a view that otherwise increased the value of the property should not be borne by the owner whose land was taken for public purposes without his permission (3).

Two examples from New York concerning road construction emphasize that the law of aesthetics is becoming more important and that intangible factors that used to be considered outside the public interest are now as important as other more easily measured factors. In *Clarance* v. *State of New York,* the court awarded the plaintiff $10,000 because a road degraded his view of Seneca Lake and denied him easy access to the lake (3). In a similar case, *Dennison* v. *State of New York,* a property owner was awarded damages because a new highway caused loss of privacy and seclusion, loss of view of forests and mountains, traffic noise, lights, and odors. The court concluded that all of these factors tended to cause damage to the plaintiff's property (3).

The Storm King Mountain dispute is a classic example of possible conflict between a utility company and a conservationist. In 1962, the Consolidated Edison Company of New York announced plans for a hydroelectric project approximately 40 miles north of New York City in the Hudson River Highlands, an area considered by many to have unique aesthetic value because nowhere else in the eastern United States is there a major river eroded through the Appalachian Mountains at sea level, giving the effect of a fjord (6). The early plans for the project called for a powerhouse to be constructed above ground, requiring a deep cut into Storm King Mountain. The project was redesigned to site the powerhouse entirely underground, eliminating the cut on the mountain (Figure 15.3). Regardless of this, the conservationist continued to oppose the project, and the issues have now been broadened to include possible damage to the fishery industries. The argument is that the high rate of water intake, 8 million gallons per minute, from the river would draw many fish larva into the plant where they would be destroyed by turbulence and abrasion. The most valuable sport fish is the striped bass, and one study showed

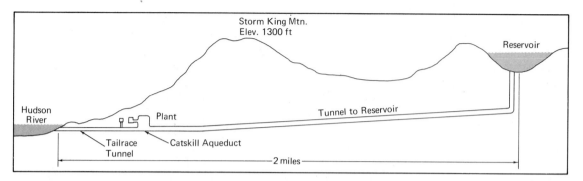

FIGURE 15.3 *Idealized diagram showing how the entire Storm King Mountain hydroelectric project might be placed underground. (Source: L. J. Carter,* Science, *vol. 184, June 1974, pp. 1353–58. Copyright 1974 by the American Association for the Advancement of Science.)*

that 25 to 75 percent of the annual bass hatch might be destroyed if the plant were operating. The fish return from the ocean to tidal water to spawn, and since the Hudson River is the only estuary north of Chesapeake Bay where the striped bass spawn, the concern for the safety of the fisheries is real. The problem is even more severe because the proposed plant is located near the lower end of the 8-mile reach in which the fish spawn. The project is still being studied, and it is the responsibility of the utility company to show that the project will not do unacceptable damage to the fishery resource (6).

After a decade of legal activity, the Storm King Mountain controversy is still unresolved. The case is interesting because it emphasizes the difficulty in making decisions about multidimensional issues. On one hand, there is a utility company trying to survive in New York City where there are unbelievably high peak-power demands accompanied by high labor and maintenance costs. On the other hand, conservationists are fighting to preserve our beautiful landscape and fishery resources. Both have legimate arguments, but in light of their special interests, it is difficult to resolve conflicts. Present law and procedures are sufficient to resolve the issues, but trade-offs will be necessary. Ultimately, an economic and environmental price must be paid for any decision, and this reflects our desired life-style and standard of living.

A recent case reported by Yannacone, Cohen, and Davison in Colorado emphasizes the significance of the Trust Doctrine and the Ninth Amendment as they pertain to land management (2). The conflict surrounded the use of 1800 acres of land near Colorado Springs. The land is part of the Florissant Fossil Beds where insect bodies, seeds, leaves, and plants were deposited in an ancient lake bed approximately 30 million years ago. Today, they are remarkably preserved in thin layers of volcanic shale. Unfortunately, the fossils are delicate and unless protected tend to disintegrate when exposed. The fossils are considered by many people to be unique and irreplaceable. At the time of the controversy, a bill had been introduced into Congress to establish a Florissant Beds National Monument. The bill had passed the Senate, but the House of Representatives had not yet acted on it.

While the House of Representatives was deliberating the bill, a land development company that had contracted to purchase and develop recreational homesites on 1800 acres of the ancient lake bed announced they were going to bull-doze a road through a portion of the proposed national monument site in order to gain access to the property they wished to develop. A citizens group formed to fight the development until the House acted on the bill. They tried to obtain a temporary restraining order, which was first denied because there was no law preventing the owner of the property from using his land in any way he wished, provided he did not break any existing law. The conservationists then went before an appeals court and argued that even though there was no law protecting the fossils, they were subject to protection under the Trust Doctrine and the Ninth Amendment. The argument was that protection of an irreplaceable, unique fossil resource was an unwritten right retained by the people under the Ninth Amendment and that further-more, since the property had tremendous public interest, it was also protected by the Trust Doctrine. An analogy used by the plaintiffs was that if someone were to find the Constitution of the United States buried on his land and wanted to use it to mop the floor, certainly he would be restrained. After several more hearings on the case, the court issued a restraining order to halt development, and shortly thereafter the bill to establish a national monument was passed by Congress and signed by the president (2).

The court order prohibiting the destruction of the fossil beds may have deprived a land owner of making the most profitable use of his property, but it does not prohibit all uses consistent with protecting the fossils. For instance, the property owners are free to develop the land for tourism or scientific research. While this might not result in the largest possible return on the property owner's investment, it probably would return a reasonable profit.*

Environmental Legislation

The ultimate goal of persons concerned with how man treats and uses his natural environment and resources is to insure that ecologically sound, responsible, socially acceptable, economically possible, and politically feasible legislation is passed (2). To attain this, we need professional people to assist the legislators at all levels of government in drafting the needed laws and regulations.

Environmental legislation has already had a tremendous impact on the industrial community. New standards in regulations limiting the discharge of pos-sible pollutants into the environment have placed restrictions on industrial activity. Furthermore, any new activity that directly or indirectly is involved with the federal government is required to complete a study to evaluate the environmental impact of their proposed activity. Beyond this, the federal legislation has set an example which many states are following in passing environmental protection legislation.

A complete discussion of all federal legislation that has environmental implications is beyond the scope of our discussion. However, a discussion of some of

---

*Yannacone et al., *Environmental Rights and Remedies* (Rochester, N.Y.: The Lawyers Cooperative Publishing Co., 1972), pp. 39–46.

the major acts is valuable in understanding the basics of such legislation. For this purpose, we will discuss the Refuse Act of 1899, the Fish and Wildlife Coordination Act of 1958, the National Environment Policy Act of 1969, and the Water Quality Improvement Act of 1970.

The Refuse Act of 1899   This act states that it is unlawful to throw, discharge, or deposit any type of refuse from any source except that flowing from streets and sewers into any navigable water. Furthermore, the act implies that it is unlawful to discharge refuse into tributaries of navigable water. For all practical purposes, this means that it is against the law to pollute any stream in the United States. However, the Secretary of the Army can, if he wishes, permit the discharge of refuse into a stream if a permit is first applied for.

The Fish and Wildlife Coordination Act of 1958   This act establishes a national policy that recognizes the important contribution of wildlife resources and specifically provides that conservation of wildlife shall be balanced with other factors in water resources development planning (2).

Wildlife resources in the act are broadly defined to include birds, fish, mammals, and all other types of animals, as well as aquatic and land vegetation upon which the wildlife depends. Furthermore, the act provides for coordination of wildlife aspects of resource development and requires that projects to develop power, control flooding, or facilitate navigation must first consult with the U.S. Fish and Wildlife Service, as well as the head of the state agency exercising administrative control of wildlife resources in the particular state where the project is planned. This applies, with the exception of impondments of less than 10 acres, to the waters of any stream or other body of water to be imponded, ditched, diverted, or otherwise modified for any purpose by any department or agency of the United States (2).

The object of the Fish and Wildlife Coordination Act is to prevent damage or loss of wildlife resources due to water resources development and, at the same time, provide for development and improvement of wildlife and necessary habitat. Reports by the Fish and Wildlife Service, along with those of state agencies, provide details of expected damage to wildlife resources by a particular project, and they include recommendations concerning how to reduce the projected damages and develop and improve the fish and wildlife resources. According to the act, agencies receiving the reports must consider these recommendations fully and include in the project plans ways and means by which wildlife conservation is to be achieved (2).

The National Environment Policy Act of 1969   The philosophical purposes of the Environmental Policy Act are to declare a national policy that will encourage harmony between man and his physical environment; to promote efforts that prevent or eliminate environmental degradation, thereby stimulating the health and welfare of man; and to improve our understanding of relations between ecological systems and important natural resources. To promote interest, research, and authority to achieve these purposes, the act establishes the Council on Environmental Quality. The council is in the Executive Office of the President and is responsible for preparation of a yearly Environmental Quality Report to the Nation. It also provides advice and assistance to the president on environmental policies.

Perhaps the most significant aspect of the act is that it requires an environmental-impact statement before major federal actions are taken that could significantly affect the quality of the human environment. This requirement extends to activities such as construction of nuclear facilities, airports, federally assisted highways, electric power plants, and bridges; release of pollutants into navigable waters and their tributaries; and resource development on federal lands, including mining leases, drilling permits, and other uses. By May of 1972, nearly 3000 environmental-impact statements had been filed with the Council on Environmental Quality (Figure 15.4).

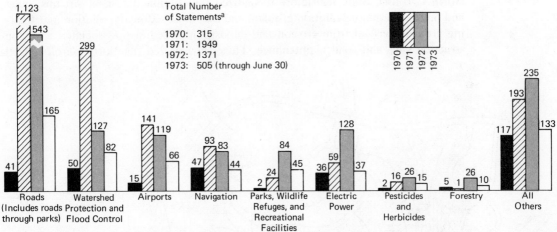

[a] Includes all final statements and draft statements on actions for which a final statement has not yet been filed.

FIGURE 15.4 *Environmental-impact statements filed with the Council on Environmental Quality from 1970 to 1973. (Source: Council on Environmental Quality.)*

The Water Quality Improvement Act of 1970 This act is a comprehensive water-pollution control law that essentially gives more power to the Federal Water and Pollution Control Act of 1956. The purpose of the latter act was to enhance the quality and value of our water resources and to establish a national policy to prevent, control, and abate pollution of the country's water resources. The 1970 Act provides for control of oil pollution by vessels and offshore and onshore oil wells; control of hazardous pollutants other than oil; control of sewage from vessels; research and development methods to control and eliminate pollution of the Great Lakes; research grants to universities and scholarships to students to train people in water quality control; and projects to demonstrate methods to eliminate and control mine drainage of acid water from both active and abandoned mines (2).

The main regulatory function of the Water Quality Improvement Act of 1970 requires that all facilities or activities that involve discharge into navigable water and that require a federal license must obtain a certificate of reasonable assurance that the proposed activity will not violate the state's water quality standards as approved by the Environmental Protection Agency.

State and Local Environmental Legislation   The far-reaching federal legislation program is a good example for state and local governments to follow. In fact, a number of states have already enacted legislation analogous to the federal laws, and it is expected that this trend will continue. Areas of particular public concern at the state and local level are environmental impact, floodway regulation, sediment control, land use, and water and air quality. The most effective programs and laws, excepting federal activity, will probably be at the local level because acceptance and enforcement, when combined with education and communication, are probably most effective where environmental degradation is experienced firsthand.

An example of important state and county legislation comes from North Carolina. State legislators recognized that sedimentation of streams, lakes, and ponds is a major pollution problem and that most sediment pollution in urbanizing areas is derived from erosion and deposition of sediment associated with construction sites and road maintenance. Thus, they passed the North Carolina Sedi-

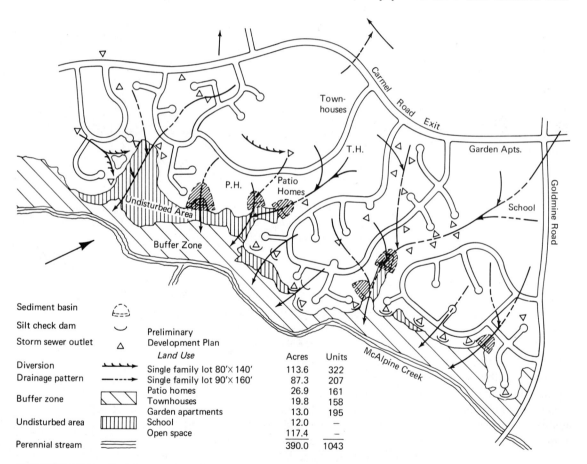

FIGURE 15.5   *Erosion- and sediment-control plan for a housing development including single-family homes, patio homes, townhouses, garden apartments, a school, and open space. (Courtesy of Braxton Williams, Soil Conservation Service.)*

ment Pollution Control Act of 1973. The act establishes that it is vital to the public interest and necessary for public health and welfare that control of erosion and sedimentation be an important goal. Furthermore, the act provides for creation, administration, and enforcement of a sediment-control program complete with minimal mandatory standards that allow future development with minimal detrimental effects from sediment pollution.

The Sediment Pollution Control Act of 1973 in North Carolina recognizes the need for local participation, and it encourages local government to draft ordinances consistent with the act. Mecklenburg County, part of the fourth largest urbanizing area in the southeastern United States, recently drafted a sediment-control ordinance that calls for erosion- and sediment-control plans to be submitted before land-disturbing activities are started, if more than one contiguous acre is to be uncovered. These plans are to be approved by the County Engineer and reviewed by the Soils and Water Conservation District Office for comments and recommendations. An example of such a plan is shown in Figure 15.5.

Communication and education are the keys to success with the sediment-control ordinance. Therefore, a significant number of activities, including urban field trips and workshops on sediment control, are part of an ongoing program. The idea is not to punish developers into sediment control, but to indicate the benefits of sound conservation practices.

Water Law

Water is so necessary to all aspects of human use and interest that water resources may be the most legislated and discussed commodity in the environmental law arena. Struggles for sufficient water by large areas such as New York City and Southern California are examples. Intrastate and legislative contracts were necessary to obtain water for New York City from the Delaware River waters in the Catskill Mountains. In California, there has been a 50-year fight between California and Arizona over Colorado River water, and it is only partially resolved (3).

There will always be problems in allocating water resources, and the problem is most severe in areas such as the southwestern United States where water is scarce (Figures 15.6 and 15.7). California is a good example of the type of conflicts that arise when a large population with accompanying industrial and agricultural activity is concentrated in an area with a natural deficiency of water. Approximately two-thirds of the state's water supply comes from the northern third of California, but the greatest need for water is in the southern two-thirds of the state where the vast majority of people live and where most of the industry and agricultural activity takes place (7).

The deficiency of water in Southern California, combined with an almost insatiable demand for water, resulted in the construction of the California aqueduct to move water from the northern and southeastern parts of the state to the Los Angeles area (Figure 15.8). The legal grounds that allowed the California citizens to vote for and pass a state bond of nearly $2 billion for the California aqueduct which now transports water from the northern part of the state to the southern (Figure 15.9) was derived from the Constitution of the State of California:

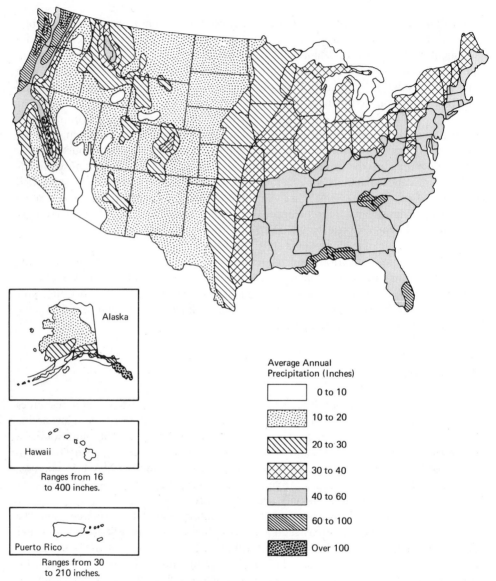

Alaska

Hawaii

Ranges from 16
to 400 inches.

Puerto Rico

Ranges from 30
to 210 inches.

Average Annual
Precipitation (Inches)

0 to 10

10 to 20

20 to 30

30 to 40

40 to 60

60 to 100

Over 100

FIGURE 15.6 *Average annual precipitation for the United States. (Source:* The Nation's Water Resources, *Water Resources Council, 1968.)*

"The general welfare requires that water resources of the state be put to beneficial use to the fullest extent of which they are capable . . . that the conservation of such water is to be exercised with a view to the reasonable and beneficial use thereof in the interest of the people and for the public welfare" (3). From a philosophical viewpoint, a question might be raised—Is the continued development in Southern California warranted? Perhaps the people should be located where the water is

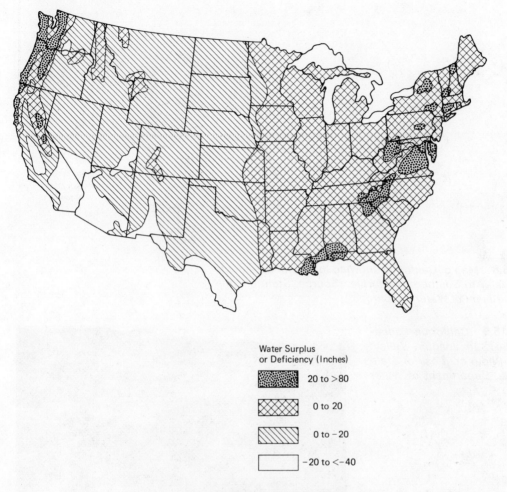

Water Surplus
or Deficiency (Inches)

20 to >80

0 to 20

0 to −20

−20 to <−40

FIGURE 15.7 *Areas of natural water surplus and deficiencies in the United States. (Source:* The Nation's Water Resources, *Water Resources Council, 1968.)*

rather than huge quantities of water moved hundreds of miles over rough terrain and active faults at tremendous costs to an area already overcrowded and suffering from pollution and other environmental problems. Value judgments in these matters have little validity, and the fact that the aqueduct was constructed indicates the price people are willing to pay to support their standard of living and life-style. Furthermore, it should be pointed out that many of the people in the Feather River and Owens River areas from where the water transported to Los Angeles is diverted are not so pleased as those receiving the water.

There is a rather elaborate framework of law surrounding the use of surface water, two major aspects of which are the Riparian Doctrine and the Appropriation Doctrine.

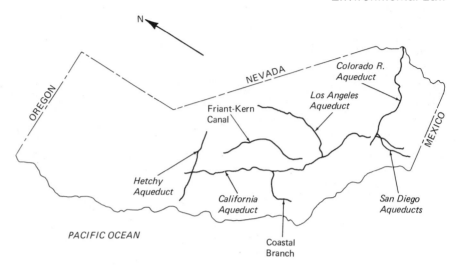

FIGURE 15.8 *Map of California showing the major aqueducts supplying water to Southern California. (Source: State of California, Department of Water Resources.)*

FIGURE 15.9 *California aqueducts in the San Joaquin Valley, California. (Photo courtesy of State of California, Department of Water Resources.)*

Riparian rights to water are restricted to owners of the land adjoining a stream of standing water. The word *Riparian* comes from the Latin word meaning *bank,* and traditional Riparian Doctrine is a common law concept. It essentially holds that each landowner has the right to make reasonable use of water on his lands provided the water is returned to the natural stream channel before it leaves his property. Furthermore, the property owner has the right to receive the full flow of the stream undiminished in quantity and quality. However, he is not entitled to make withdrawals of water that infringe upon the rights of other Riparian owners (8).

The Riparian Doctrine was the water law used by most states before 1850, and it is still used in all the states east of the Mississippi and in the first tier of states immediately west of the river (Figure 15.10). The right to use water is considered as real property, but the water itself does not belong to the property owner. Riparian water rights are considered as natural rights and property that enters into the value of land. They may be transferred, sold, or granted to other people (9).

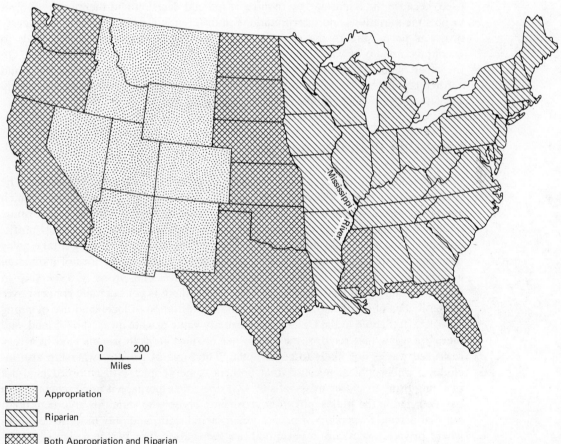

FIGURE 15.10 *Surface water laws for the conterminous United States. (Data from New Mexico Bureau of Mines, Circular 95, 1968.)*

The Appropriation Doctrine in water law holds that prior usage is a significant factor. That is, the first to use the water for beneficial purpose is prior in right. Furthermore, this right is perfected by use and is lost if beneficial use ceases (9).

Appropriation water law is common in the western part of the United States, and, generally, states with the poorest water supply manage their water most closely. Arizona is a good example. With an average precipitation of less than 15 inches per year, Arizona must, of necessity, manage its water very closely. The state constitution states that Riparian water rights are not authorized, and the state's comprehensive water code declares that all water is subject to appropriation. Preferred uses are domestic, municipal, and irrigation. Colorado also has a limited water supply and has declared that all streams are to be considered as public property subject to appropriation (9).

Comparison of the two doctrines suggests that management of water resources is considerably more effective when the principles of appropriation are used, because the Riparian law requires a judicial decision and therefore is subject to possible variations and interpretations in different courts. As a result, a property owner is never sure of his position. Furthermore, the Riparian system tends to encourage nonuse of water and thus is counterproductive in times of shortage. On the other hand, states with an appropriation system have the power to make and enforce regulations based upon sound hydrologic principles, and this is more likely to lead to effective management of water resources (9).

Land-Use Planning and Law

Few environmental topics are as controversial as land-use planning legislation. The controversy is a result of several factors. First, unlike air and water pollution, which can be measured, evaluated, and possibly corrected, it is very difficult to determine the "highest and best use" of the natural environment rather than the "most profitable use." Second, landowners are fearful that land-use planning would take away their right of deciding what to do with their property; that is, the idea of individual ownership of property would be transferred into a social property ownership in which the individual has a caretaker role. Third, there is considerable concern over the basic idea of the federal or state government initiation of local land-use planning.

People in the United States greatly value private ownership of land, and therefore a law that requires the landowner in rural areas to use his land in a very restricted way is not likely to be accepted. The topic of private ownership also includes a philosophical question that bothers some people: Will enforced land-use planning bring us closer to socialism? If it does, then perhaps it might encourage a basic change in the philosophy of government. Those who state this argument point out that private ownership of property is a sacred right, and only in those countries where private ownership is permitted are there also other personal rights. Everything considered, it is extremely unlikely that private ownership in the United States will be abolished. On the other hand, it is increasingly apparent that private ownership of land does not mean that the owner has the right to deliberately or inadvertently degrade the environment. To be effective, land-use planning must leave the property owner certain alternatives from which he is free to choose how to best manage his land.

It seems inevitable that a federal land-use planning act will eventually become law. The purpose of the act will probably be to assure that all the land in the nation is to be used in such a way as to facilitate harmonious existence between man and his natural environment, such that environmental, economic, social, and other requirements of present or future generations are not degraded. Furthermore, the legislation will probably require individual states to establish land-use planning agencies to initiate planning or to approve local government planning.

The argument in favor of a federal land-use planning act stresses that such planning is urgently needed if degradation of land resources of statewide or national importance are to be minimized. Those in favor of such an act point out that land use is the most important aspect of environmental quality control that remains to be stated as national policy (10).

Alternative federal roles to a rigidly controlled land-use planning act might include the following: increased federal money to sponsor continued research and dissemination of information programs in the area of land-use planning; federal funds to underwrite selected experiments in guiding growth and land use; and, finally, if Congress should determine that it is federal responsibility to recommend land-use planning at the state and local levels, then perhaps a special revenue-sharing measure might be appropriate (11).

A number of states are now considering, or have passed, land-use planning legislation. Table 15.1 summarizes land-management programs in the United States. For example, in 1970, the Maine legislature enacted a site law, the objective of which is to maintain the highest and best use of the natural environment. The law was drafted to regulate large developments in such a way that the burden of proof was shifted to the developer; that is, the developer must show that his project will not degrade the land, water, or air. A recent test of the law in court upheld the legislature's policy to regulate land use. This decision was significant because it strongly reaffirmed the right of states to limit property use in such a way as to preserve the environment (12).

In 1970, Vermont also passed a state land-use program that required developers, before they could receive a development permit, to prove to a district environment commision that their project first would not cause undue air or water pollution; second, could obtain sufficient water; third, would not result in excessive soil erosion, highway congestion, or municipal service and educational burden; and, last, would not have undue detrimental effects on the natural scenery, historical sites, or rare and irreplaceable natural areas. In addition, the development would have to conform with existing local or regional land-use plans as well as any subsequently adopted plan (11). The law also required establishment of an environmental board to hear appeals in the permit procedure and to prepare a state land-use plan consisting of a map showing broad categories of the proper use of all the land in the state. In discussing the Vermont law, one author states that the environmental problems that lead to the law (that is, shoddy construction, defective sewage systems, poorly designed roads, etc.) could have been solved at the local level, and that the land-use law is unnecessary and, furthermore, produces many more problems (11). This may be correct, for as the author points out, the Vermont legislature in 1974

TABLE 15.1   *State land-use programs as of September, 1974.*

	Statewide Land-Use Planning and Control[a]	Coastal Zone Management[b]	Wetlands Management[c]	Power-Plant Siting[d]	Surface Mining[e]	Designation of Critical Areas[f]	Land-Use Tax Incentives[g]	Flood-Plain Management[h]
Alabama	P	—	—	—	yes	—	—	—
Alaska	—	—	—	yes	—	—	yes	—
Arizona	P	NA	—	yes	—	—	—	yes
Arkansas	—	NA	—	yes	yes	—	yes	yes
California	P	yes	—	yes	—	—	yes	yes
Colorado	P	NA	—	yes	yes	yes	yes	yes
Connecticut	P	—	yes	yes	—	—	yes	yes
Delaware	P	yes	yes	—	—	—	yes	—
Florida	P and R	yes	yes	yes	—	yes	yes	—
Georgia	P	—	yes	—	yes	—	—	—
Hawaii	P and R	yes	—	yes	—	yes	yes	yes
Idaho	—	NA	—	—	yes	—	—	—
Illinois	—	—	—	yes	yes	—	yes	—
Indiana	—	—	—	—	yes	—	yes	yes
Iowa	—	NA	—	—	yes	—	yes	yes
Kansas	—	NA	—	—	yes	—	—	—
Kentucky	—	NA	—	yes	yes	—	yes	—
Louisiana	—	—	yes	—	—	—	yes	—
Maine	P and R	yes	—	yes	yes	yes	yes	yes
Maryland	P and R	—	yes	yes	yes	—	yes	yes
Massachusetts	—	—	yes	yes	—	—	—	—
Michigan	P	yes	—	—	yes	—	—	yes
Minnesota	P and R	yes	yes	yes	yes	yes	yes	yes
Mississippi	—	yes	—	—	—	—	—	—
Missouri	—	NA	—	—	yes	—	—	—
Montana	—	NA	—	yes	yes	—	yes	yes

SOURCE: The Council of State Governors, Lexington, Kentucky.
NOTES: Indications that a state has a program in one of the above categories does not constitue an evaluation of the effectiveness of the program; nor does it indicate that the program is based on specific enabling legislation.
NA means not applicable.
[a]P indicates that the state has a land-use planning program under way. R indicates that the state has authority to review local plans or has direct control.
[b]The state has authority to plan or review local plans or the ability to control land use in the coastal zone.
[c]The state has authority to plan or review local plans or the ability to control land use in the wetlands.
[d]The state has authority to determine the siting of power plants and related facilities.
[e]The state has authority to regulate surface mining.

stripped the environmental board of responsibility to develop a state land-use plan and created a land-use study commission to pursue the matter. The law might have worked better had it involved more local planning at the county and city level; that is, local planning tends to be more detailed than a broad-brush state plan that may be difficult to interpret.

California citizens concerned about coastal areas (Figure 15.11) voted in 1972 to require a permit for any development taking place in a zone extending

TABLE 15.1, continued

	Statewide Land-Use Planning and Control[a]	Coastal Zone Management[b]	Wetlands Management[c]	Power-Plant Siting[d]	Surface Mining[e]	Designation of Critical Areas[f]	Land-Use Tax Incentives[g]	Flood-Plain Management[h]
Nebraska	—	NA	—	yes	—	yes	yes	yes
Nevada	P and R	NA	—	yes	—	—	—	—
New Hampshire	—	—	yes	yes	—	—	yes	—
New Jersey	—	—	yes	yes[i]	—	—	yes	yes
New Mexico	—	NA	—	yes	yes	—	yes	—
New York	P	—	yes	yes	yes	yes	yes	—
North Carolina	P	yes	yes	—	yes	—	yes	yes
North Dakota	—	NA	—	—	yes	—	yes	—
Ohio	—	—	—	yes	yes	—	yes	—
Oklahoma	—	NA	—	—	yes	—	—	yes
Oregon	P and R	yes[j]	—	yes	yes	yes	yes	—
Pennsylvania	P	—	—	yes	yes	—	yes	—
Rhode Island	P	yes	yes	yes	—	—	yes	—
South Carolina	—	—	—	yes	yes	—	—	—
South Dakota	—	NA	—	—	yes	—	yes	—
Tennessee	—	NA	—	yes[j]	yes	—	—	—
Texas	—	yes	—	—	—	—	yes	—
Utah	—	NA	—	—	—	yes	yes	—
Vermont	P and R	NA	yes	yes	—	—	yes	yes
Virginia	—	yes	yes	—	yes	—	yes	—
Washington	—	yes	yes	yes	yes	—	yes	—
West Virginia	—	NA	—	—	yes	—	—	—
Wisconsin	P	yes	yes	—	yes	yes	yes	yes
Wyoming	—	NA	—	—	yes	—	yes	—
Guam	P and R	yes	—	yes	yes	—	—	yes
Puerto Rico	P and R	—	—	yes	yes	—	—	yes

[f]The state has established, or is in the process of establishing, rules, regulations, and guidelines for the identification and designation of areas of critical state concern (for example, environmentally fragile areas and areas of historical significance).

[g]The state has adopted tax inducements to withhold or delay development of open space (for example, a tax on present use, a rollback penalty, or a contract between the state and landowners to provide preferential tax for commitment to open-space usage).

[h]The state has authority to regulate the use of floodplains.

[i]Only in coastal zone areas.

[j]Tennessee Valley Authority.

from 1000 yards inland from the sea's high tide to 3 miles seaward. The law also calls for a master plan for coastal development to be prepared and adopted as state policy.

North Carolina, recognizing the importance of planning for a fragile environment likely to be heavily developed, passed the Coastal Area Management Act of 1974. This act recognizes the need to preserve for the people an opportunity to enjoy the aesthetic, cultural, and recreational quality of the natural coastlines. The first two objectives of the act are to provide a management system that allows

FIGURE 15.11 *Citizens concerned for coastal areas prompted legislation calling for planning in the coastal environment of California extending from 1000 yards inland. This oblique aerial photograph is of Morro Bay, California. (Photo courtesy of the California Division of Beaches and Parks.)*

the preservation of the estuaries, Barrier Islands, sand dunes, and beaches such that their natural productivity and biological, economic, and aesthetic value are safeguarded; and to insure that development in the coastal area does not exceed the capabilities of the land and water resources.

The North Carolina act establishes a cooperative program of coastal management between local and state governments such that local government has the initiative in the planning. The role of state government is to set standards and to review and support local government in its planning program. The planning has three main steps: development of state planning guidelines for coastal areas; development and adoption of a land-use plan for each county in the coastal area; and use of plans as criteria for issuing or denying permits to develop land or water resources in the coastal area.

SUMMARY AND CONCLUSIONS

A land ethic is emerging in our national environmental policy. Productive and enjoyable harmony between man and his environment is now encouraged, and legal theories based upon the Trust Doctrine or the Ninth Amendment are being used to argue in court for the right of this generation and future generations to an unspoiled land.

The term *environmental law* has only recently gained common usage, but the subject is becoming increasingly popular among lawyers and promises to be more significant in the future. Topics of particular concern to individuals and society are degradation of air, water, scenery, and other natural resources.

Environmental legislation is having a tremendous impact on the industrial community. In addition to regulations controlling emissions of possible pollutants, perhaps the most significant and far-reaching piece of environmental legislation is the National Environmental Policy Act of 1969. The act requires that before any environment-affecting activity that is directly or indirectly involved with federal government can begin, a statement evaluating the environmental impact must be completed. Other important legislation includes the Refuse Act of 1899, the Fish and Wildlife Coordination Act of 1958, and the Water Quality Improvement Act of 1970. Many states are following the federal government's leadership and are developing environmental and land-management legislation.

Areas of particular concern at the state level are environmental impact, floodway regulation, sediment control, land use, and air and water quality.

Water law remains a significant issue and is of particular importance in regions with deficiencies of water. Eastern states generally have what is known as *Riparian rights,* whereby owners of land that adjoins water have the right to reasonable use. In the western states, water law is generally governed by the Appropriation Doctrine, which holds that prior beneficial usage is the key to water rights. In some states, such as Arizona and Colorado, where water is especially valuable, all water is appropriated on the basis of prior and preferred uses. In comparing the two systems, one can conclude that appropriation of water is superior because it leads to better management of water resources.

Perhaps the most controversial and potentially significant of all environmental legislation is land-use planning. Conflicts result because it is often difficult to determine the highest and best use of land. Also, landowners are fearful that land-use planning will cause property owners to lose the right to control their property, and there is considerable concern over the basic idea that federal and state governments should initiate local planning.

The argument in favor of land-use planning legislation stresses that planning is urgently needed if degradation of the land is to be controlled, and that land use is the remaining important aspect in environmental quality control that is not stated as national policy.

The significant effect of land-management legislation is that the burden of proof that an activity will not degrade the land is shifted to the developer. Furthermore, we will have to live with the concept that the most profitable use of land is not always necessarily the best use. This comes directly from the Trust Doctrine and an emerging land ethic.

References

1. Landan, N. J., and Rheingold, P. D. 1971. *The environmental law handbook*. New York: Ballantine Books.
2. Yannacone, V. J., Jr.; Cohen, B. S.; and Davison, S. G. 1972. *Environmental rights and remedies* 1. San Francisco: Bancroft-Whitney.
3. Coates, D. R. 1971. Legal and environmental case studies in applied geomorphology. In *Environmental Geomorphology*, ed. D. R. Coates, pp. 223–42. Binghamton, New York: State University of New York.

4. Juergensmeyer, J. C. 1970. Control of air pollution through the assertion of private rights. *Environmental Law:* 17–46. Greenvale, New York: Research and Documentation Corporation.

5. Murphy, E. F. 1971. *Man and his environment: law.* New York: Harper & Row.

6. Carter, L. J. 1974. Con Edison: endless Storm King dispute adds to its troubles. *Science* 184: 1353–58.

7. Cargo, D. N., and Mallory, B. F. 1974. *Man and his geologic environment.* Menlo Park, California: Addison-Wesley.

8. Private Remedies for Water Pollution. 1970. *Environmental Law:* 47–69. Greenvale, New York: Research and Documentation Corporation.

9. Legal Approach to Water Rights. 1972. In *Water quality in a stressed environment,* ed. W. A. Pettyjohn, pp. 255–76. Minneapolis: Burgess.

10. Healy, M. R. 1974. National land use proposal: land use legislation of landmark environmental significance. *Environmental Affairs* 3: 355–95.

11. McClaughry, J. 1974. The land use planning act—an idea we can live without. *Environmental Affairs* 3: 595–626.

12. Wainwright, J. K., Jr. 1974. Spring Valley: public purpose and land use regulation in a "taking" context. *Environmental Affairs* 3: 327–54.

APPENDIX

Common Conversion Factors
(Computed to Two Significant Figures)

LENGTH
1 yard = 3 ft, 1 fathom = 6 ft, 1 rod = 16.5 ft

	in.	ft	mi	cm	m	km
1 inch (in.) =	—	0.08	—	2.54	0.03	—
1 foot (ft) =	12	—	—	30.48	0.30	—
1 mile (mi) =	—	5280	—	—	1609	1.61
1 centimeter (cm) =	0.39	—	—	—	0.01	—
1 meter (m) =	39.37	3.28	—	100	—	—
1 kilometer (km) =	—	3281	0.62	—	1000	—

AREA
1 square mi = 640 acres, 1 acre = 43,560 ft^2

	in.^{2}	ft^2	mi^2	cm^2	m^2	km^2
1 in.^{2} =	—	—	—	6.45	—	—
1 ft^2 =	144	—	—	929	0.09	—
1 mi^2 =	—	—	—	—	—	2.59
1 cm^2 =	0.16	—	—	—	—	—
1 m^2 =	1550	10.76	—	10,000	—	—
1 km^2 =	—	—	0.39	—	1,000,000	—

VOLUME

	in.^{3}	ft^3	yd^3	m^3	qt	liter	barrel	gal
1 in.^{3} =	—	—	—	—	—	0.02	—	—
1 ft^3 =	1728	—	—	0.03	—	28.3	—	7.48
1 yd^3 =	—	27	—	0.76	—	—	—	—
1 m^3 =	61,020	35.31	1.3	—	—	1000	—	—
1 quart (qt) =	—	—	—	—	—	0.95	—	0.25
1 liter (ℓ) =	61.02	—	—	—	1.06	—	—	0.26
1 barrel (oil)	—	—	—	—	168	159.6	—	42
1 gallon (U.S.) =	231	0.13	—	—	4	3.8	0.02	—

MASS AND WEIGHT

1 pound = 453.6 grams = 0.45 kilogram = 16 ounces

1 short ton = 2000 pounds = 900 kilograms

1 long ton = 2240 pounds = 1008 kilograms

1 metric ton = 2205 pounds = 1000 kilograms

ENERGY AND POWER

1 kilowatt-hour = 3413 Btus = 860,421 calories

1 Btu = 0.000293 kilowatt-hour = 252 calories

1 watt = 3.413 Btu/hr = 14.34 calories/min

1 calorie = the amount of heat necessary to raise the temperature of 1 gram (1 cm³) of water 1 degree Celsius

TEMPERATURE

$F = \dfrac{9}{5} C + 32$ F is degrees Fahrenheit.
C is degrees Celsius (centigrade).

Fahrenheit		Celsius
	Freezing of H_2O	
32	(Atmospheric Pressure)	0
50		10
68		20
86		30
104		40
122		50
140		60
158		70
176		80
194		90
212	Boiling of H_2O (Atmospheric Pressure)	100

OTHER CONVERSION FACTORS

1 ft³/sec = 0.03 m³/sec = 7.48 gal/sec

1 acre-foot = 43,560 ft³ = 1233 m³ = 325,829 gal

Strength of Common Rock Types

	Rock Type	Range of Compressive Strength (Thousands of psi)	Comments
Igneous	Granite	23 to 42.6	Finer-grained granites with few fractures are the strongest. Granite is generally suitable for most engineering purposes.
Igneous	Basalt	11.8 to 52.0	Brecciated zones, open tubes, or fractures reduce the strength.
Metamorphic	Marble	6.7 to 34.5	Solutional openings and fractures weaken the rock.
Metamorphic	Gneiss	22.2 to 36.4	Generally suitable for most engineering purposes.
Metamorphic	Quartzite	21.1 to 91.2	Very strong rock.
Sedimentary	Shale	Less than 1 to 33.5	May be a very weak rock for engineering purposes; careful evaluation is necessary.
Sedimentary	Limestone	5.3 to 37.6	May have clay partings, solution openings, or fractures that weaken the rock.
Sedimentary	Sandstone	4.8 to 34.1	Strength varies with degree and type of cementing material, mineralogy, and nature and extent of fractures.

SOURCE: Data primarily from *Handbook of Tables for Applied Engineering Science,* eds. R. E. Bolz and G. L. Tuve (Cleveland, Ohio: CRC Press, 1973.)

Geologic Time Scale

Era	Approx. Age in Millions of Years (Radio-activity)	Period (System)[a]
Cenozoic		Recent (Holocene)
		Pleistocene
		} Man ? Neogene
	13	Pliocene
	25	Miocene }
	36	Tertiary { Oligocene
	58	Eocene } Paleogene
		Paleocene } Mammals
	65	
Mesozoic	135	Cretaceous
	181	Jurassic · Flying reptiles · First bird
		Triassic · Dinosaurs
	220–230	
	280	Permian · First reptiles
		Pennsylvanian } Carboniferous
		Mississippian }
Paleozoic	345	Devonian · First amphibian · First insect fossils
	405	Silurian · First land plant fossils
	425	Ordovician · First vertebrate fossil fish
	500	Cambrian
	570–600	
	700	First multi-celled organisms
	3400	First one-celled organisms
Precambrian	4000	Approximate age of oldest rocks discovered
	4500	Approximate age of the earth and meteorites

[a]*Period* refers to a time measure; *system* refers to the rocks deposited during a period.

GLOSSARY

A soil horizon   Uppermost soil horizon sometimes referred to as the **zone of leaching.**

Absorption   Taking up, incorporating, or assimilating of a material.

Adsorption   Attachment of molecules of gas or molecules in solution to the surface of solid materials with which they come in contact.

Aerobic   Characterized by the presence of free oxygen.

Aesthetics   Originally a branch of philosophy; defined today by artists and art critics.

Age element   Elements that characteristically accumulate in tissue with age.

Aggregate   Any hard material such as crushed rock, sand, gravel, or other material that is added to cement to make concrete.

Alkaline soils   Soils in arid regions that contain a large amount of soluble mineral salts (primarily of sodium) which in the dry season may appear on the surface as a crust or powder.

Alluvium   Unconsolidated sediments, including sand, gravel, and silt, deposited by streams.

Anaerobic   Characterized by the absence of free oxygen.

Angle of repose   The maximum angle that loose material will sustain.

Anhydrite   Evaporite mineral ($CaSO_4$) calcium sulfate.

Anthracite   A type of coal characterized by a high percentage of carbon and low percentage of volatiles, providing a high heat value. Anthracite often forms as a result of metamorphism of bituminous coal.

Anticline   A type of fold characterized by an upfold or an arch. The oldest rocks are found in the center of the fold.

Appropriation doctrine, water law   Holds that prior usage of water is a significant factor. The first to use the water for beneficial purposes is prior in right.

Aquiclude   Earth material that retards the flow of groundwater.

Aquifer   Earth material containing sufficient groundwater that the water can be pumped out. Highly fractured rocks and unconsolidated sands and gravels make good aquifers.

Area (strip) mining   Type of strip mining practiced on relatively flat areas.

Artesian   Refering to a groundwater system in which the groundwater is isolated from the surface by a confining layer and the water is under pressure. Groundwater that is under sufficient pressure that it will flow freely at the surface from a spring or a well.

Asbestos   Fibrous mineral material used as insulation. It is suspected of being either a true carcinogen or a carrier of carcinogenic trace elements.

Ash fall   Volcanic ash eruption that blows up into the atmosphere and then rains down on the landscape.

Ash flow   Mixture of volcanic ash, hot gases, and fragments of rock and glass that flows rapidly down the flank of a volcano. May be an extremely hazardous event.

Ash, volcanic   Unconsolidated volcanic debris, less than 4 mm in diameter, physically blown out of a volcano during an eruption.

Avalanche   A type of landslide involving a large mass of snow, ice, and rock debris that slides, flows, or falls rapidly down a mountainside.

Azonal soil   Recent surface materials such as floodplain deposits that do not have distinctive soil layering.

B soil horizon   Intermediate soil horizon sometimes known as the **zone of accumulation.**

Balancing doctrine   Asserts that public benefits and the importance of a particular action should be balanced with the potential injury to certain individuals. The method of balancing is changing now, and courts are considering possible long-range injury caused by certain activities to large numbers of citizens other than the immediate complainants.

Barrier island   Island separated from the mainland by a salt marsh. It generally consists of a multiple system of beach ridges and is separated from other barrier islands by inlets that allow the exchange of seawater with lagoon water.

Basalt   A fine-grained extrusive igneous rock. It is one of the most common igneous rock types.

Basaltic   Engineering geology term for all fine-grained igneous rocks.

Bauxite   A rock composed almost entirely of hydrous aluminum oxides. It is a common ore of aluminum.

Bed material   Sediment transported and deposited along the bed of a stream channel.

Bedding plane   The plane that delineates the layers of sedimentary rocks.

Bentonite   A type of clay that is extremely unstable. Upon wetting, it expands to many times its original volume.

Biosphere   The zone adjacent to the surface of the earth that includes all living organisms.

Biotite   A common ferromagnesian mineral, a member of the mica family.

Bituminous coal   Very common type of coal characterized by relatively high carbon content and low volatiles. Sometimes called **soft coal.**

Blowout   Failure of an oil, gas, or disposal well resulting from adverse pressures that may physically blow part of the well casing upward. May be associated with leaks of oil, gas, or, in the case of disposal wells, harmful chemicals.

BOD   Biological oxygen demand, a measure of the amount of oxygen necessary to decompose organic materials in a unit volume of water. As the amount of organic waste in water increases, more oxygen is used, resulting in a higher BOD.

Braided river   A river channel characterized by an abundance of islands that continually divide and subdivide the flow of the river.

Breccia   A rock or zone within a rock composed of angular fragments. Sedimentary, volcanic, and tectonic breccias are recognized.

Breeder reactor   A type of nuclear reactor that actually produces more fissionable (fuel) material than it uses.

Brine   Water that has a high concentration of salt.

British thermal unit (Btu)   A unit of heat defined as the heat required to raise the temperature of one pound of water one degree Fahrenheit.

Brittle   Material that ruptures before any plastic deformation.

Bulk Element   Common elements that make up the bulk of living material.

C soil horizon   Lowest soil horizon sometimes known as the **zone of partially altered parent material.**

Cadmium   A metallic element having an atomic number of 48 and an atomic weight of 112.4 As a trace element, it has been associated with serious health problems.

Calcite   Calcium carbonate ($CaCO_3OH$). Common carbonate mineral that is the major constituent of the rock limestone. It weathers readily by solutional processes, and large cavities and open weathered fractures are common in rocks containing the mineral calcite.

Caliche   A white to gray irregular accumulation of calcium carbonate in soils of arid regions.

Calorie   The quantity of heat required to raise the temperature of one gram of water from 14.5 to 15.5 degrees Celsius.

Capillary action   The rise of water along narrow passages, facilitated and caused by surface tension.

Carbonate   A compound or mineral containing the radical (CO_3^{--}). The common carbonate is calcite.

Carcinogen   Any material that is known to produce cancer in man or other animals.

Cesium-137   A fission product produced from nuclear reactors and having a half-life of 33 years.

Channelization   An engineering technique to straighten, widen, deepen, or otherwise modify a natural stream channel.

Circum-Pacific belt   One of the three major zones where earthquakes occur. This belt is essentially the Pacific plate. It is also known as the **ring of fire,** as many active volcanoes are found on the edge of the Pacific plate.

Clay   May refer to a mineral family or to a very fine-grained sediment. It is associated with a good deal of environmental problems such as shrinking and swelling of soils and sediment pollution.

Coal   A sedimentary rock formed from plant material that has been buried, compressed, and changed.

Columnar jointing   System of fractures (joints) that break rock into polygons typically with five or six sides. These polygons form columns. This type of fracturing is common in basalt and is most likely caused by shrinking during cooling of the lava.

Common excavation   Excavation that can be accomplished with an earth mover, a backhoe, or a dragline.

Composite volcano   Steep-sided volcanic cone produced by alternating layers of pryoclastic debris and lava flows. It characteristically has magma that is intermediate in silica content (60 percent).

Conchoidal fracture   A shell-like or fan-shaped fracture characteristic of the mineral quartz and natural glass.

Cone of depression   A cone-shaped depression in the water table caused by withdrawal of water table at rates greater than the rates at which the water can be replenished by natural groundwater flow.

Conglomerate   A detrital sedimentary rock composed of rounded fragments, 10 percent of which are larger than 2 mm in diameter.

Connate water   Water that is no longer in circulation or in contact with the present water cycle. Generally saline water trapped during the deposition of sediments.

Contact metamorphism   Type of metamorphism produced when country rocks are in close contact with a cooling body of magma below the surface of the earth.

Continental drift   The movement of continents in response to sea-floor spreading. The most recent episode of continental drift supposedly started about 200 million years ago with the breakup of the supercontinent Pangaea.

Continental shelf   Relatively shallow ocean area between the shoreline and the continental slope that extends to approximately a 600-foot water depth surrounding a continent.

Contour (strip) mining   Type of strip mining used in hilly terrain.

Convection   The transfer of heat involving the movement of particles; for example, the boiling of water in which hot water rises to the surface and displaces cooler water which moves toward the bottom.

Convergent plate boundary   Boundary between two lithospheric plates in which one plate descends below the other (subduction).

Corrosion   A slow chemical weathering or chemical decomposition that proceeds from the surface in. Objects such as pipes experience corrosion when buried in soil.

Cost-benefits analysis   A type of site selection that compares benefits and costs of a particular project. The most desirable projects are those for which the benefits-to-cost ratio is greater than one.

Creep   A type of downslope movement characterized by slow flowing sliding, or slipping of soil and other earth materials.

Crystal settling   A sinking of earlier-formed crystals to the bottom of a magma chamber.

Crystalline   A material with a definite internal structure such that the atoms are in an orderly, repeating arrangement.

Crystallization   Processes of crystal formation.

Debris flow   Rapid downslope movement of earth material often involving saturated, unconsolidated material that has become unstable due to torrential rainfall.

Deep-well disposal   Method of waste disposal that involves pumping waste into subsurface disposal sites such as fractured or otherwise porous rocks.

Detrital   Mineral and rock fragments derived from preexisting rocks.

Diamond   A very hard mineral composed of the element carbon.

Dilatancy of rocks   Inelastic increase in volume of a rock that starts after the stress on the rock has reached one-half of the breaking strength of the rock.

Discharge   The quantity of water flowing past a particular point on a stream. Generally is measured in cubic feet per second (cfs).

Disseminated mineral deposit   A mineral deposit in which ore is scattered throughout the rocks. Examples include diamonds in kimberlite and many copper deposits.

Divergent plate boundary   Boundary between lithospheric plates characterized by the production of new lithosphere. Found along oceanic ridges.

Dose dependency   Relates to the fact that effects of a certain trace element on a particular organism depend upon the dose or concentration of the element.

Drainage basin   The area that contributes surface water to a particular stream network.

Drainage net   The system of stream channels that coalesce to form a stream system.

Dredge spoils   Solid material, such as sand, silt, clay, rock or other material deposited from industrial and municipal discharges, that are removed from the bottom of water bodies to improve navigation.

Driving forces   Those forces that tend to make earth material slide.

Ductile   Material that ruptures following elastic and plastic deformation.

Earthquake   Natural shaking or vibrating of the earth in response to the breaking of rocks along faults. The earthquake zones of the earth generally correlate very well with lithospheric plate boundaries.

Ecology   Branch of biology that treats relationships between organisms and their environments.

Economic geology   The application of geology to locating and evaluating mineral materials.

Effluent   Any material that flows outward from something. Examples include waste water from hydroelectric plants and water discharged into streams from waste-disposal sites.

Effluent stream   Type of stream where flow is maintained during the dry season by groundwater seepage into the channel.

Elastic deformation   Type of deformation where the material returns to its original shape after the stress is removed.

Engineering geology   The application of geologic information to engineering problems.

Environment   That which surrounds an individual or a community; both physical and cultural surroundings. **Environment** is also sometimes used to denote a certain set of circumstances surrounding a particular occurrence, for example, environments of deposition.

Environmental geology   The application of geologic information to environmental problems.

Environmental geology map   A map that combines geologic and hydrologic data expressed in nontechnical terms to facilitate a general understanding by a large audience.

Environmental-impact statement   A written statement that assesses and explores possible impacts associated with a particular project which may affect the human environment. The statement is required by the National Environmental Policy Act of 1969.

Environmental law   A field of law that is growing rapidly and becoming a significant part of our jurisprudence.

Environmental resource unit (ERU)   A portion of the environment with a similar set of physical and biological characteristics. Supposedly a natural division characterized by specific patterns or assemblages of structural components such as rocks, soils, vegetation, etc., and natural processes such as erosion, runoff, soil processes, etc.

Ephemeral   Temporary or very short-lived. Characteristic of beaches, lakes, and some stream channels that change rapidly (geologically).

Epicenter   The point on the surface of the earth directly above the focus (area of first motion) of an earthquake.

Evaporite   Sediments deposited from water as a result of extensive evaporation of seawater or lake water. Dissolved materials left behind following evaporation.

Exponential growth   A type of compound growth in which a total amount or number increases at a certain percentage per year, and each year's rate of growth is added to the total from the previous year. Characteristically stated in terms of a particular doubling time, that is, the time in years it will take the original number to double. Commonly used in reference to population growth.

Extrusive igneous rocks   Igneous rock that forms when magma reaches the surface of the earth. A volcanic rock.

Fault   A fracture or fracture system that has experienced movement along opposite sides of the fracture.

Feldspar   The most abundant family of minerals in the crust of the earth. They are silicates of calcium, sodium, and potassium.

Ferromagnesian mineral   Minerals containing iron and magnesium. These minerals are characteristically dark in color.

Fertile materials   Materials such as uranium-238, which is not naturally fissionable but upon bombardment by neutrons is converted to plutonium-239, which is fissionable.

Fission   The splitting of an atom into smaller fragments with the release of energy.

Floodplain   Flat topography adjacent to a stream in a river valley that has been produced by the combination of overbank flow and lateral migration of meander bends.

Floodway district   That portion of a channel and floodplain of a stream designated to provide passage of the 100-year regulatory flood without increasing the elevation of the flood by more than one foot.

Floodway fringe district   Land located between the floodway district and the maximum elevation subject to flooding by the 100-year regulatory flood.

Fluorine   Important trace element essential for nutrition.

Fluvial   Concerning or pertaining to rivers.

Fly ash   Very fine particles (ash) resulting from the burning of fuels such as coal.

Focus, earthquake   The point in the earth where an earthquake originates.

Folds   Bends that develop in stratified rocks due to tectonic forces.

Foliation   Property of metamorphic rock characterized by parallel alignment of the platy or elongated mineral grains. Foliation is environmentally important because it can affect the strength and hydrologic properties of rock.

Formation   Any rock unit that can be mapped.

Fossil fuels   Fuels such as coal, oil, and gas that have formed by the alteration and decomposition of plants and animals from a previous geologic time.

Fracture zone   A fracture system that may or may not be active and may or may not have an alteration zone along the fracture planes. Fracture zones are environmentally important because they greatly affect the strength of rocks.

Fumarole   A natural vent from which fumes or vapors are emitted; for example, geysers and hot springs characteristic of volcanic areas.

Fusion, nuclear   Combining of light elements to form heavy elements with the release of energy.

Gaging station   Location at a stream channel where the discharge of water is measured.

Gasification   Method of producing gas from coal.

Geochemical cycle   Migratory paths of elements during geologic changes and processes.

Geologic cycle   A group of interrelated cycles known as the hydrologic, rock, tectonic, and geochemical cycles.

Geomorphology   The study of landforms and surface processes.

Geothermal energy   The useful conversion of natural heat from the interior of the earth.

Glacier   A landbound mass of moving ice.

Gneiss   A coarse-grained, foliated metamorphic rock in which there is banding of the light and dark minerals.

Gravel   Unconsolidated, generally rounded fragments of rocks and minerals greater than 2 mm in diameter.

Groin   A structure designed to protect shorelines and trap sediment in the zone of littoral drift. The structure is generally constructed perpendicular to the shoreline.

Groundwater   Water found beneath the surface of the earth within the zone of saturation.

Grout   A mixture of cement and sediment that is sufficiently fluid to be pumped into open fissures or cracks in rocks, thereby increasing the strength of a foundation for an engineering structure.

Gypsum   An evaporite mineral, $CaSO_4 \cdot 2H_2O$.

Half-life   The amount of time necessary for one-half of the atoms of a particular radioactive element to decay.

Halite   A common mineral, NaCL (salt).

Hematite   An important ore of iron, a mineral (Fe_2O_3).

High-value resource   Materials such as diamonds, copper, gold, and aluminum. These materials are extracted wherever they are found and transported around the world to numerous markets.

Humus   Black organic material in soil.

Hydrocarbon   Organic compounds consisting of carbon and hydrogen.

Hydroconsolidation   Consolidation of earth materials upon wetting.

Hydrofracturing   Pumping of water under high pressure into subsurface rocks to fracture the rocks, thereby increasing their permeability.

Hydrograph   A graph of the discharge of a stream with time.

Hydrologic cycle   Circulation of water from the oceans to the atmosphere and back to the oceans by way of evaporation, runoff from streams and rivers, and groundwater flow.

Hydrology   The study of surface and subsurface water.

Hydrothermal ore deposit   A mineral deposit derived from hot water solutions of magmatic origin.

Igneous rocks   Rocks formed from the solidification of magma. They are extrusive if they crystallize on the surface of the earth and intrusive if they crystallize beneath the surface.

Impermeable   Earth materials that retard greatly or prevent the movement of fluids through them.

Infiltration   The movement of surface water into rocks or soil.

Influent stream   Type of stream that is everywhere above the groundwater table and flows in direct response to precipitation. Water from the channel moves down to the water table, forming a recharge mound.

Intrazonal soil   Soil characteristically determined by local conditions such as excess water due to flooding or high groundwater table.

Island arc   A curved group of volcanic islands associated with a deep-oceanic trench and subduction zone (convergent plate boundary).

Itaiitai disease   Extremely painful disease that attacks bones, causing them to become very brittle so that they break easily. Associated with heavy metals, especially cadmium, in concentrations of a few parts per million in the soil or in food consumed by the victims of the disease.

Juvenile water   Water derived from the interior of the earth that has not previously existed as atmospheric or surface water.

Karst topography   A type of topography characterized by the presence of sinkholes, caverns, and diversion of surface water to subterranean routes.

Kimberlite pipe   An igneous intrusive body that may contain diamond crystals disseminated (scattered) throughout the rock type.

Land ethic   Ethic that affirms the right of all resources, including plants, animals, and earth materials, to continued existence and, at least in some locations, continued existence in a natural state.

Landslide   Specifically, rapid downslope movement of rock and/or soil. Also used as a general term for all types of downslope movement.

Land-use planning   Complex process involving development of a land-use plan to include a statement of land-use issues, goals, and objectives; a summary of data collection and analysis; a land-classification map; and a report that describes and indicates appropriate development in areas of special environmental concern. An extremely controversial issue.

Laterite   A soil formed from intense chemical weathering in tropical or savanna regions.

Lava   Molten material that is produced from a volcanic eruption, or a rock that forms from solidification of molten material.

Leachate   Obnoxious liquid material capable of carrying bacteria produced as surface water or groundwater comes into contact with solid waste.

Leaching   Process of dissolving, washing, or draining earth materials by percolation of groundwater or other liquids.

Lignite   A type of low-grade coal.

Limestone   A sedimentary rock composed almost entirely of the mineral calcite.

Limonite   Rust; hydrated iron oxide.

Liquefaction   Extracting oil from coal.

Lithosphere   Outer layer of the earth approximately 100 kilometers thick of which the plates that contain the ocean basins and the continents are composed.

Littoral   Pertaining to the near-shore and beach environments.

Loess   Angular deposits of windblown silt.

Low-value resource   Resources such as sand and gravel that have primarily a place value. They are extracted economically because they are located close to where they are to be used.

Magma   A naturally occurring silica melt, a good deal of which is in a liquid state.

Magma tap   The attempt to recover geothermal heat directly from magma. Feasibility of such heat extraction is unknown.

Magnetite   A mineral and important ore of iron, Fe_3O_4.

Magnitude, earthquake   A number on a logarithmic scale which refers to the amount of energy released by an earthquake.

Manganese oxide nodule   Nodules of manganese, iron with secondary copper, nickel, and cobalt, which cover vast areas of the deep-ocean floor.

Marble   Metamorphosed limestone.

Marl   Unconsolidated clays, silts, sands, or mixtures of these materials that contain a variable content of calcareous material.

Meanders   Bends in a stream channel that migrate back and forth across the floodplain, depositing sediment on the inside of the bends, forming point bars, and eroding the outsides of bends.

Meteoric water   Water derived from the atmosphere.

Methane   A gas, CH_4, the major constituent of natural gas.

Mica   A common rock-forming silicate mineral.

Mudflow   A mixture of unconsolidated materials and water that flows rapidly downslope or down a channel.

Myth of superabundance   The myth that assumes that land and water resources are inexhaustible and therefore management of resources is unnecessary.

National Environmental Policy Act of 1969 (NEPA)   Act declaring that it is national policy that harmony between man and his physical environment be encouraged. Established the Council on Environmental Quality. Established requirements that an environmental-impact statement be completed prior to major federal actions significantly affecting the quality of the human environment.

Neutron   A subatomic particle having no electric charge and found in the nuclei of atoms. Neutrons are crucial in sustaining nuclear fission in a reactor.

Nonrenewable resource   A resource that is cycled so slowly by natural earth processes that once used, it is essentially not going to be made available in any useful time framework.

Nuclear reactor   A device in which controlled nuclear fission is maintained. The major component of a nuclear power plant.

Oil shale   An organic-rich shale containing substantial quantities of oil that may be extracted by conventional methods of destructive distillation.

Ore   An earth material from which a useful commodity can be extracted profitably.

Osteoporosis   Disease characterized by a reduction in bone mass.

Outcrop   A naturally occurring or man-caused exposure of rock at the surface of the earth.

Overburden   Earth materials (spoil) that overlie an ore deposit. Most frequently refers to material overlying or extracted from a surface (strip) mine.

Oxidation   Chemical process of combining with oxygen.

P wave   One of the seismic waves produced by an earthquake. The fastest of the seismic waves, it can move through liquid and solid materials.

Pathogen   Any material that may cause disease; for example, microorganisms, including bacteria and fungi.

Pebble   A rock fragment between 4 and 64 mm in diameter.

Pedology   The study of soils.

Pegmatite   A coarse-grained igneous rock that may contain rare minerals rich in elements such as lithium, boron, fluorine, uranium, and others.

Percolation test   A standard test used to determine the rate at which water will infiltrate into the soil. Primarily used to determine the feasibility of the septic-tank disposal system.

Permafrost   Permanently frozen ground.

Permeability   A measure of the ability of an earth material to transmit fluids such as water or oil.

Petrology   The study of rocks and minerals.

Physiographic determinism   A type of site selection that utilizes the philosophy of designing with nature.

Physiographic province   A region characterized by a particular assemblage of landforms, climate, and geomorphic history.

Placer deposit   A type of ore deposit found in material transported and deposited by agents such as running water, ice, or wind; for example, gold and diamonds found in stream deposits.

Plastic deformation   A type of deformation involving a permanent change of shape without rupture.

Plate tectonics   A model of global tectonics which suggests that the outer layer of the earth known as the **lithosphere** is composed of several large plates which move relative to one another. Continents and ocean basins are passive riders on these plates.

Plutonium-239   A radioactive element produced in a nuclear reactor. It has a half-life of approximately 24,000 years.

Point bar   Accumulation of sand and other sediments on the inside of meander bends in stream channels.

Pollution   Any substance, biological or chemical, in which an identified excess is known to be detrimental to desirable living organisms.

Pool   Common bed form produced by scour in meandering and straight channels with a relatively low channel slope. Characterized at low flow by slow-moving, deep water. Generally but not exclusively found on the outside of meander bends.

Porosity   The percentage of void (empty space) in earth material such as soil or rock.

Potable water   Water that may be drunk safely.

Pyrite   Iron sulfide, a mineral, commonly known as fool's gold. Environmentally important because in contact with oxygen-rich water, it produces a weak acid that may pollute water or dissolve other minerals.

Pyroclastic activity   Type of volcanic activity characterized by eruptive or explosive activity in which all types of volcanic debris, from ash to very large particles, are physically blown from a volcanic vent.

Quartz   Silicon oxide, a very common rock-forming mineral.

Quartzite   Metamorphosed sandstone.

Quick clay   Type of clay which when disturbed, such as by seismic shaking, may experience a spontaneous liquefaction and lose all shear strength.

Radioactive waste   Type of waste produced in the nuclear fuel cycle. Generally classified as high-level or low-level.

Radon   A colorless, radioactive, gaseous element.

Reclamation, mining   Restoring of land used for mining to other useful purposes, such as agriculture or recreation, after mining operations are concluded.

Recycling   The reuse of resources reclaimed from waste.

Regional metamorphism   Wide-scale metamorphism of deeply buried rocks by regional stress accompanied by elevated temperatures and pressures.

Renewable resource   A resource such as timber, water, or air that is naturally recycled or recycled by man-induced processes within a time framework useful for man.

Reserves   Known and identified deposits of earth materials from which useful materials can be extracted profitably with existing technology and under present economic and legal conditions.

Resisting forces   Those forces that tend to oppose the downslope movement of earth materials.

Resistivity   A measure of an earth material's ability to retard the flow of electricity. The opposite of conductivity.

Resources   Includes reserves plus other deposits of useful earth materials that may eventually become available.

Riffle   A section of stream channel characterized at low flow by fast, shallow flow. Generally contains relatively coarse bed-load particles.

Riparian rights, water law   Each landowner has the right to make reasonable use of water on his land, provided the water is returned to the natural stream channel before it leaves his property. Furthermore, the property owner has the right to receive the full flow of the stream undiminished in quantity and quality.

Rippable excavation   Type of excavation that requires breaking up the soil before it can be removed.

Riprap   A layer or an assemblage of broken stones emplaced to protect an embankment against erosion by running water or breaking waves.

Riverine environment   Land area adjacent to and influenced by a river.

Rock   **Geologic**—an aggregate of a mineral or minerals. **Engineering**—any earth material that has to be blasted in order to be removed.

Rock cycle   A group of processes that produce igneous, metamorphic, and sedimentary rocks.

Rocksalt   Rock composed of the mineral halite.

Rotational landslide   Type of landslide that develops in homogeneous material. The movement is likely to be rotational along a potential slide plane.

S wave   Secondary wave, one of the waves produced by earthquakes.

Saline   Salty; characterized by a high salinity.

Salinity   A measure of the total amount of dissolved solids in water.

Salt dome   A structure produced by the upward movement of a mass of salt. Frequently associated with oil and gas deposits on the flanks of a dome.

Sand   Grains of sediment having a size between 1/16 and 2 mm in diameter. Often sediment composed of quartz particles of the above size.

Sand dune   A ridge or hill of sand formed by wind action.

Sandstone   A detrital sedimentary rock composed of sand grains that have been cemented together.

Sanitary landfill   A method of solid-waste disposal not producing a public health problem or nuisance. Confines and compresses the waste and covers it at the end of each day with a layer of compacted, relatively impermeable material such as clay.

Scarp   A steep slope or cliff commonly associated with landslides or earthquakes.

Scenic resources   The visual portion of an aesthetic experience. Scenery is now recognized as a natural resource with varying values.

Schist   A coarse-grained metamorphic rock characterized by a foliated texture of the platy or elongated mineral grains.

Schistosomiasis   Snail fever, a debilitating and sometimes fatal tropical disease.

Sea wall   An engineering structure constructed at the water's edge to minimize coastal erosion by wave activity.

Secondary enrichment   A weathering process of sulfide ore deposits which may concentrate the desired minerals.

Sedimentology   The study of environments of deposition of sediments.

Seismic   Refering to vibrations in the earth produced by earthquakes.

Seismograph   An instrument that records earthquakes.

Selenium   An important nonmetallic trace element with an atomic number of 34.

Septic tank   A tank that receives and temporarily holds solid and liquid waste. Anaerobic bacterial activity breaks down the waste. The solid wastes are separated out, and liquid waste from the tank overflows into a drainage system.

Serpentine   A family of ferromagnesian minerals. Environmentally important because they form very weak rocks.

Sewage sludge   Solid material that remains after municipal waste-water treatment.

Shale   A sedimentary rock composed of silt- and clay-sized particles. The most common sedimentary rock.

Shield volcano   A broad, convex volcano built up by successive lava flows. The largest of the volcanoes.

Silicate minerals   The most important group of rock-forming minerals.

Silt   Sediment between 1/16 and 1/256 mm in diameter.

Sinkhole   A surface depression formed by the solution of limestone or the collapse over a subterranean void such as a cave.

Sinuous channel   Type of stream channel (not braided).

Slate   A fine-grained, foliated metamorphic rock.

Slump   A type of landslide characterized by the downward slip of a mass of rock, generally along a curved slide plane.

Soil   **Soil science**—earth material modified by biological, chemical, and physical processes such that the material will support rooted plants. **Engineering**—earth material that can be removed without blasting.

Soil horizons   Layers in soil (A, B, C) that differ from one another in chemical, physical, and biological properties.

Soil survey   A survey consisting of a detailed soil map and descriptions of soils and land-use limitations. Generally prepared by the Soil Conservation Service in cooperation with local government.

Solar energy   Collecting and using energy from the sun directly.

Solid waste   Material such as refuse, garbage, and trash.

Spoils, mining   Banks or piles that are accumulations of overburden removed during mining processes and discarded on the surface.

Storm surge   Wind-driven oceanic waves.

Strain   Change in shape or size of a material as a result of applied stress. The result of stress.

Stress   Force per unit area. May be compression, tension, or shear.

Strip mining   A method of surface mining.

Subduction   A process in which one lithospheric plate descends beneath another.

Subsidence   A sinking, settling, or otherwise lowering of parts of the crust of the earth.

Subsurface water   All of the waters within the lithosphere.

Surface water   Waters above the surface of the lithosphere.

Surface wave   One of the types of waves produced by earthquakes. These waves generally cause most of the damage to structures on the surface of the earth.

Suspended load   Sediment in a stream or river carried off the bottom by the fluid.

Syncline   A type of fold in which younger rocks are found in the core of the fold. Rocks in the limbs of the fold dip inward toward a common axis.

System   Any part of the universe that is isolated in thought or in fact for the purpose of studying or observing changes that take place under various imposed conditions.

Tar sands   Naturally occurring sand, sandstone, or limestone that contains a very viscous petroleum.

Tectonic   Refering to rock deformation.

Tectonic creep   Slow, more or less continuous movement along a fault.

Tectonic cycle   A group of processes which collectively produce external forms on the earth, such as ocean basins, continents, and mountains.

Tephra   Any material ejected and physically blown out of a volcano; mostly ash.

Texture, rock   The size, shape, and arrangement of mineral grains in rocks.

Tidal energy   Electricity generated by tidal power.

Till   Unstratified, heterogeneous material deposited directly by glacial ice.

Toxic   Harmful, deadly, or poisonous.

Transform fault   Type of fault associated with oceanic ridges. May form a plate boundary such as the San Andreas fault in California.

Translation (slab) landslide   Type of landslide in which the movement takes place along a definite fracture plane such as a weak clay layer or bedding plane.

Tropical cyclone   Severe storm generated from a tropical disturbance. Called **typhoons** in most of the Pacific Ocean and **hurricanes** in the Western Hemisphere.

Tsunami   Seismic sea wave generated by submarine volcanic or earthquake activity. Characteristically has very long wave length and moves very rapidly in the open sea. Incorrectly referred to as *tidal wave*.

Tuff   Volcanic ash that is compacted, cemented, or welded together.

Unconfined aquifer   Type of aquifer in which there is no impermeable layer restricting the upper surface of the zone of saturation.

Unconformity   A buried surface of erosion representing a time of nondeposition. A gap in the geologic record.

Unified soil classification system   A classification of soils, widely used in engineering practice, based upon the amount of coarse particles, fine particles, or organic material.

Uniformitarianism   Concept which states that the present is the key to the past. That is, we can read the geologic record by studying present processes.

Volcanic breccia, aggolomerate   Large rock fragments mixed with ash and other volcanic materials cemented together.

Volcanic dome   Type of volcano characterized by a very viscous magma with high silica content. Activity is generally explosive.

Water table   The surface that divides the zone of aeration from the zone of saturation. The surface below which all the pore space in rocks is saturated with water.

Weathering   Changes that take place in rocks and minerals at or near the surface of the earth in response to physical, chemical, and biological changes. The physical, chemical, and biological breakdown of rocks and minerals.

Zinc   An important trace element necessary in life processes.

Zonal soil   Soil in which the profile is in adjustment to the climatic zone.

Zone of aeration   The zone or layer above the water table in which some water may be suspended or moving in a downward migration toward the water table or laterally toward a discharge point.

Zone of saturation   Zone or layer below the water table in which all of the pore space of rock or soil is saturated.

INDEX